OUR EVOLVING UNIVERSE

About The Cover

The cover displays some southern circumpolar constellations that include main sequence stars similar to the sun. Some of these stars may possibly have planets similar to Earth upon which life could evolve. Among these constellations is Reticulum, The Net, named by Nicolas Louis de Lacaille in the mid-eighteenth century. This constellation emerged from obscurity during the Betty and Barney Hill incident in the early 1960s. Betty and Barney Hill were allegedly picked up and examined by beings on a UFO. During the incident Zeta Reticuli, a pair of stars in the constellation just visible to the unaided eye, was indicated as the point of origin of the UFO. The constellation Reticulum is located just west of the Large Magellanic Cloud.

OUR EVOLVING UNIVERSE

WALTER H. HESSE   California State Polytechnic University, Pomona

WADSWORTH PUBLISHING COMPANY, INC.   Belmont, California

ISBN: 0-8221-0184-X

Library of Congress Catalog Card Number: 76-27551

Printed in the United States of America

Printing (last digit): 9 8 7 6 5 4 3 2

Interior and cover art and design by Bert Johnson, Graphics Two.

Constellations appearing on title page and chapter opening pages
reproduced from woodcuts from Hyginus' *Poetica Astronomica* (Venice,
1482). Courtesy of The Granger Collection.

This book is dedicated to my wife, Helen, who suffered with me through the many stages necessary to bring the work to fruition.

CONTENTS

PREFACE

The study of astronomy has evolved through the centuries from the simple observation of celestial activity for the purpose of telling time and direction to an attempt to understand the development of the universe from its origin and the part life plays in the evolutionary scheme of the universe. Is life an accident that occurred only on Earth or does life exist throughout the universe? Is life an extension of the process whereby galaxies, stars, and planets are formed or is life an event in the history of the universe unique to Earth? These are some of the questions dealt with in this text, which was written for students who are not majoring in the sciences. Hopefully the book will appeal to the future business executive, poet, recreation director, and historian.

The text is designed for the nonscience student who has had minimal exposure to science and mathematics. A knowledge of mathematics beyond high school algebra and geometry should not be required to cope with the few mathematical equations and examples included in the text.

Some historical material is included to provide a view of how science progresses. In this way it can be shown that theories are not "hatched out of thin air" but rather in most instances are painfully slow in evolving toward the ultimate truth. These historical discussions usually occur at the introduction of new topics. In this way it is hoped that the principles of astronomy are imparted by means of a historical development of the ideas. The exploration of space during the past fifteen years, up to and including the Viking landing, has been covered by means of a discussion of the discoveries made. There has been a minimum reference to the hardware used.

Two chapters are included on matters pertaining to life in the universe. Chapter 1, for the most part, deals with aspects of this subject not usually found within the realm of science but rather as part of the myth and lore relating to extraterrestrial life. Although no scientific knowledge may be derived from this sort of discussion, it does provide some background and peripheral information on the thinking of the lay person on these matters. These topics were developed from questions frequently asked by students in an astronomy course. Therefore, a brief treatment of these subjects should be of value in anticipating some of the questions and at the same time providing a basis for further discussion. Also included in this chapter is a brief account of Project Cyclops, a technique that has been proposed for use in the search for extraterrestrial life. Chapter 19 deals more directly with the scientific aspects of the search for extraterrestrial life. Included are discussions on the possibility of its occurrence, the manner in which the search for evidence is being undertaken, and some of the possible ramifications of making contact with extraterrestrial life.

The balance of the text follows the format of topics more often than not used by teachers of astronomy. Where appropriate the development of life or the formation of the precursors of life are described. This material is then summarized in the last chapter.

Controversy in astronomy is not avoided but is described to show that astronomy, or for that matter any science, is subject to differences of opinion. Such topics as heliocentric versus geocentric universe, wave versus particle theory of light, and evolutionary versus steady state universe are used to point this up.

Questions are provided at the end of each chapter. Also included are some projects which are designed to lead students into a more active role in searching out the mysteries of the universe with a minimum of equipment. A glossary is included at the end of the book as well as tables and charts useful in the study of astronomy.

A professional supplement is provided that is designed to assist the instructor in using the text to the best possible advantage. Included in the professional supplement is a discussion of the mathematical equations that will enable the instructor to pursue these aspects of astronomy in more depth. Also there are answers to chapter-end questions, examination questions, suggested films and some supplementary material that hopefully will be useful in enhancing interest in the course.

The text may be used in a one-semester or two-quarter course. With the judicious use of supplementary material the text can cover a full year. In the two-quarter sequence, chapters 1 through 3 and 6 through 12, covering material pertaining to the solar system, can be used in the first quarter, and chapters 4, 5, and 13 through 19, on stellar astronomy, can be covered in the second quarter. The material has been used in this manner by the author and has proved to be a satisfactory division of topics.

The publishing of a text is not the work of one person but is the combined effort and interaction of a number of people. The author wishes to acknowledge the input by the many students whose questions helped to guide the direction of the text. A thank you is due to the reviewers of the manuscript who pointed out its strengths and weaknesses and made many helpful suggestions for its improvement. These include Professor Igor Alexandrov, California State University, Long Beach; Dr. Sune Engelbrektson, Pace College; the late Professor Virginia Larsen, Los Angeles Pierce College; Professor John M. Samaras, University of South Carolina; Professor Michael Stewart, San Antonio College; Dr. Peter D. Usher, Pennsylvania State University; Professor James C. Vogt, Santa Rosa Junior College; and Professor Lester Winsberg, University of Illinois, Chicago Circle. A thank you also to Jenny Donner for the preparation of the manuscript, Elaine Linden for copy editing, and Janet Greenblatt who did a magnificent job in handling the many details in the production of the book.

W.H.H.

OUR EVOLVING UNIVERSE

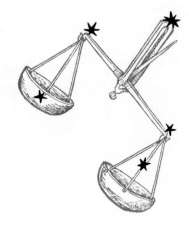

Libra

CHAPTER 1   QUEST FOR EXTRATERRESTRIAL LIFE

> I do not know what I may appear to the world; but to myself I seem to have
> been only like a boy playing on the seashore and diverting myself, now and
> then finding a smoother pebble or a prettier shell than ordinary, whilst the
> great ocean of truth lay all undiscovered before me.
>
> —*Sir Isaac Newton*

1.1 LIFE IN THE UNIVERSE

Astronomy has been traditionally regarded as the science of the stars, a
science dealing with motion, shape, size, and energy relationships of
objects in the universe. Until recently little serious thought was given to
the possibility of life existing in parts of the universe other than the earth.
In the past the great men of astronomy, including Aristotle, Hipparchus,
Ptolemy, Copernicus, Kepler, Galileo, and Newton, asked such questions
as: Is the earth flat or round? Does the earth move or do all other objects in
space move? What is the cause of the lunar and solar eclipses? What is a
star? What is the Milky Way? The answers to these questions are now
available, and it is possible to see how simple the problems really were.
Despite the apparent simplicity (in hindsight) of the questions, gaining the
answers has ranked among the great triumphs of human inquiry. Yet we
can see that these questions relate only to the physical properties of the
universe. Nowhere is there a serious question relating to the prospect of
other forms of life in the universe. It would appear that astronomy was
limited to the study of a physical cosmology, and that biology was
restricted to study of life forms on earth, excluding inquiry into the
possibility of extraterrestrial life.

In modern times increasingly complex problems in astronomy have
been posed. The questions being asked are: How was the universe
created? What if anything existed before the creation of the universe? Are
there undiscovered forces and energies at work? More recently the
following questions have been included: Are there intelligent beings
elsewhere in the universe? Will we be able to communicate with them?

Let us begin by thinking about one of the more intriguing of these questions to which answers are now being sought. Is there intelligent life or, for that matter, any form of life in the universe beyond that which is found on earth? Life, in myriad forms, may be found virtually everywhere on earth—from tropical rain forests to barren ice caps of the polar regions, from high mountain slopes to the abyssal depths of the ocean. On the highest mountain peaks and the driest desert, we find a wide variety of organisms, some quite simple in form, even microorganisms. Some forms of life have the ability to adapt to widely changing conditions and survive. For example a tardigrade (Figure 1.1) or moss animal can be dehydrated, shriveled to almost crystalline appearance, and then, when provided with water, begin to crawl around and reproduce. This indicates the ability of life to survive and develop under extreme environmental conditions. Considering the varied circumstances under which life can exist on earth, is it not possible for life to occur on similar planets orbiting other stars?

There is as yet no positive answer to the question of the existence of life in the universe, and before exploring such possibilities (see chapter 19) we may inquire into the reasons for making such a search.

A human being needs contact with others to fulfill his or her own humanity. This desire to communicate with others through the interchange of information and impressions or ideas is a thoroughly human characteristic. We may apply the same need to humankind—the totality of all individuals—seeking companionship in the broader universe to answer questions about itself. We seek other societal entities to find the answers to such questions as: Where do we come from? Where are we going? What is our place and meaning in the cosmos? The scientist wants to learn if the biochemistry of life as we recognize it on earth is common in the universe or if there are alternate forms of life, how life evolved on distant planets and what societal organization exists there, how that planet evolved

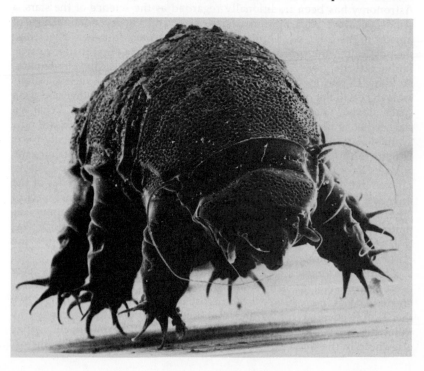

Figure 1.1 The tardigrade (Echiniscus arctomys) can revert to an almost deathlike state (cryptobiosis) by becoming almost completely dehydrated and, upon the addition of moisture, revive again. (Courtesy of John H. Crowe)

geologically as well as sociologically. This may lead to an understanding of whether life is a stage in the evolution of the universe or just a chance occurrence here on Earth. Are we unique in the universe—a freak accident of nature on a tiny world orbiting an average star in a fairly remote portion of a galaxy? Or are we a part of a grand scheme in the evolutionary process that pervades the entire universe?

The desire for contact with other intelligent forms and the suppressed need to feel we are not alone in the universe occurs on several levels. Not only is the scientist interested in seeking such contact but also the lay person has, in many ways, expressed the wish for companionship in the universe. Some people view extraterrestrial life as an intelligence that may resolve all problems on Earth and relieve us from the continuous cycle of war and social and economic upheavals. In a sense this is a larger version of the "cargo cults" that still exist on some of the islands in the Pacific Ocean. The island people saw the arrival of ships and planes during World War II as an unending source of riches and the means of a life of continual ease. The belief still persists that special agents will, in the future, divert cargo to cult members, and as a result crude airstrips and warehouses have been built on some islands to receive the expected largess. There is some manifestation of this type of wish fantasy in our own society.

In the past people have imaginatively expressed in literature their need to feel they are not alone in the universe. Now the feeling that there are other populated worlds is leading to the inevitable step—a search for extraterrestrial life. The purpose of this chapter is to briefly explore the means whereby we attempt to prove to ourselves that other beings exist.

The fact that we want to believe we are not alone in the universe has made possible such occurrences as the "great moon hoax" perpetrated by Richard Locke in 1835. Richard Locke was a New York journalist who, upon hearing that the noted English astronomer Sir John Herschel was going to South Africa to view the southern skies, agreed to write a series of articles for a New York paper on the progress of the expedition. Although Locke never went to Africa and was never in contact with Sir John, he wrote very colorful and imaginative accounts of the proceedings. The articles culminated in a description of Sir John's viewing of the moon, where he allegedly saw beautiful lush gardens and people walking about. Of course Locke invented the entire contents of the articles, but his stories were readily accepted because his readers wanted to believe in the existence of other-world beings.

When Giovanni Schaparelli announced sighting channel-like structures (which he called canali) on Mars in 1877, people easily accepted the idea of canals, which helped create the climate that led to the acceptance of the ideas of Percival Lowell more than a quarter of a century later. Lowell, an American astronomer, said he had seen the dividing of the canals, the spread of vegetation, and the geometry of the Martian irrigation systems. These were accepted by some as signs that intelligent creatures inhabited Mars. More recently, the worldwide enthusiasm for the Apollo program was partly prompted by the very slim possibility that man exploring the moon might, among other things, discover life thereon, or at least would find that life had appeared there at some time during the moon's evolution. No signs of life were found during the Apollo project, but the search has been continued by means of various probes to Mars and the other planets.

Some indications of interest in the existence of alien life-forms are related to astronomy and are of scientific predisposition while others are not. Science fiction, the UFO controversy, and other more-or-less imaginative speculations are examples of the latter. It is these aspects, although not a part of astronomy nor even considered scientific, that we will review briefly in the following pages.

1.2 SCIENCE FICTION

Science fiction is a form of writing in which unusual plots are made believable by utilizing scientific explanations for an imaginative theme. The plot may have little relationship to astronomy other than that some of the action occurs within the astronomical realm. By no means do all science fiction stories deal with life-on-other-world plots, but those that do, speculate on life-forms that range from pure energy to anthropomorphic forms to monsters.

Science fiction, leading us through our imaginations far beyond the boundaries of the earth, was first identified as an art-form by Hugo Gernsbeck in 1929. Gernsbeck is considered the father of modern science fiction, having published in 1926 the world's first science fiction periodical, *Amazing Stories*. However, science fiction may be traced back to the second century A.D. when Lucien of Greece wrote in his *Vera Historica* of a flight of men to the moon. In this tale he told of voyagers who were picked up by a whirlwind and set down on the moon. There they assisted the moon men in a battle against the men from the sun over the colonization of the morning star. This first science fiction story, dealing with life on other worlds, was a forerunner to others of a similar theme, many by well-known writers such as H. G. Wells *(War of the Worlds)*; Edgar Rice Burroughs *(Pirates of Venus)*; Ray Bradbury *(The Martian Chronicles)*; and Isaac Asimov *(The Foundation)*.

Motion pictures have made good use of imaginative techniques and special effects to depict the science fiction theme of life-from-other-worlds. Television's popular *Star Trek* series had as part of its introduction ". . . to explore strange new worlds, to seek out new life and new civilizations, to boldly go where no man has gone before." The popularity of science fiction in any form does seem to indicate something of people's disinclination to feel that they are alone in the universe. It appears that, to humankind, the existence of other beings of any type, whether malevolent or benevolent, is preferable to the alternative of existing alone as a unique phenomenon in the universe.

1.3 ANCIENT ASTRONAUTS?

While science fiction buffs enjoy stories of imaginary extraterrestrial life, there are those who consider other-worldly beings as a reality. These are the individuals who accept the factuality of certain biblical incidents or the presence of massive ancient structures as evidence of past visitations by intelligent life from other worlds. Following are a few examples of such incidents and how they have been interpreted as being related to ancient visits, as chronicled by Eric Von Daniken, one of the principal proponents of these theories.

A typical example of a biblical passage that is interpreted as a visit by extraterrestrial life is Gen. 6:2.

> The sons of God seeing the daughters of man, that they were fair, took
> to themselves wives of all which they chose.

The suggestion was made that the "sons of God" were actually astronauts, although no proof of this interpretation is offered. A broader reading of the same passage seems to indicate that the descendants of Seth and Enos were called the "sons of God" due to their piety and virtuous ways, whereas the descendants of Cain were called the "children of man" because of their evil ways. In making the statement that the ancient astronauts were the "sons of God," Von Daniken assumes that the paths of evolution on another world were such as to produce beings identical to humans and therefore capable of reproducing with them. The odds against such an event are fantastic.

Proponents of the ancient astronaut theory cite the incident described by Ezekiel (Ezek. 1:4–28) while standing by the river Chobar as a landing of a spacecraft. Furthermore, it is claimed, Ezekiel also described the words of "one who spoke"—this supposedly being the voice of one of the occupants of the spacecraft. In Ezek. 10:1–22 a vehicle is described that has "wheels," which have been construed as helicopter-like blades of an atmospheric-entry vehicle, possibly atomic powered. Such interpretations should only be accepted after first considering the symbolism in statements and writings of the ancient prophets. Their teachings were often in allegorical form and accepting them in the literal sense could lead to erroneous conclusions.

Another biblical event, the destruction of Sodom and Gomorrah, has been explained by those subscribing to the ancient astronaut hypothesis as an event involving a nuclear explosion. This incident, described in Gen. 19:1–28, tells of a meeting between Lot and two angels (interpreted as astronauts) who strongly urge him to take his family and flee Sodom. As he leaves he is warned not to look back, but his wife does not heed the warning and is turned to "salt." The end of Sodom and Gomorrah is described as follows:

> The Lord rained on Sodom and Gomorrah brimstone and fire from the
> Lord out of heaven, and He overthrew those cities, and all the valley,
> and all the inhabitants of the cities and what grew on the ground.

A further description of the destruction of the cities is given through the eyes of Abraham:

> And he looked down toward Sodom and Gomorrah, and toward all the
> land of the valley, and he saw, and behold, the smoke of the land
> ascended like the smoke of a furnace.

This is interpreted by Von Daniken as the mushroom cloud of an atomic explosion. However, there is no residual evidence of past atomic blasts in the Dead Sea area. It has been determined that the lack of vegetation is the result of high salt content in the soil and not the effect of ancient radiation damage, thus removing an atomic explosion as the cause of the destruction of Sodom and Gomorrah. There is some evidence of earthquake and volcanic activity in the area, and it is more probable that an earthquake was responsible for the demise of these cities. We can assume that structures were not earthquake-proof and even a moderate earthquake could have caused considerable damage.

It appears unusual that, despite the many supposed alien visits to Earth, never has an authenticated artifact, a piece of hardware, or any remnant of such a visit been exhibited or found, although claims of such finds have been made. From the manner in which many items were left on

Figure 1.2 The Pyramid of Cheops, or the Great Pyramid, as it is also known, is among the most celebrated of such structures of antiquity. (Bettmann Archive)

the moon by several teams of astronauts, it would seem that some articles would have been discarded by these extraterrestrial visitors. If, as it is claimed, the people of biblical times venerated such alleged visitors from space as angels or messengers from heaven, would not the discarded items have been kept as sacred relics? This does not seem to be the case, and in fact structures that are pointed out as examples of those built with the aid of aliens could readily have been built by earthlings.

Proponents of the early extraterrestrial visitation theory point to the statues on Easter Island, the Egyptian obelisk, Stonehenge, and the monoliths at Tiahuanaco as examples of structures that were raised or moved in ancient times. The claim is made that the massive nature of the rock involved made it impossible for ancient Earth people to move the stones, and that they could only have been moved by some technically advanced beings. This is merely an assumption and one not entirely based on fact, for it has been shown that ancient people were capable of moving large masses of rock great distances and were technologically capable of building the ancient structures without extraterrestrial assistance. We must also remember that a structure like Stonehenge is not an isolated monument. There are approximately 900 such structures in Great Britain alone, although none as well known as Stonehenge. Those who oppose the "ancient astronaut" theory might point out that it is unlikely extraterrestrial visitors would spend their time constructing memorials in such great numbers to commemorate their visits.

Up to this time, despite the many claims to the contrary, no substantive scientific evidence exists to show that alien beings have visited the earth in the past. The prospects for such a visit are extremely remote (not impossible but remote) due to the tremendous distances and almost insurmountable problems involved in space travel as we now view it. Some of these problems will be discussed in later chapters (Figure 1.2).

1.4 UNIDENTIFIED FLYING OBJECTS OR UFO'S

Whenever the question of extraterrestrial life is discussed within the astronomical context, questions relating to UFO's inevitably arise. Individuals will ask: Do you believe in UFO's or in flying saucers? Do they really exist? Are they visitors from outer space? These questions cannot be answered in the affirmative because the answers are not known. However, it is felt that a brief discussion of this rather controversial subject may be of value here. Unidentified flying objects, or UFO's as they are more familiarly known, are not really a modern phenomenon since they have been observed for many decades all over the world. Generally they went unreported or when reported were ignored. But some of the reports stimulated a great deal of interest. One of these was the "airship mystery" of 1896 and 1897 when witnesses in the western United States at different times and from different locations reported observing a huge cigar-shaped craft flying overhead.

The confusion about UFO's in recent times has been so great and has persisted for so long it is not surprising that the subject is a matter of ridicule and derision among many scientists. A part of the confusion arises from the interchangeable use of the terms UFO and "flying saucer," when in fact they have different meanings. "Flying saucer" was originally used as a journalistic expression for objects in flight which could not be identified. The term was later used by certain groups to indicate vehicles whereby aliens visited Earth. Members of these groups claimed contact with the aliens aboard flying saucers, and some even insisted they had traveled on board such vehicles to Mars and Venus. To a certain extent it was these groups, although small in number, that contributed greatly to the skepticism of the entire subject held by scientists and many laymen. The claims by these so-called contactees of trips to Venus or Mars on flying saucers cast severe doubt on their existence as well as on the sanity of the reporters. Sincere as these individuals may have been, they generally contributed UFO reports that were easily categorized with the 80 percent of those reports considered invalid.

The remaining 20 percent of the sightings have generated a great deal of interest. These sightings may be defined as a light or object seen in the sky which by its action or appearance defies logical explanation, even by technically qualified persons, and therefore *remains unidentified*. These are the real unidentified flying objects that have been reported generally by reliable, sensible people, including airline pilots, control tower operators, housewives, policemen, and even astronauts.

Typical of such reports was an incident that occurred on November 2, 1957. A patrolman on duty at the Levelland, Texas (pop. 10,000), police station received ten calls between 11:00 P.M. and 1:30 A.M. regarding a strange craft seen by a number of people on the outskirts of the town. All callers, some quite agitated, reported seeing a glowing object sitting on or near the highway or flying low overhead. The observations occurred at different locations in the vicinity of Levelland and at slightly different times. In each case, when the object was approached by car it was reported that the car engine invariably stopped and its lights went out. One observer attempted to approach the object, but it immediately rose vertically into the air and disappeared from sight. In each case, as the object disappeared the car engine would start and the lights would function normally. No explanation was ever given for the phenomenon, but the report is real and cannot be ignored in view of the number of witnesses that saw the object. There are a sufficient number of such reported sightings on file (over 12,000 of all types held on file by the Air Force and many more held by NICAP* and similar organizations) to make

*NICAP—National Investigations Committee on Aerial Phenomena.

Figure 1.3 Photograph of a UFO taken by Mr. and Mrs. Paul Trent of McMinville, Oregon, on May 11, 1950. The photograph was subjected to extensive analysis by experts who concluded that no tampering with the negative had occurred. The object was described by the Trents as a disk-shaped metallic-appearing object and to this day remains unidentified (Courtesy of NICAP, Kensington, Md.)

the valid UFO's a significant number. As previously stated, it has been estimated that 80 percent of reported sightings are misconceptions of natural phenomena such as ball lighting, planets, meteors, or are imaginary sightings or outright hoaxes. However, 20 percent of the reports have been from reliable persons who provided accounts that did not fall into the above categories.

The source of UFO's is unknown, although a number of hypotheses have been suggested. In the late 1940s and 1950s, when great public interest was exhibited in UFO's, it was suggested that they were secret aircraft being tested by the United States, by Russia, or by some power or organization of whose technology we were unaware. The "extraterrestrial life" hypothesis was not long in appearing. In addition, it was suggested that the vehicles were time machines from the future, that they were ball lighting or plasma or swamp gas or other rare atmospheric phenomena. Of course none of these suggested hypotheses satisfy all circumstances, nor is there agreement on the possibility of any being valid UFO sources (Figure 1.3).

After World War II, when sightings became common, there was considerable concern that the unidentified objects could be a danger to national security, and the Air Force was given the responsibility for investigating their origins and determining whether or not they represented a risk. The investigative program was originally named Project Sign in 1947, renamed Project Grudge in 1949, and in 1951 the name was changed to Project Bluebook. Criticism leveled at the Air Force UFO research program noted that it was severely understaffed for this type of investigation. Generally, an officer below field grade rank was in charge, assisted by a junior officer or a sergeant. Reports were accumulated and classified only in chronological order of receipt. Evaluations were sketchy, often being made by a sergeant with little or no scientific experience or training. Recommendations to cross-index or computerize the data were rejected, and the data remained essentially unprocessed.

In 1966, the Air Force sponsored a scientific study of the UFO problem—a study which was conducted by a committee directed by Dr. E. U. Condon at the University of Colorado. The Condon Report, which included a conclusion and recommendations as to future courses of action, was the end result. Project Bluebook was terminated in 1969 primarily upon the recommendation of the Condon Report. Officially the UFO problem became a closed matter, and the military and civilian authorities planned no further investigations. This action was based on the conclusions in the Condon Report which do not categorically state that UFO's do not exist nor that they are all explainable on the basis of natural phenomena. Rather it states that the available record leads to the conclusion that ". . . further extensive study of UFO's probably cannot be justified in the expectation that science will be advanced thereby."

The study, report, and conclusions have been both praised and damned by many in the scientific community. The National Academy of Sciences endorsed the report of the Condon Committee and approved its methodology. However, disapproval was expressed by others who felt that a number of well-documented but unexplained cases, some of which were included in the Condon Report, should not be ignored. They suggested a continued effort be expended with the first step an improvement in data collection and data processing. No responsible scientist believes that UFO's are of extraterrestrial origin, because there is not a shred of evidence to support such a hypothesis. Yet it is paradoxical that some in the scientific community are intensely interested in the search for life in the universe and at the same time ridicule the UFO experience as nonsense. It would appear that for them the only adequate proof, as someone stated, would be "the landing of a UFO on the White House lawn." The last word has probably not been heard.

1.5 COMMUNICATION WITH EXTRATERRESTRIAL LIFE

Some scientists are seriously exploring the possibilities of extraterrestrial life, and many believe there is a high probability that life of some sort exists elsewhere in the universe despite the fact that there is very little direct evidence to support the idea (see chapter 19). One of the problems is the great distances to even the nearest stars. The stars, possibly with planets upon which life could develop, are so distant that the prospect of making physical contact is at present impossible with our technology. The only means available to us for making contact is by radio, but because of the vastness of space the prospect of receiving a recognizable signal from some form of intelligent life, or of a signal of ours being received and recognized, is extremely small.

Although radio is currently thought to be the only feasible means of communication, other rather interesting but improbable methods which predate the invention of the radio have been suggested. The discovery of channel-like markings on Mars by Schaparelli (1877) caused many to consider Mars a prime candidate for the abode of intelligent life. One of the suggestions made for communicating with the Martians was to plant gigantic windbreaks on the steppes of Russia and in Siberia to form an enormous geometric display of the Pythagorean Theorem because it was assumed that the principle of the triangle would be recognized by other intelligent beings.

Those who seriously thought about the problem realized that we are unable to see canals on Mars viewed from Earth because, at a minimum distance of over 50 million kilometres, the canals would be too narrow. Percival Lowell suggested that what we were seeing was the vegetation growing along the canals which were used for irrigation. The sides of the triangle would therefore have to be made up to several hundred kilometres wide to form lines that would be visible on Mars.

Another fanciful idea proposed that a series of canals be dug in the Sahara Desert to form a huge triangle with a square on each side, again suggesting the Pythagorean principle. Then it was suggested that kerosene or some similar combustible be poured in to coat the water in the canals, and on a dark clear night this was to be ignited. The Martians would then see a flaming manifestation of life on Earth.

These imaginative schemes were never carried out because radio was developed, which was obviously a much more plausible method of communication. After World War II radio astronomy evolved, and with this came the development of the giant antenna systems capable of detecting even faint signals over enormous distances. One such antenna, the 27-metre-diameter dish at Green Bank, West Virginia, was in the late 1950s used in a plan developed by Frank Drake of the National Radio Astronomy Observatory (NRAO) in an attempt to detect signals from possible planets orbiting nearby stars. The plan was not received with wholehearted enthusiasm by many astronomers but nevertheless was approved by the Observatory.

One difficult problem Drake faced was determining which frequency this extraterrestrial intelligence would be transmitting on and the need to find an obvious guidepost that an extraterrestrial being would also recognize. Drake realized that such a mutually distinguishable focal point in the radio spectrum was the 21-centimetre line produced by radiation from neutral hydrogen atoms drifting in space (see chapter 14). The atoms of hydrogen, the most abundant element in space, spinning freely, occasionally collide with each other and as a result increase in energy or become "excited." The energy emitted by the atoms corresponds to the 21-centimetre wavelength, and in a huge cloud of such atoms a steady signal would be transmitted at the 21-centimetre band. Drake proposed that the 27-metre antenna be aimed at two nearby stars, Epsilon Eridani and Tau Ceti, both within 12 light years of the earth, to ascertain if some form of organized signal was emanating from them.

On April 6, 1960, the 27-metre antenna was trained on Tau Ceti in the first serious scientific effort to detect signals from another world. This was on the 21-centimetre band, but no signal was detected. When the antenna was turned to Epsilon Eridani a definite signal was received. After considerable excitement the source was later determined to be from a radar countermeasure experiment being conducted in the vicinity by the Air Force.

In 1967, British astronomers at Cambridge detected rhythmic signals with extremely regular pulse from several objects. At first they were thought to be signals from other planets, but eventually it was determined that they were neutron stars (pulsars) rotating at high speeds. Other more recent endeavors to detect signals have been made in Russia and Great Britain and have yielded negative results. In 1972, G. L. Verschuur of the NRAO attempted to detect signals on the 21-centimetre band. Ten stars were included in the study, two of which showed evidence of planetary presence. Even with use of the much larger antennae now available at Greenbank—the 50-metre dish and the 100-metre dish—his results were negative. Although thus far yielding negative results, these efforts are continuing, even to the extent of attempting to detect signals from nearby galaxies.

In 1971, the National Aeronautics and Space Administration (NASA) sponsored a study conducted jointly by Stanford University and Ames Research Center. The purpose of *Project Cyclops*, as the study was called, was to determine the requirements in hardware, time, manpower, and funding necessary to detect extraterrestrial life. The proposal suggests that the cost of construction be spread over a period of 10 to 15 years, with the detection of extraterrestrial life as a possible long-term goal of decades or even centuries. This would require a long-term commitment in funds and also the construction of highly automated equipment to monitor space over long intervals of time.

Initially, it is suggested, a small complex should be built that could be added to year by year and in the meantime could be used to gather radio astronomy data of current interest. The nature of the instrumentation could be such as to greatly enhance existing equipment used for deep space probes and radar astronomy. As now envisioned, the design would include a series of antennae, each approximately 100 metres in diameter. There would ultimately be thousands of these units providing a receiving area of from 7 to 20 square kilometres. Each of the units would be able to focus on the same area of the sky, and all of the input fed simultaneously into a computer capable of correcting for the geometry of the system. The computer could extract from the data any evidence of narrow band signals or coherent signals from distant objects (Figure 1.4).

What would be the benefits from such a project? Bernard M. Oliver, a co-director of Project Cyclops, thinks that communication between

Figure 1.4 Artist's concept of low-aerial view of portion of Cyclops system antenna array, showing central control and processing building (NASA-Ames Research Center)

worlds may already have been accomplished by other intelligent beings. He feels that 4 to 5 billion years ago the first civilizations in our galaxy came into contact with each other and began to build up a great intellectual community within the galaxy by an exchange of ideas and histories. If such a heritage exists, he suggests, interstellar communication would provide not only an exciting discovery for the human race but access to conditions in the remote past as well. There might be available pictures of the early forms of our galaxy and the universe taken over a span of 5 billion years as well as the geologic history of countless planets, some of which may be quite similar to our own. Even more intriguing would be information whose nature we cannot even imagine at our present stage of development.

"Is anyone trying to communicate with us?," is another question that has been raised in speculating about extraterrestrial life. One suggestion is that "they" know we are here but leave us alone and maintain the earth much as we preserve wilderness areas. "They" keep out of sight and allow us to evolve as an interesting phenomenon to be sustained. This is the so-called "zoo hypothesis."

The "messenger idea" is another speculation about alien world contact. Satellites or messengers are thought to be sent to orbit candidate stars in an attempt to detect narrow band signals from planets associated with that star. The signals periodically are relayed to the parent planet, indicating the existence of a form of life capable of communicating.

Now that we have contemplated several ways whereby people have expressed the need and the will to explore the universe for life, we can look at the requirements for life, how it may come about, and the prospects of finding life on alien worlds. But first it would be well to view the physical universe—how it appeared to us in the past and how we see it today.

1.6 SUMMARY

Most of the questions asked by early-day astronomers have been answered, and the answers appear relatively simple compared to questions currently asked, such as, Is there life in the universe? There is no real evidence of the existence of life beyond Earth, but nevertheless we seek life because we do not wish to be alone. Evidence of this occurs in science fiction literature, in attempts to assign the existence of ancient relics and structures, and more recently of UFO's, as evidence of the appearance of extraterrestrial beings on Earth. The UFO phenomenon has created considerable controversy, not so much as to its nature but because of the manner in which the entire subject has been investigated.

Many scientists believe there is a high probability that intelligent life exists in the universe. The scientist is challenged by the prospect of life existing on other worlds and is attempting to develop techniques for communicating with other beings. Some fanciful suggestions have been made but it appears that communication by radio will be the most practical. With the development of radio astronomy and the giant antennae utilized by radio astronomers, it is possible to train these instruments on stars that possibly have a planetary system associated with it and listen for narrow band signals that may indicate the presence of intelligent life. A number of attempts using this technique have thus far yielded negative results. Currently there is a study, Project Cyclops, underway to determine the feasibility of constructing a gigantic antenna which will ultimately cover up to 20 square kilometres. The purpose of this antenna will be to search for intelligent life through the detection of coherent signals. This is proposed as a long-term project which may extend for decades or even centuries.

QUESTIONS

1. What were some of the important questions asked by early astronomers? Without further reading, can you answer any of these questions?
2. What are some of the questions asked by astronomers today? Have you any thoughts about the answers to these questions?
3. What do scientists hope to learn about extraterrestrial life?
4. In what way has man expressed his belief in the existence of extraterrestrial life?
5. Describe the Project Cyclops installation and its intended function.
6. What benefits are expected to be derived from Project Cyclops?

EXERCISES AND PROJECTS

1. Evaluate a segment of the Von Daniken theory (for example, biblical evidence, monuments, surface markings on Nazca Plain) and provide alternate reasons for the existence of these features other than evidence of ancient extraterrestrial visits. This project may take the form of a term paper in which Von Daniken's views are cited along with alternative views, your evaluation of these views, and your conclusions.

FOR FURTHER READING

CONDON, E. U., Sci. Dir., and D. S. GILLMOR, Ed., *Scientific Study of Unidentified Flying Objects.* New York: E. P. Dutton & Co., 1968.

FULLER, JOHN G., *The Interrupted Journey.* New York: Dial Press.

HYNEK, J. A., *The UFO Experience.* Chicago: Henry Regnery Company, 1972.

LORE, G. I. R., and H. H. DENEAULT, *Mysteries of the Skies.* Englewood Cliffs, N.J.: Prentice-Hall, 1968.

Project Cyclops, A Design Study of a System for Detecting Extraterrestrial Intelligent Life, Stanford University, Ames Research Center, 1971.

SULLIVAN, W., *We Are Not Alone.* New York: McGraw-Hill, 1964.

VON DANIKEN, E., *Chariots of the Gods.* New York: Bantam Books, 1971.

WILSON, C., *Crash Go The Chariots.* New York: Lancer Books, 1972.

Draco, Ursa Major, Ursa Minor

CHAPTER 2   **ANCIENT ASTRONOMY**

Merely to realize there are more things in heaven and Earth than are dreamed
of in one's philosophy is hardly an end in itself. The end should be to expand
one's philosophy so as to include them.

—*Lord Rayleigh*

The astronomers of antiquity, viewing the sky, saw the physical universe
quite differently from the way it is seen by modern astronomers. Ancient
astronomers saw the sun's daily rising and setting, the moon's waxing and
waning, and the annual changing of the stellar pattern. Their observations
led them to describe a universe which had as the central figure a
motionless, spherical Earth.

The ancient astronomers were able to accomplish much in their
science even without the aid of sophisticated instruments available to the
present-day astronomers. By the time man had developed a tool-making
capability, he was also able to measure the passage of time by, for
example, counting the number of sunrises occurring between one full
moon and the next. These types of achievements seemed to occur
worldwide, since there are records of similar activities by the ancient
Chinese, Egyptians, Babylonians, Greeks, and Mayans. Although early
people in China, Egypt, and Central America presumably had little contact
with each other, their accomplishments in what may be considered the
oldest of sciences followed a parallel course.

2.1 ASTRONOMY OF ANCIENT CHINA

Early civilizations were obsessed with the need to develop a satisfactory
calendar, and to this end observations of the movements of planets and the
phases of the moon were organized into months and years. The ancient
Chinese were no exception and may actually have been among the earliest
societies to use astronomy for practical purposes. They maintained
excellent records of astronomical phenomena, some of which go back over
4000 years.

The Chinese recognized the lunar cycle with the moon moving
eastward in the sky against the background of more stationary stars, and

they calculated that one revolution of the moon around the Earth required about 29 days. They also determined the length of the celestial cycle (the year) by watching the pattern of the stars repeating itself every 365 days. From these observations the Chinese were able to fashion a calendar 2000 years before the birth of Christ.

The Chinese made star charts showing, quite accurately, the positions of the more prominent stars. By observing the motions described by these stars, they were able to recognize the polar star, and its position was used to align all the important buildings and temples in China in an exact north-south direction. The emperor's throne as well as the front door of each important house was made to face exactly south.

The Chinese were skilled in the ability to predict lunar and solar eclipses, quite possibly having been the first culture with this capability. Records are available from as early as 2200 B.C. showing the predictions made at that time. The Chinese considered an eclipse a celestial event of great importance: they believed it represented an attempt by a dragon to swallow the sun. The predictions were therefore considered essential so that the people could be marshalled for noise-making demonstrations to frighten the dragon away. So important was this that the emperors held astronomers in high esteem and maintained them in grand style. However, neglect of their duty could be fatal. About 2200 years ago court astronomers Ho and Hsi were beheaded for not accurately predicting the occurrence of an eclipse.

The Chinese kept accurate records of "guest stars," now known as novae and supernovae. Records of the passage of comets were also maintained, and from these records the earliest known passage of Halley's comet around the sun has been established for the year 240 B.C. There is an even earlier record for the year 467 B.C., but the observations made were not clear enough to establish the orbit accurately (Figure 2.1).

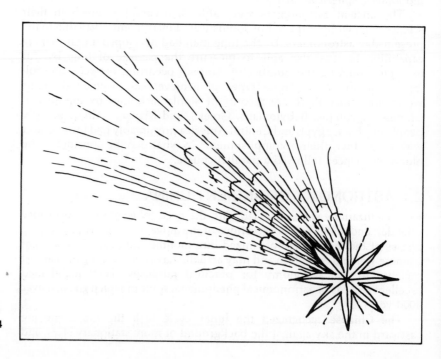

Figure 2.1 The oldest known drawing of Halley's comet from the Nuremberg chronicle 684 A.D.

Astronomy was greatly stressed by the Chinese, who used it to establish a system of references to the four cardinal points (NSEW) of the sky, with their center the polar star. China was placed in the center of the Earth—"the middle kingdom"—around which the Chinese conceived a square world. From this cosmology was derived a comprehensive world view, which in the 1000 years before Christ dominated Chinese thinking in all matters and to a limited degree affects Chinese thought today.

2.2 ASTRONOMY OF ANCIENT EGYPT

The Egyptians, like the Chinese, had a highly developed astronomy which became an important adjunct to their society. Whereas the Chinese emperor was called the Son of Heaven, the pharaoh was called the Son of the Sun in religious ceremonies. The ability to orient structures, such as pyramids and temples, with respect to celestial objects may have reached a higher degree of precision than that achieved by the Chinese. The 5000-year-old Great Pyramid of Cheops, for example, varied only slightly from the cardinal points, and its entrance shaft was pointed directly at the celestial pole marked at that time approximately by the star Alpha Draconis (Figure 2.2).

Time was of great importance to the Egyptians, especially the passage of the seasons. The first calendar, adopted in 4246 B.C., was composed of 12 months of 30 days each, using as a basis the lunar cycle and the annual passage of the stars. The priests who maintained the calendar noted that this measure of the year was short by 5 days, so they were added at the conclusion of the twelfth month. It was considered bad luck to work on the

Figure 2.2 Cross section of the Pyramid of Cheops looking west. The entrance shaft is pointed toward Alpha Draconis.

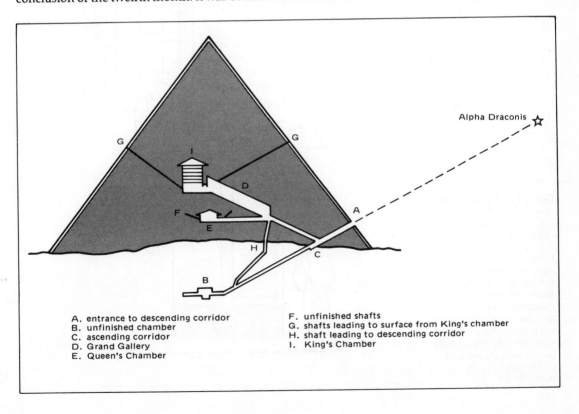

A. entrance to descending corridor
B. unfinished chamber
C. ascending corridor
D. Grand Gallery
E. Queen's Chamber
F. unfinished shafts
G. shafts leading to surface from King's chamber
H. shaft leading to descending corridor
I. King's Chamber

extra days; therefore, they were generally an occasion for feasting. A second calendar was maintained for agricultural purposes. Its purpose was to keep track of the annual flooding of the Nile, but this calendar was variable because the flooding did not occur exactly on schedule each year. The flooding of the Nile loosely coincided with the appearance in the spring of the bright star Sirius in the eastern horizon just before sunrise. For this reason the appearance of Sirius at this hour was used to predict the annual flooding of the Nile; however, the predictive accuracy was so poor, this may have been purely symbolic. Egyptians also subdivided the day into hours, using the sundial during the day and a water clock at night. The sundial was marked in uniform divisions, and since the sun passes overhead in an arc its shadow passed over the gradations of the dial, causing early morning and evening hours to be shorter than those hours in midday. This concept of varying lengths in hours as shown on the sundial became so fixed in the minds of Egyptians that they arranged their water clocks to register hours of varying duration (Figure 2.3).

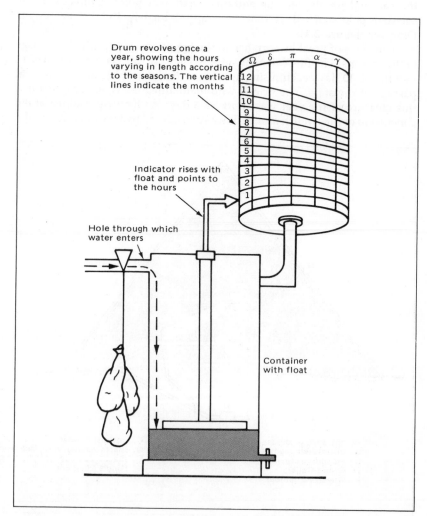

Figure 2.3 An Egyptian water clock. Water flowing into the cylinder raised the float holding the figure that pointed to hours on the drum. The hours varied in length according to the seasons. The vertical lines indicated the months, and the drum was rotated each month by a slave who was also responsible for the daily recharging of the water supply that operated the float.

2.3 STONEHENGE

Stonehenge, a collection of huge stone monoliths looking like a lonely and silent sentinel on the Salisbury Plain in southern England, is an enigma. Many legends as to its origin and purpose exist, and now another theory has been proposed which states that Stonehenge was built by early British inhabitants as a prehistoric astronomical observatory. Few records of Stonehenge exist, but the painstaking work by many specialists in archaeology and other sciences has gradually unraveled some of the mystery surrounding the monument. Stonehenge, it seems, was built between 3500 and 4000 years ago. The period of its construction spanned approximately 300 years and occurred in three distinct stages. The first stage of building was done by late British Stone Age people thought to have descended from farmers who had migrated from the Continent. The second stage was built by the Beaker people, so-called from their custom of burying pottery with their dead. And the final stage was built in the early Bronze Age by the Wessex people, who may have traded and communicated with civilizations in the Mediterranean region.

Construction of the first stage of Stonehenge did not involve the moving and placement of large stones, with the exception of the 35-ton "heel" stone. Initially, a circular ditch was dug, with the soil therefrom forming the banks on both sides of the ditch. Inside the inner embankment a series of 56 holes were dug and refilled. The "Aubrey" holes, as they are now called (after J. Aubrey who discovered them) are very regularly spaced, making a circle approximately 87¼ metres (288 feet) in diameter. Outside the entrance and about 30 metres from the circle the "heel" stone was erected (Figure 2.4).

The second constructive stage included the placement of a double circle of stones, with the stones of the inner circle aligned with those in the outer circle, resembling the spokes of a wheel. At least 82 bluestones, as they were called, were used in the arrangement which, for no apparent reason, was left unfinished. The bluestone, each weighing about 5 tons, appears to have come from a site approximately 225 kilometres from the monument. The original ditch was widened somewhat, as was the entrance, to facilitate the movement of these large stones into the monument area. An ancient roadway of sorts has been traced for some distance to a point near the Avon River, and may represent the route over which the stones were moved to Stonehenge.

In the third and final stage of construction the bluestone circles were dismantled, to be replaced by the large sarcen (sandstone) boulders seen today. Five "trilithons" were placed close to the center in a horseshoe-shaped arrangement. The trilithons, meaning three stones, consisted of

Figure 2.4 Stonehenge on the Salisbury Plains, England, as seen from the heel stone. (Bettmann Archiev)

two uprights capped by a lintel, each weighing in the vicinity of 40 to 50 tons. Some of the discarded bluestones were placed inside the arrangement of trilithons to approximate the horseshoe shape, but no lintels were used. Surrounding this was a circle of 30 upright stones completely capped with lintels approximately 25 tons each. Between this circle of stone and the Aubrey holes, two concentric circles of holes called the Y and Z holes were dug.

A variety of legends as to its origin surrounds Stonehenge. The Druids, a religious order, have laid claim to Stonehenge as a temple, and some believe it was built by early Druids. It has also been suggested that Stonehenge was the burial place of an ancient British king who had the monument built in much the same manner as the Egyptians built the pyramids. In mythology Merlin has been credited with stealing the monument from Ireland and by means of his magical skill spiriting it off to Salisbury Plain. The first known reference to Stonehenge being built as an astronomical observatory was by William Stukeley, a contemporary of Isaac Newton. Stukeley deduced that the monument was oriented toward the sunrise in midsummer, an idea supported by Dr. John Smith later in the eighteenth century. Now in the latter half of the twentieth century, Gerald S. Hawkins, with the aid of a computer, has been able to establish the distinct possibility that Stonehenge was used by the ancient Britons to keep track of important astronomical positions of the sun and moon and thereby forecast the occurrence of solar and lunar eclipses.

2.4 ASTRONOMY OF THE MAYANS

Half a world away from Egypt and Stonehenge, the Mayan civilization rose and flourished and died. All that is known about the Mayans is through inference. From their vague beginnings to their decline in 900 A.D., they left no record of their civilization except in their buildings, murals, pottery, and statues. This kind of evidence has shown that the Mayans expended great amounts of time on astronomical studies. Astronomy was the basis for the measurement of time and for their religious activities.

The Mayans believed that because time was cyclical, historical events would constantly be repeated. Careful analysis of the cycle, they thought, would enable them to predict the recurrence of some events. They also felt that time had no beginning and no end, although one of their calendars began at a point equivalent to 3111 B.C.

As did the Egyptians, the Mayans had several calendars. But, unlike the Egyptians, the temporal measurements made by the Mayans were quite precise. One calendar measured the passage of the traditional year and was composed of 18 periods of 20 days, each year having a terminal period of 5 days. Like the Egyptians, the Mayans believed that anyone who labored on those days would be inflicted with misfortune. The Mayans recognized the quarter-day period in addition to the 365 days which made a full year.

The second calendar was evidently a religious calendar because the 260-day period has no evident astronomical significance. The 260 may have been a base number similar to our 100. The third or long calendar was calculated from a mythical starting point of the Mayan era (3111 B.C.), although there is no archaeological evidence to indicate that the Mayans as a people existed at that remote time.

The calendar was a central part of Mayan life, playing a major role in their art and religion. Since the planet Venus was the second most

Calendar	Starting dates	
Jewish	3761 B.C.	Presumed date of Earth's creation
Greek	776 B.C.	Date of first Olympiad
Roman	753 B.C.	Date of founding of Rome
Islamic	622 A.D.	Year Mohammed left Mecca
Christian	Birth of Christ	
Japanese	660 B.C.	
Byzantine	5509 B.C.	

important deity after the sun, Venus's position was noted with great precision and human sacrifices were made when the planet appeared. Lunar eclipses, which they believed were caused by ants eating the moon, could also be predicted, as could the appearance of the planets. The planets represented minor gods to the Mayans which, if not properly appeased, could end the world.

2.5 ASTRONOMY IN ANCIENT BABYLON

The ancient Babylonians, Assyrians, and Chaldeans lived in what is now Iraq between the Tigris and Euphrates rivers, and their cultures were closely related. The Babylonians are considered to have been among the best astronomers of antiquity. Star charts made by the Babylonian astronomer-priests had names for *constellations* (stars grouped together) that are still used, having been passed to us from the Greeks who borrowed them from the Babylonians. In making the star charts the Babylonian astronomers noted that several celestial bodies, including the sun and the moon, had motions that differed from and were more complex than what were then considered the fixed stars. Because of this motion, these objects were later called *planets,* meaning "wanderers," by the Greeks. The Babylonians became adept at forecasting the location of the planets for any particular time. The sun and the moon were also considered planets. It was therefore possible for the astronomers to forecast the coming of an eclipse, an ability which, although not perfect, made them seem powerful forecasters of events in the future. The astronomers would frequently relate earthly events to events in the sky, and the movements and positions of the planets became important to some in the destiny of individuals and nations. For example, a tablet dated around 2000 B.C. states:

> When Ishtar [Venus] is visible in the east on the sixth of Abu there will
> be rain and destruction. Until the tenth of Nisan it remains easterly. On
> the eleventh it vanishes and remains invisible for three months. If then
> it glows in the west again on the eleventh of Du'uzu there will be
> fighting in the land but the fruit of the field will prosper.

In this manner astrology had its beginning—partly science and partly mysticism. Astrology in modern times still strongly influences the actions of many people.

2.6 OBJECTIVES OF ANCIENT ASTRONOMY

In even a brief review of ancient astronomy it is possible to see the pattern of astronomical activity and the emphasis placed upon celestial events by ancient peoples. Although relatively isolated from one another and having primitive communication methods, these societies developed their sciences along remarkably similar lines. They applied the knowledge gained toward practical purposes and needs, which may be enumerated as follows:

Time. The alternate daylight and darkness, the regularity of the lunar cycle, and the yearly passage of the stars were recognized and used to mark the passage of time. This was a valuable means of arranging human affairs on an orderly basis. It permitted merchants to meet commitments, governments to collect taxes at regular intervals, and farmers to time the planting of their crops and the breeding of their animals.

Religion. Objects in the sky were looked upon with religious reverence by ancient peoples. In Babylonia, the sign for God was a star. For many civilizations, such as the Egyptians and the Mayans, the sun was recognized as the principal source of light and heat, and for this reason it was worshipped. Many ancient gods and goddesses, prominent in mythology, were represented by various star constellations in the sky.

Direction. The rising and setting of the sun and the location of the polar star permitted travelers to maintain a sense of direction and to reach distant destinations by sea as well as by land. Navigation was dependent upon knowledge of the position of certain stars and still is today.

Prognostication. The ability to forecast the future stemmed from the success of the astronomer-priest in predicting events in the heavens such as eclipses. Celestial events of this nature were viewed by ancient people with superstitious awe and forecasting them gave the priests a very high standing in the society.

2.7 GREEK ASTRONOMY

The Greeks, being close to the Babylonian and Egyptian cultures, adapted much from their astronomies. However, they added another objective to the study of astronomy, not considered by previous societies, that of seeking knowledge for its own sake. Twenty-five hundred years ago the Greeks began to seek answers to astronomical questions for other than practical reasons. Mathematics and physics were applied in search of the true nature of the structure and form of the universe. To the Greeks the acquisition of knowledge was an end in itself.

A complete picture of the universe needs a version of the origin of the world. This was supplied by Hesiod (800 B.C.), a Greek poet, who described the Earth, formed from chaos, giving birth to the sky and the sea. Following this, Anaximander (600 B.C.), who no doubt was influenced by Hesiod, described the Earth as occupying a central position in the universe. He observed the movements of the constellations and concluded that the Earth was a cylinder with a height one-third of its diameter, floating freely in the void with the sky, spherical in shape, equidistant from the Earth on all sides. This was one of the most momentous conclusions of antiquity and probably gave rise to the *geocentric* or earth-centered hypothesis of the universe, a concept that was to influence astronomical thought for 2000 years.

Pythagoras

Pythagoras (580–500 B.C.), one of the principal Greek astronomers, was the founder of the Pythagorean Brotherhood, a religious order devoted to mathematical and religious contemplation. He is credited as being the first to suggest that the Earth was a sphere, a concept in keeping with the circular form of the universe. He was also no doubt influenced by the curvature of the Earth's shadow on the moon during a lunar eclipse.

The Pythagorean Brotherhood saw the universe as a perfect harmonious structure wherein the Earth was a stationary sphere, composed of the elements earth, water, fire, and air, sitting in the midst of the entire system. In orbits around the central Earth were the planets, including the moon, Mercury, Venus, the sun, Mars, Jupiter, and Saturn, each occupying its own sphere. The planets exhibited motions that differed from those of the stars, a fact previously determined by the Babylonians. According to the geocentric concept the planets revolved around the Earth in perfect circular orbits, each at its own particular velocity which was uniform throughout its entire orbit. The Pythagoreans considered the distance of each planet from the Earth to be a function of the planet's velocity—the lower the velocity of the planet the greater its distance from the Earth. The stars, it was thought, were fixed on an outer sphere, the celestial sphere, with the entire sphere revolving around the Earth (Figure 2.5).

This was the view of the universe at approximately 500 B.C., but continuing observations soon revealed certain apparent motions which were inconsistent with the supposedly harmonious arrangement of the system.

These motions or *inconsistencies* (inconsistent only with the established

Figure 2.5 Greek concept of the spherical universe. The motionless Earth is in the center, with the moon, Mercury, Venus, the sun, Mars, Jupiter, Saturn, and the stars moving in successive spheres around the Earth.

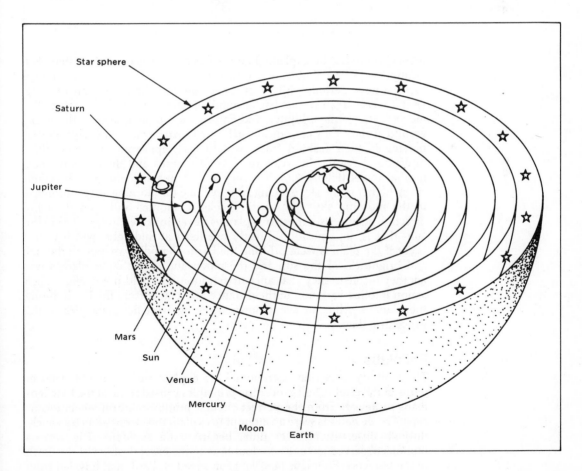

Figure 2.6 Retrograde
motion, an apparent
motion, is the result of
the difference in orbital
speeds between the
Earth and the outer
planets when these are
viewed from the Earth.

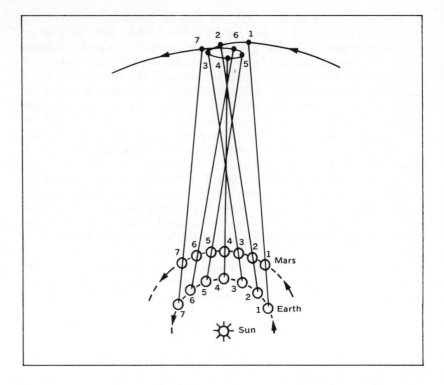

Figure 2.6 Retrograde motion, an apparent motion, is the result of the difference in orbital speeds between the Earth and the outer planets when these are viewed from the Earth.

theory) could not be explained on the basis of the geocentric system. For example, all celestial objects moved across the sky from east to west once each 24-hour period, while over longer periods the planets also moved west to east against the background of stars. In addition, when the movements of Mars, Jupiter, and Saturn were plotted against the background of stars, they periodically exhibited a *retrograde motion*. The planets, moving in their fashion from west to east, appeared to reverse their direction at regular intervals for several months and then resume their normal travel (Figure 2.6). Mercury and Venus also changed direction, rising higher in the sky evening after evening (or morning after morning if in the morning sky) for several weeks or months and then reversing direction again. Also, these two planets never appeared in *opposition* to the sun—that is, in a position directly overhead at midnight—which would occur if the planets were orbiting the Earth. Periodic variation in the size and brightness of the sun and the moon and of the other planets was another inconsistency that ancient astronomers could not explain. Such action implied variation in the distance of a planet from the Earth during the planet's orbit, an event not in keeping with the principles of the geocentric theory.

Philolaus

The inability of ancient astronomers to explain such inconsistencies in terms of the established geocentric principles caused some of the Pythagoreans to deviate from the perfect circular motion-uniform velocity concept. The deviations were an attempt to explain the inconsistencies which, through observations over time, began to be apparent. The various motions observed gave rise to doubts as to the exact position of the Earth in the universe. Philolaus (450 B.C.) conceived an Earth that traveled from

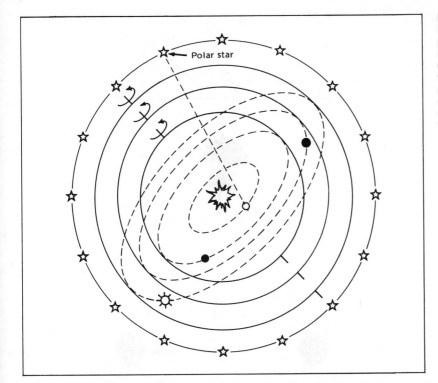

Figure 2.7 The Pythagorean Philolaus pictured the Earth revolving around a central fire each twenty-four hours. In this way he accounted for the daily east to west movement of celestial objects across the sky.

east to west around a "central fire." The central fire (never seen, according to Philolaus, since one side of the Earth constantly faced the fire) was on the side opposite to that on which Greece was located. Philolaus considered the central fire to be in the center of the universe and hidden from the Earth by a "counter earth" which kept pace with the Earth as it moved in its orbit (Figure 2.7). This arrangement of an Earth in motion accounted for the daily movement of the stars from east to west and placed the Earth with the other planets in a west to east orbit around a common center.

Heraclides

Heraclides of Pontus (350 B.C.) attempted to improve upon Philolaus's design by proposing that the Earth rotated on its axis. Such a motion would permit the Earth to remain at the center of the universe and still account for the east to west motion of the stars. Heraclides also grappled with another imperfection in the geocentric scheme. Mercury and Venus were always seen close to the sun and never seen traveling across the night sky as did the other planets. Heraclides suggested that these two planets revolved around the sun, thereby removing the Earth from its central position of all bodies in the universe.

Aristarchus

This idea was expanded upon by Aristarchus of Samos (280 B.C.), who saw the sun instead of the Earth as the stationary center of the universe. Accordingly, the Earth and the other planets orbited around the sun and the fixed stars were on a sphere surrounding the whole system. Aristarchus placed the Earth third from the sun. Those who objected to this arrangement argued that if the Earth were so far from the center of the system the stars would display parallactic displacement. *Parallactic displace-*

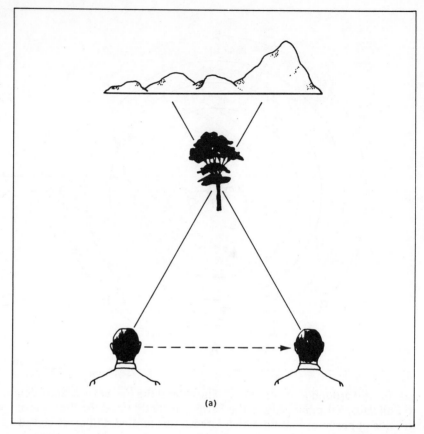

Figure 2.8 (a) An observer moving from one position to another will see a nearby object change position with respect to the more distant background. This is parallactic displacement, or parallax. (b) Stellar parallax results from the movement of the Earth around the sun, causing the apparent displacement of nearby stars with respect to more distant stars. Aristarchus was unable to demonstrate stellar parallax because of the great distances to stars.

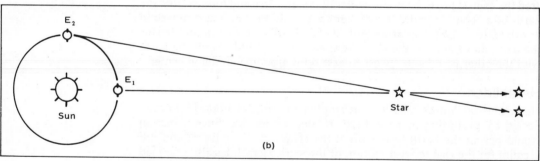

ment or *parallax* results when an object is viewed from two different positions (Figure 2.8). The object displays an apparent displacement against a more distant background. The fact that parallax could not be shown was not to Aristarchus's discredit but, rather, was due to the unavailability at that time of instruments and techniques of sufficient accuracy to measure extremely small angles.

Plato

Relatively few of the Greek philosophers were inclined to favor a deviation from the basic Pythagorean geocentric concept. Plato (427–347 B.C.) accepted the Pythagorean view of the cosmos in its entirety, consid-

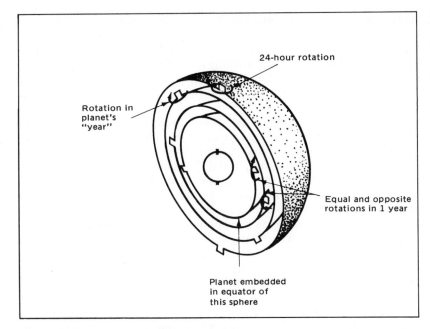

24-hour rotation

Rotation in
planet's
"year"

Equal and opposite
rotations in 1 year

Planet embedded
in equator of
this sphere

Figure 2.9 Eudoxus
utilized four concentric
spheres to simulate the
motion of a planet. In this
sketch the two innermost
spheres spun in equal
but opposite directions in
one year to produce
retrograde motion; the
third sphere rotated once
in the planet's year; and
the outermost sphere
turned once each twenty-
four hours to provide the
planets' motion around
the Earth.

ering nature as a perfect entity. Since the sphere is a perfect three-dimensional shape, he reasoned that the universe must be spherical and all motions uniform and circular. He suggested to his students that they find an arrangement whereby uniform circular motion would explain the apparent and somewhat irregular motions seen in the universe.

Eudoxus

Eudoxus (370 B.C.), perhaps the first authentic Greek astronomer, devised a mathematical model of the system wherein each planet was controlled by a series of concentric spheres. Each sphere, centered on the stationary Earth, rotated on an independent axis, thereby providing the planet with its apparent motion (Figure 2.9). Each of the five planets required four spheres, the moon and the sun each required three, and one sphere accounted for the movement of the stars. The 27 spheres imitated the motion of the planets quite well, although the mathematics involved was far from simple. However, with more accurate observations it soon became apparent that the arrangement was not yet perfect, and Eudoxus's pupil, Callipus, refined the system by adding an additional sphere for each of the five planets and the sun and moon for a total of 34 spheres. This system, which Eudoxus did not consider real but only a mathematical model, was simple and mathematically correct, and conformed to the observed motions of the planets as prescribed by the Pythagoreans.

Aristotle

Aristotle (384–322 B.C.) perhaps more than any other scholar during the classical period did much to influence thinking in favor of the geocentric concept of the universe. His influence may have been largely responsible for Aristarchus's sun-centered idea being rejected.

Aristotle favored the Pythagorean view of a spherical Earth, both on theoretical and observational considerations. He believed that a sphere had perfect symmetry and the Earth, in order to conform to the symmetrical nature of the universe, must be a sphere. Also, he reasoned, the Earth's component pieces would fall naturally towards the center and would be pressured into a round form. Aristotle was able to observe that the Earth's shadow on the moon was circular during a lunar eclipse, and found that even relatively short distances north or south changed the position of star patterns.

Aristotle accepted the organization of the system as viewed by Eudoxus, but believed the spheres to be real physical entities. Aristotle was concerned that the motion of the spheres of one planet would be transferred to another, so he posited "unrolling spheres" to negate the motion between one planet and the next. This entire system was, according to Aristotle, controlled by a prime mover located at the edge of the universe and by lesser movers which regulated the motion of the individual planets. The prime mover moved the entire system east to west and the lesser movers were responsible for moving the individual planets in the opposite direction. Aristotle stated that the influence of the prime mover was such that the outer planets moved slowly and the inner planets, being farther away from the prime mover and therefore less influenced, moved more rapidly in orbit.

Aristotle's tremendous authority as a scholar and influence as a teacher were important factors in elevating the geocentric theory to prominence during the classical period. The theory was accepted as true for about 2000 years, along with Aristotle's many other scientific and philosophical ideas. He made a distinction between theories devised to "save the phenomena" and theories based on true physical causes. Unfortunately, it appears that Aristotle subscribed to the former where the geocentric concept was concerned.

Hipparchus

The inconsistencies inherent in the Pythagorean system were not solved even with the addition of mathematical spheres by Eudoxus or by their elevation to reality by Aristotle. The spinning concentric spheres made the model more complex than previous models, and further attempts were made to simplify the system. Hipparchus (2nd century B.C.), noted as an astronomer and mathematician, found that the sun's uneven motion in orbit could be easily explained by assuming the sun moved around a circular orbit, the center of which was fixed a short distance from the Earth. He also worked out epicyclic systems for the sun and the moon. The *epicycle* is a small circle, whose center moves around the circumference of a larger circle called the *deferent*. Hipparchus was also responsible for making a very accurate star chart, and when he compared his work with earlier efforts he discovered the precession of the equinoxes (see chapter 7), a motion unexplained until the time of Isaac Newton.

Ptolemy

Ptolemy (2nd century, A.D.) set himself the task of reviewing all the records of planetary motion. Using Hipparchus's works and with many observations of his own, he devised a machine that used mechanical principles for duplicating planetary motion. He abandoned the use of spheres and achieved a more efficient representation of planetary motion by use of mechanical contrivances that described eccentric circles and epicycles. Ptolemy basically followed the Pythagorean model, with the earth as a

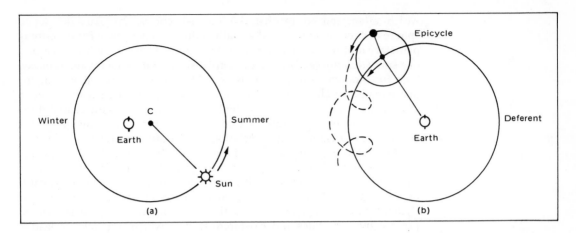

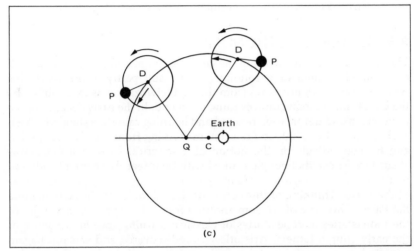

Figure 2.10 (a) Variation in distance of the sun from the Earth could be explained by means of an "eccentric," a device in which the center of the sun's circular orbit was placed a short distance from Earth. (b) In the epicycle scheme a planet moves around the epicycle, a smaller circle whose center follows the deferent. The deferent is a larger circle with its center located at the Earth. This device enabled Hipparchus to show retrograde motion. (c) Ptolemy's device had a planet P revolve around D at uniform speed with respect to Q. Q is the equant which maintained a constant distance from C. The motionless Earth appeared opposite C from the equant.

sphere at the center of the system and the sun and moon shown moving around the Earth with relatively simple epicyclic motion. The five planets required more complex schemes, and it was necessary not only to remove the Earth from the center of the system but also move the center of planetary motion an equal distance out on the other side to a point called the "equant" (Figure 2.10). Although complicated, the arrangement functioned to provide accurate predictions of planetary motion and position for years into the future and accounted for those long back into the past. Ptolemy did not consider that his device duplicated real motion but only that it was a mechanical means of describing the workings of the universe.

Ptolemy's great work, *The Almagest,* was the most comprehensive astronomical treatise of antiquity. In it he described in detail the functioning of his devices and provided tables which gave the motion of each of the planets. The work became the basic astronomical authority for the next 1400 years, being translated from Latin to Arabic and back into Latin again as the centers of learning moved east and then returned to Europe in the Middle Ages.

From the foregoing it is possible to see that the Pythagorean concept, although basically maintained, was subject to a series of modifications which attempted to resolve the inconsistencies. Eudoxus developed spheres which he considered a mathematical convenience for describing planetary motion, whereas Aristotle later judged the spheres to be real and of a soft, frictionless material. Ptolemy discarded the spheres and replaced their function with mechanical devices. His system was based on complex mathematics and had for him no physical reality. Aristarchus attempted to change the system by replacing the Earth with the sun at the center of the system. This was generally found unacceptable, since the geocentric concept of Pythagoras was psychologically more attractive.

Although to the modern mind it is full of absurdities, the Pythagorean concept was a valuable contribution to the early development of cosmology. It expressed the idea that the universe is an orderly structure capable of being grasped and expressed in the language of mathematics. The Pythagorean concept also paved the way for other hypotheses as expressed by Heraclides and Aristarchus, ideas which were later to lead to the epoch-making work of Copernicus in the sixteenth century.

2.8 MEDIEVAL ASTRONOMY

The medieval period, covering the centuries from about 400 to 1500 A.D., contributed nothing to astronomy. As a result partly of the successive waves of barbarian invasions from the east, civilization as exemplified by the Greek and Roman cultures came to an end. In the early decades of the period ruin and devastation reigned and learning came to a standstill. Men such as Lactantius (300 A.D.) rejected everything from the classical period and heaped ridicule on the notion of a spherical Earth. He felt it was absurd to believe that people walked with their feet above their heads and that mountains, seas, and fields hung without support from the other side of the Earth. Thinking of this sort contributed to the medieval belief in a flat Earth. This type of wisdom endured for centuries but with the rise of the monasteries as repositories of classical learning and the influence of the early church fathers, Aristotle was rediscovered and was once again accepted as an infallible guide to knowledge and wisdom. During this period the Arabic world, while not adding greatly to the store of knowledge, also did much to preserve what had existed, and this ultimately was passed on to an awakening intellectual community in Europe. However, new insights as to the form of the universe did not arise with a renewal in the interest for learning. The Pythagorean geocentric concept prevailed. The need for spheres to accomplish planetary motion as advanced by Aristotle competed with Ptolemaic eccentrics and epicycles, but the Aristotelian authority was sustained.

Thus it was that the Italian poet Dante, in the fourteenth century, followed the Aristotelian prescription in describing the universe. The stationary Earth was composed of the four elements earth, water, air, and fire. Earth and water possessed gravity and had a tendency to fall, while air and fire were characterized by levity and had a tendency to rise. The fact that air and fire never escaped the Earth was due to their being filled with impurities which prevented them from achieving their proper sphere.

Dante described the skies as being formed of an incorruptible fifth element not subject to change and decay as were the four earthly elements. Ten skies or spheres surrounded the Earth, each transparent but sufficiently solid and real to contain one of the celestial bodies as it rotated about the stationary Earth. The innermost sphere, nearest the Earth, contained the moon. The next six spheres carried the sun and the five planets and the eighth sphere held the stars. A ninth sphere contained no planets or stars and had no visible signs to indicate it existed. However, it was considered existent because it was the *primum mobile* or prime mover— the motivating force that turned all the other spheres or skies in the heavens. Surrounding this was the tenth sphere—the Empyrean Heaven— the abode of God. This sphere was stationary. Thus cosmological thinking in the fourteenth century had reverted back to the Aristotelian geocentric concept of spheres to explain the structure of the universe, and it was this concept that Copernicus challenged in the first half of the sixteenth century.

2.9 SUMMARY

Astronomy is probably one of the oldest of sciences, having been of interest to ancient man throughout historical times. Each civilization as it came into being developed some form of cosmological theory, in part for religious purposes, for maintaining some form of calendar, for navigation, or for the purpose of foretelling the future. Despite the relative isolation of the early peoples—Chinese, Egyptians, Mayans—their uses of astronomy were quite similar.

The Greeks appeared to be the first to organize the universe into some semblance of order, providing it with a structure which had as its center a spherical, stationary Earth. The organization was such that the motion and positions of the celestial objects could be expressed in mathematical terms. The geocentric system was imperfect and resulted in inconsistencies that could not be explained on the basis of such a cosmological arrangement. Some of the ancient astronomers attempted to resolve the inconsistencies by deviating from the earth-centered concept, and one astronomer went so far as to replace the Earth with the sun as the central object. Others made use of artificial devices such as spheres (which some of the ancient astronomers came to believe were real), eccentrics, and epicycles. These devices were useful for the most part in mathematically describing apparent motions but did not resolve the problem of the inconsistencies.

No significant knowledge was added to astronomy during the medieval period. It was a period ushered in by chaos followed by the gradual restoration of order and the pursuit of learning. By the fourteenth century astronomical knowledge and theories had advanced little beyond what was taught during the time of Aristotle 1700 years earlier.

QUESTIONS

1. What were some of the astronomical phenomena recognized by ancient astronomers prior to the rise of the Greek culture (circa 600 B.C.)?
2. In summary, what were the objectives of ancient people in studying astronomy?
3. What additional objective did the Greeks have for studying astronomy that was not considered by previous astronomers?

4. Describe the evolution of the geocentric concept of the universe as developed by the Greeks.
5. What was the view of the universe as seen by the Pythagoreans?
6. What inconsistencies in celestial motion were inherent in the geocentric system as viewed by ancient Greeks?
7. Trace the evolution of the sun-centered concept of the universe that culminated in Aristarchus's view of the system.
8. How did the Greek astronomers attempt to resolve the inconsistencies seen in the geocentric system?
9. What was the fourteenth-century view of the geocentric system as described by Dante?

EXERCISES AND PROJECTS

Using your imagination and ingenuity as the ancient Egyptians must have done, build a sundial that will give a reasonable estimation of the correct time. As you work on the project make note of the problems that arise and what you did to overcome them. Be prepared to present this in class. The construction and report may be a group effort with input from all participants.

FOR FURTHER READING

HAWKINS, G. S., *Stonehenge Decoded.* New York: Delta, 1965.
LOCKYER, J. N., *Dawn of Astronomy.* Cambridge, Mass.: M.I.T. Press, 1964.
THIEL, R., *And There Was Light.* New York: Mentor Books, 1957.
VON HAGEN, V. W., *Ancient Sun Kingdoms of the Americas.* New York: World Press, 1961.
YOUNG, L. C., ed., *Exploring the Universe.* New York: McGraw-Hill, 1963.

Aries

CHAPTER 3

MODERN ASTRONOMY

Open wide the door for us, so that we may look out into the immeasurable
starry universe; show us that other worlds like ours occupy the ethereal
realms; make clear to us how the motion of all worlds is engendered by
forces; teach us to march forward to greater knowledge of nature.

—*Giordano Bruno (1548–1600)*

Classical astronomy and the study of astronomy as it extended into the
Middle Ages was based essentially on three suppositions: the earth-
centered concept of the universe, the principle that the universe was
limited by spheres of fixed stars, and a dualistic physics wherein there was
a fundamental distinction made between the nature of terrestrial matter
and celestial matter. These beliefs came under increasing attack in the
latter part of the Middle Ages as observational evidence and a new
approach to viewing the cosmos showed them to be erroneous. New
discoveries were made and fresh conceptual schemes emerged which laid
the groundwork for the modern view of astronomy. It was in such an
atmosphere that Copernicus, in the early sixteenth century, challenged the
old ideas and loosened the hold of a geocentric cosmology upon philosophical thought.

3.1 COPERNICUS (1473–1543)

There is never a clear distinction between one epoch in science and the
next. However, as a convenience we may consider that Copernicus
ushered in the modern era of astronomy when he introduced the sun-
centered or heliocentric theory of the universe.

Copernicus—the latinized form for Niklas Koppernigk—was born in
Poland during an era of awakening intellectualism, a condition stimulated
by the invention of the printing press in the mid-fifteenth century. It was
an era when such men as Columbus, Michelangelo, and Leonardo da Vinci
were extending the physical and intellectual boundaries of the world. To
be a student in those times was an exciting experience, and Copernicus,
orphaned at ten and raised by an uncle who planned a career in the church
for him, became what in modern parlance is known as a professional
student. He studied at Crakow, Bologna, Rome, Padua, and Ferrara for
fifteen years before the death of his uncle in 1512 required Copernicus to

35

assume his duties as the canon of Frauenberg. During his stay at the University of Crakow Copernicus became interested in astronomy and mathematics, and as he carried out the many duties connected with his clerical office he began to think very deeply about the organization of the universe.

Copernicus reviewed the many works of astronomy going back to the ancient Pythagoreans and found, to his dismay, that numerous adjustments and changes were required through the centuries to keep the earth-centered system in phase with observational evidence. He found that errors, which supposedly had been corrected or accounted for by use of such devices as the equant and epicycle, were constantly recurring. Copernicus expressed the opinion that "... *a system of this sort seemed neither sufficiently absolute nor sufficiently pleasing to the mind.*" Referring to classical works, Copernicus found that a number of astronomers in the past had ascribed motions to the Earth which could account for some of the observed movements in the sky. Philolaus and Heraclides had suggested deviations from the geocentric concept, and Aristarchus had removed the Earth from the center of the system and replaced it with the sun. Copernicus felt that such an arrangement might have merit since motion, he must have reasoned, would occur regardless of whether the observed object moves or whether the observer moves with respect to the object. Thus the Earth moving around the sun would give the same appearance as the sun moving around the Earth.

Further reasoning and observation suggested to Copernicus that a *heliocentric* or sun-centered concept would represent a much simpler arrangement than the geocentric system. Copernicus's contribution here was not to a new cosmological theory but rather to a theory of planetary motion. He placed the sun in the center and arranged the planets in the order of their periods of revolution. The moon, in this system, lost its status as a planet and took up a position as a satellite of the Earth. Mercury, with the shortest period of revolution, was closest to the sun with Venus next in order and the Earth, now a planet, placed third from the sun. Beyond the Earth, Mars, Jupiter, and Saturn followed in that order. Surrounding this arrangement, and at great distance, was the motionless sphere containing all the stars. The nightly movement of the stars across the sky and the journey of the sun from east to west was attributed by Copernicus to the rotation of the Earth on its axis (Figure 3.1).

Copernicus did not completely discard all the trappings of the geocentric theory. He maintained the motions of the planets in perfect circular orbits and at uniform velocities around the sun. For this reason it was necessary for him to retain a few of the epicycles, but on a much reduced scale from that used by Ptolemy. Copernicus did find that the heliocentric arrangement would explain the inconsistencies experienced by the Greeks with the geocentric system. The daily motion from east to west of all objects in the sky could be ascribed to the rotation of the Earth, and the slower west to east motion of the planets was laid to the movement of the planets in orbit around the sun. The fact that Mercury and Venus never appeared in opposition could now be seen to result from their position between the Earth and the sun, and the retrograde motion of the outer planets could be viewed as an apparent motion resulting from the difference in velocity with which the Earth and the outer planets traveled around the sun.

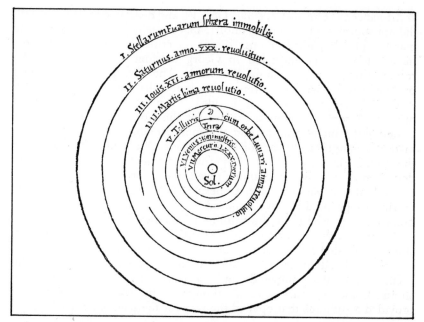

Figure 3.1 The heliocentric system as described by Copernicus in which all observed celestial motions could be accounted for by substituting the sun for the Earth at the center of the system. (Yerkes Observatory photograph)

The simplicity of the new system and the resolution of the inconsistencies did not result in the immediate acceptance of the heliocentric theory. On the contrary, there was much resistance from both the academic and religious community to the Copernican ideas, primarily because they did not conform to the Aristotelian concept of the universe—a concept that had become an integral part of intellectual thought during the ensuing centuries. Furthermore, Copernicus could not prove that his system was correct and that of Aristotle and Ptolemy incorrect. He could only claim that his version represented a simpler arrangement capable of resolving some of the difficulties experienced with the geocentric system. Opponents of Copernicus also argued that placing the Earth at such a distance from the sun would permit the nearby stars to exhibit parallactic displacement—a fact not in evidence. Copernicus's explanation that parallax could not be demonstrated because of the enormous distances to the stars was not acceptable to people who were generally incapable of visualizing such a distant Heaven while Hell was so near underfoot.

Other objections to the heliocentric system were based on the belief that a rotating Earth would fly apart as a result of centrifugal force. Copernicus argued that the far vaster sphere holding the fixed stars would be more susceptible than the Earth to such a force. However, Aristotelian thought ascribed to the heavens special properties which were immune to such earthly forces as centrifugal force. Ptolemy, in reviewing the rotating-earth idea suggested earlier by Heraclides, stated that such a motion would result in a high wind blowing constantly from the east or that objects thrown in the air would never return to the same place. The same argument was used against the Copernican system, and Copernicus could only counter that the air was an integral part of the Earth and rotated with it.

Such reasoning delayed the general acceptance of the Copernican system for over a century after the theory was first published. The entire work describing the system (six books), entitled *Nicolai Copernici Torimensis de Revolutionibus Orbium Coelestium Libra VI*, and dedicated to Pope Paul III, did not appear in printed form until 1543. For many years prior to this the manuscript was passed among friends for criticism, and late in Copernicus's life it was entrusted to Georg Joachim who undertook the responsibility for publishing the work. The first printed copy was presented to Copernicus a few hours before he died. Another friend, Andreas Osiander, afraid that the idea of a moving earth might offend many philosophers, wrote a preface for the work. In it he stated that Copernicus's intent was to find the most suitable manner for mathematically computing the motion of the planets. Opponents later pointed to this preface claiming that even Copernicus did not believe that his theory described the real world. Although well-intentioned, Osiander's preface was actually at variance with Copernicus's beliefs. The original manuscript disappeared for two hundred years, but when rediscovered, the preface was proved to be the addition of another author.

Ultimately the importance of the Copernican system was recognized, since it provided a truer picture of the universe than did the geocentric concept and led to an explanation of other phenomena such as the occurrence of the tradewinds and the cause of tides. It did much to stimulate and revitalize scientific thought in the sixteenth century and provided a stimulus that opened the way for the scientific revolution.

3.2 TYCHO BRAHE (1546–1601)

Tycho Brahe, born in Denmark of noble parentage, aspired to a career as an astronomer at an early age. After studying mathematics and astronomy in Switzerland and Germany, he accepted an appointment as astrologer to the Court of King Frederick II of Denmark and was made governor of Hven, a Danish island in the Baltic Sea. It was on Hven that he built his magnificent observatory, a monument to his organizational ability and his patron's gold.

Tycho Brahe was not a theoretician. Rather his genius lay in his ability to make observations and plot the motions of the planets to a degree of accuracy never before achieved. He was able to measure the positions of the stars to within four minutes of arc (0.067°) of the modern values as compared to twenty minutes of arc (0.33°) achieved by Ptolemy or ten minutes of arc (0.167°) by Ulugh Beigh, astronomer of Samarkand in the fifteenth century. Brahe accomplished this without the benefit of the telescope (invented after his death) and without the development of new instruments. He reproduced instruments used by the ancient Greeks but improved their accuracy by ingenious methods of scaling and sighting and added innovations which further improved their performance (Figure 3.2).

Realizing that the Copernican tables were computed from old observations made with instruments of less accuracy than his own, Tycho Brahe set himself the task of plotting anew the positions of the important stars, and in 1592 published a catalogue of stars which became the standard reference for that time. With the star charts he was able to observe the motions of the sun, moon, and planets and calculate the elements of their orbits with a high degree of accuracy. Early in his efforts Brahe discovered that previous observations suffered from an error induced by the refraction (bending) of light as it passed through atmospheric layers of differing densities. He corrected for this, thereby greatly improving the precision of his astronomical data.

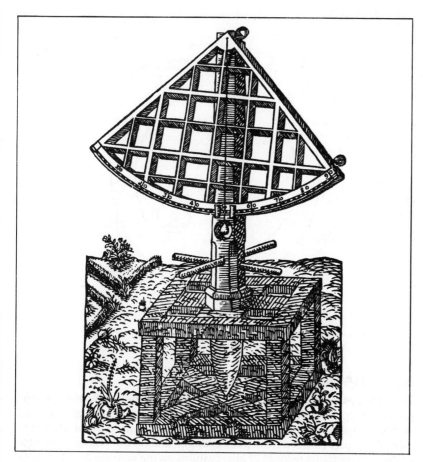

Figure 3.2 A quadrant reproduced from Tycho Brahe's *Astronomiae Instauratae Mechanica*, published in 1598. (Bettmann Archive)

Tycho Brahe was dissatisfied with the Copernican theory, and part of his purpose in accumulating accurate data on planetary motion was to bolster his geostatic system (Figure 3.3). In the geostatic system the sun orbited around a stationary Earth. The moon was the Earth's satellite and the planets revolved around the sun. In this way Brahe was able to retain the Earth in its central position as prescribed by the ancient Greeks, and also satisfied the conditions of observed planetary motion. Brahe did not believe in the existence of the planetary spheres which were so much a part of the Pythagorean system. From his study he was able to show that comets were celestial objects, and that the spheres could not physically exist because the comets passed without interference through that region of space supposedly occupied by spheres.

In 1597, Christian IV, who succeeded Frederick II, ousted Brahe from his favored position in court, and Brahe moved to Prague to become court astrologer to Emperor Rudolph II. It was here that Brahe met Johannes Kepler, a mathematician and theoretician, and took him on as an assistant. Kepler was principally interested in the great amount of accurate data on planetary motion that Brahe had accumulated over many years. This data, perhaps Brahe's most significant contribution to astronomy, enabled Kepler to establish his new celestial mechanics.

Figure 3.3 Tycho Brahe's geostatic system with all planets except Earth orbitting the sun and the sun revolving around a stationary Earth.

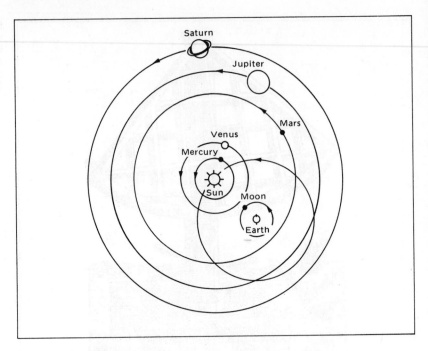

3.3 JOHANNES KEPLER (1571–1630)

For Johannes Kepler, joining forces with Tycho Brahe was a stroke of good fortune. Kepler believed in the Copernican order of the planets, but demonstrating it mathematically, a task he set for himself, would be very difficult. Kepler suffered from poor eyesight, and for this reason he needed Brahe's voluminous records of observations on the motions of the planets, particularly of Mars.

While a student at Tubingen, Kepler wrote his *Mysterium Cosmographicum* (Mysterious Universe) wherein he described his views on the relationship of the planets by fitting the five Pythagorean figures between the orbits of the six planets (Figure 3.4). The Pythagorean figures, recognized in antiquity, included the cube, tetrahedron, dodecahedron, icosahedron, and octahedron and were the only solid figures possible with identical, equilateral faces. Kepler found that by placing the figures one inside the other he could demonstrate the relative spacing of the planets. He felt that any differences could be accounted for by observational error, and that by using Tycho Brahe's data he could show this relationship to be true.

One of the first problems Kepler encountered was that the values Brahe had obtained for the orbit of Mars would not fit the uniform circular motion prescribed by the Copernican theory. There was a discrepancy of eight minutes of arc between the data and a perfect circle, an amount too large to be accounted for by observational error. After years of tedious work, Kepler came to the realization that the planets did not orbit the sun but rather a point offset from the sun—a fact recognized earlier by Copernicus but largely ignored. Kepler eventually hit upon the ellipse as the proper shape of the orbits and from this was able to state his first law of motion.

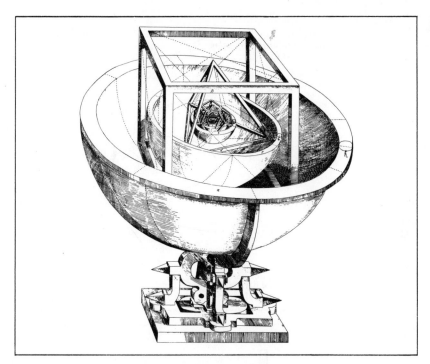

Figure 3.4 The five Pythagorean figures were used by Kepler to space the orbits of the six planets (1596).

First Law: The planets revolve around the sun following elliptical orbits with the sun at one of the focal points of the ellipse.

An *ellipse* is a conic section, that is, a geometric figure formed by a plane cutting a cone at an angle to the axis of the cone (Figure 3.5). The curve of the ellipse is such that the sum of the distances from the focal points to any point on the curve is a constant value. The degree of eccentricity of the elliptical orbits of planets is quite small in most cases, and if the orbits were drawn to scale it would be difficult to distinguish them from circles.

Continuing this line of reasoning, Kepler came to the conclusion that the planets did not follow their orbital paths at uniform speeds. Rather the planets speeded up when coming closest to the sun at *perihelion* and slowed down when farthest from the sun at *aphelion*. Combining this idea with that of elliptical orbits, Kepler was able to formulate his second law of motion.

Second Law: An imaginary line from the center of the sun to the center of a planet sweeps over equal areas in equal time periods.

This means that a planet travels from A to B in the same period of time required for it to travel from C to D (Figure 3.6). However, the imaginary line will cover equal areas because of the differences in the planetary distances from the sun at these two locations.

Kepler's first two laws, supported by observational data and worked out with mathematical rigor, were announced in 1609 and confirmed the validity of the Copernican system. The laws permitted the description of planetary motion without the aid of mechanical devices such as the epicycles used by Ptolemy. At the same time the laws effectively elimi-

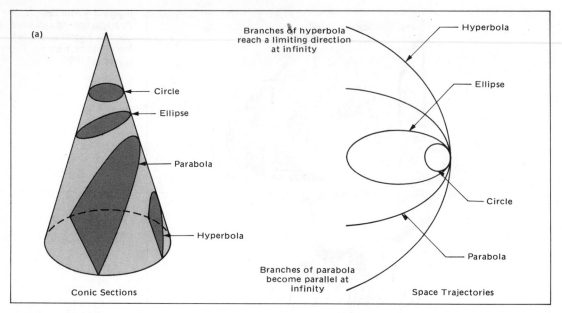

(a)

Circle

Ellipse

Parabola

Hyperbola

Conic Sections

Branches of hyperbola reach a limiting direction at infinity

Hyperbola

Ellipse

Circle

Parabola

Branches of parabola become parallel at infinity

Space Trajectories

Figure 3.5 (a) Conic sections define the orbits of planets and satellites permissible under the law of gravity. (b) An ellipse may be constructed with tacks, string, and pencil. An ellipse is a conic section in which the sum of the distances from the focal points to any point on the circumference of the ellipse is a constant value.

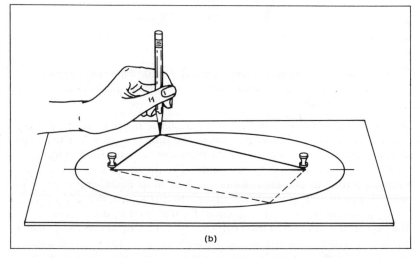

(b)

nated the ancient concepts of circular orbits and uniform motions of the planets, the last vestige of a now thoroughly discredited geocentric theory.

For the next decade Kepler labored to develop a relationship between a planet's period of revolution and its distance from the sun. He had no evidence, only an intuitive belief that such a relationship existed. To accomplish his purpose it was first necessary to determine the distance of the several planets from the sun. He used the distance of the Earth to the sun as the basic unit of measure (called *astronomical unit* by Cassini at the end of the seventeenth century). Kepler knew that Mars returned to the same position in orbit every 687 days. This is the sidereal period of Mars, or the length of time it takes a planet to make a complete 360-degree orbit. Kepler could not measure this directly but he could measure the *synodic period*, which is the time it takes for an inner and faster planet to gain one lap on an outer planet. In other words, if the Earth and Mars were to be

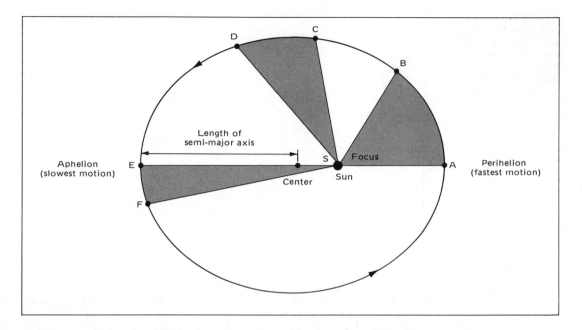

aligned with the sun, the Earth would make two complete laps plus a small portion of an additional lap before they would again reach this position of alignment. The synodic period can therefore be measured directly, and from this it is possible to calculate the sidereal period by means of the following equations.

Figure 3.6 Kepler's law of areas. An imaginary line from a planet to the sun will sweep over equal areas in equal time periods. Therefore the area *SBA* = *SCD* = *SEF*.

For a planet between the Earth and the sun (inferior planet):

$$\frac{1}{P_{sid}} = \frac{1}{E_{sid}} + \frac{1}{P_{syn}}$$

For a planet beyond the Earth from the sun (superior planet):

$$\frac{1}{P_{sid}} = \frac{1}{E_{sid}} - \frac{1}{P_{syn}}$$

where P_{sid} is the sidereal period of the planet, P_{syn} is the synodic period of the planet and E_{sid} is the sidereal period of Earth (365.25 days). We can now calculate the sidereal period for Mars whose synodic period is 780 days. Since Mars is a superior planet (see chapter 6)

$$\frac{1}{P_{sid}} = \frac{1}{365.25} - \frac{1}{780} \quad \text{or} \quad \frac{1}{P_{sid}} = 0.001456$$

Thus P_{sid} equals 687 days for the sidereal period of Mars. During this period, Earth had completed one revolution plus the major portion of a second.

If lines are drawn from the two positions of the Earth (Figure 3.7) E_1 and E_2 to the sun S and Mars M, the relative distance of Mars to the sun

Figure 3.7 After first calculating the sidereal period of Mars, Kepler was able to determine the relative distance (Earth's distance to sun equals unity) of Mars from the sun.

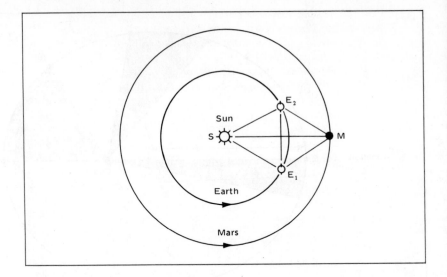

and the Earth to the sun can be geometrically calculated. The difference between two Earth years of 730½ days and one Martian year of 687 days is 43½ days, which is the segment of the orbit represented by arc E_1E_2. Angle E_1SE_2 is equal to arc E_1E_2, and the distance E_1S is equal to the distance E_2S, since these are the radii of the Earth's orbit and are approximately equal. With this kind of information Kepler was able to solve the triangles and calculate the relative distances of the planets to the sun. Now he could determine the relationship between distances and periods that he sought and he constructed his third law of motion.

Third Law: The squares of the orbital periods (P) of any two planets are proportional to the cubes of their mean distances from the sun (a).

The third law may be expressed in the following form.

$$\left(\frac{P_1}{P_2}\right)^2 \quad \propto \quad \left(\frac{a_1}{a_2}\right)^3$$

If the Earth is one of the two planets being considered (the usual case), then the following relationship exists

$$P^2 = Ka^3$$

where K is a numerical constant. Where the Earth's period of one year and its distance of one astronomical unit is used, K will have a value of 1 yr²/1 AU³.

The significance of Kepler's three laws of planetary motion can be seen when it is realized that the laws continue to be acceptable even after 350 years. For example, not only is it still possible to calculate planetary motion with Kepler's laws, but with minimal modification, the laws are used to determine the orbits of manmade satellites. If one wishes to know the period of a satellite in orbit around the Earth, it would first be necessary to recalculate K, from Kepler's third law, for an Earth-bound orbit. This can be accomplished by using the moon's period in days and its distance in kilometres, since the moon may be considered a satellite of the Earth with orbital characteristics similar to that of an artificial satellite. The constant K would then have a value in terms of days^2/kilometres^3. If the distance of the satellite from the Earth's center (radius of the Earth plus distance of the satellite above the Earth's surface) is known, the period of the satellite's orbit can be determined.

The importance of the laws goes beyond mere durability, however. With the knowledge of Kepler's third law, Isaac Newton was able to derive his law of universal gravitation. This was one of the most substantial and far-reaching discoveries in astronomy, and Kepler's contribution may have been one of the things Newton had in mind when he stated: "If I have seen further than others, it was because I stood upon the shoulders of giants." Kepler's three laws are also important because they mark the first time in the history of science that physical laws were constructed from data collected previous to the formulation of the laws.

3.4 GALILEO GALILEI (1564–1642)

Galileo, one of the few scientists known by his first name, was born and educated in Pisa, Italy, and later became a professor of mathematics at the University of Padua. He was the envy of his colleagues because of the large groups of students that he attracted to his lectures. Galileo was an innovator in the art of introducing students to new concepts through visual demonstration. For example, he showed that sound travels at measurable speeds by firing a cannon from a distant hill. The flash was seen instantaneously. The length of time required for the sound to reach the observer was measured and divided into the distance to give the speed of sound. Also, by using animal bones, he showed that hollow pipe-construction saved materials without sacrificing strength. The demonstration approach to teaching was far superior to the rote learning that was the normal method of instruction and to the theorizing for or against the ideas of Aristotle that prevailed at the time.

In 1609, Galileo learned of the development of the telescope by Hans Lippershey of the Netherlands and immediately set to work building his own instruments. By experimenting with lenses of different power he was eventually able to reach a magnification of thirty diameters. At this magnification the church tower which he sighted through his telescope would jump out of his field of view with the slightest movement.

Galileo has the distinction of being the first man credited with using the telescope for celestial observation, and the first object he viewed was the moon. He found it not smooth, as was previously thought, but mountainous, marked with great craters and smooth dark areas which he thought were seas. From shadows cast he was able to calculate that lunar mountains were comparable in height to those on Earth. He looked at the stars and found that these were not magnified because of the great distances, but that a great many more were visible. For example, instead of six stars visible in the Pleiades, a group of stars seen in the winter skies, he found more than fifty, and when he viewed the Milky Way he found a profusion of tiny stars.

The planets differed from the stars when viewed through the telescope, appearing as small round discs rather than pinpoints of light. Venus, he found, went from a round disc to a crescent shape in a matter of a few months, while at the same time varying in size (Figure 3.8). This, he felt, was additional evidence that Venus orbited the sun in accordance with Copernican theory. Galileo also discovered four satellites revolving around Jupiter, which to him made Jupiter appear like a miniature solar system (Figure 3.9). During the first viewing he observed three starlike objects in a straight line—two on the left and one on the right of Jupiter. This did not greatly impress him, for he assumed the objects to be stars a great distance in the background. However, on the next night he was astonished to find all three of the starlike objects to the right of Jupiter and concluded that these must be moons orbiting the large planet. A week after the first sighting he discovered a fourth satellite in approximately the same line as the other three. Galileo continued to observe and found the positions of the satellites changing with respect to Jupiter but at all times accompanying the large planet during its direct and retrograde movements in orbit. Saturn was somewhat of a surprise to Galileo. It was disc-shaped as he expected, but the planet appeared to have appendages on either side. He was unable to determine the nature of these appendages because of the limited ability of his telescope to resolve small details. In the middle of the seventeenth century Christian Huygens, with better equipment, was able to resolve this mysterious characteristic of Saturn into rings of matter which revolved around the planet.

Galileo's direct viewing of the sun through his telescope was a near tragedy which left him temporarily blind and permanently impaired his vision. As a result of Galileo's experience, Fr. Christopher Scheiner, a Jesuit priest, viewed the sun first through clouds and then through lenses made of dark-colored glass. He discovered dark spots on the sun's surface, spots which later served as evidence from which Galileo concluded that

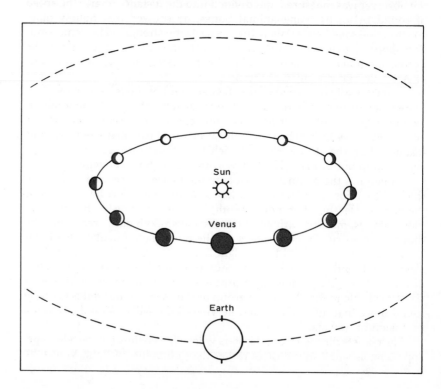

Figure 3.8 Through the use of his telescope, Galileo discovered that Venus went through phases as it revolved around the sun.

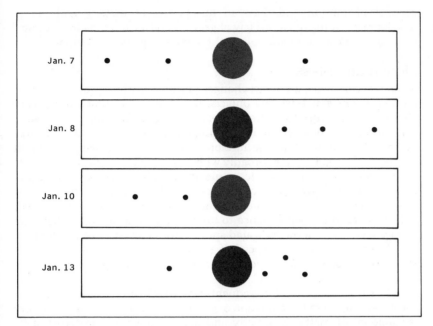

Figure 3.9 Positions of the four large satellites of Jupiter discovered by Galileo in 1610 as they appeared on the dates given.

the sun rotated on its axis. Galileo projected the sun's image on a sheet of paper in a darkened room and watched the spots appear, move across the face of the sun for several weeks, and then disappear around the opposite edge.

These discoveries were made by Galileo in the relatively short period of a few years. Never before had one man made so many facts available to astronomy, nor has this record been repeated even in the present. Galileo received much acclaim from all segments of society, and he lectured on the meanings of his discoveries which provided considerable substantiation of the Copernican theory. When someone quoted Aristotle, Galileo would state that Aristotle was a man with eyes and ears but that he, Galileo, was a man with eyes, ears, and a telescope. Galileo felt that his ability to see more clearly the workings of nature gave him a decided advantage over Aristotle. However, Galileo did have difficulty with the leaders of the Church over his support of the Copernican concept. Galileo was brought before the Inquisition where, after a trial, he was enjoined from teaching that the sun was the center of the planetary system.

3.5 ISAAC NEWTON (1642–1727)

Newton, a physicist and mathematician, was one of the greatest figures in the history of modern science. He entered Cambridge at the age of nineteen, where he performed so brilliantly that when he received his doctorate the professor under whom he worked retired to allow Newton to replace him. Quite early in his career Newton discovered the binomial theorem and invented calculus, accomplishments which alone would have assured him a place in history. He invented the reflecting telescope which utilized mirrors instead of lenses to focus light, and he experimented with prisms with which he was able to show that white light was composed of various color components. Newton also derived the laws of motion and the law of universal gravitation, but before discussing these we must first examine the concepts of motion and mass. Galileo had considered the concept of motion and mass in his studies of motion, and in the development of the universal law of gravitation Newton took into account Galileo's work as well as Kepler's three laws of planetary motion.

Motion and Mass

Motion occurs when an object changes position. In physics the concept of motion involves at least two objects, the object whose motion is being studied and the object with reference to which the first object is moving. Thus when we say a vehicle is traveling down the road at 100 kilometres per hour, we imply that the motion of the vehicle is being measured relative to the road or more broadly with respect to the Earth. The simplest type of motion is that which occurs in a straight line covering equal intervals at equal increments of time, or *speed*. The 100 kilometre-per-hour designation is the speed of the car, and can be considered *constant* speed. When the ratio of total distance to total time is considered, we have the *average* speed. When the indicated speed is for that instant of time only, it may be called *instantaneous* speed. If a direction is included with speed, as for example 100 kilometres per hour in a west direction, we then have the car's *velocity*.

Mass is a concept that is frequently confused with weight. *Weight* is a force resulting from the attraction of gravity upon a mass. *Mass* is basically the amount of matter in a body and may be measured in terms of its resistance to change of motion or its *inertia*. Mass may be found by measuring the acceleration produced by a known force on a body.

The fundamental unit of mass is the kilogram or 1000 grams. Thus if we weighed an object representing the fundamental unit of mass on the surface of Earth we would obtain a value of 1000 grams. However, if we transport the object to the moon, the object's weight would register only 166⅔ grams because the force of gravity on the moon is only one-sixth that on Earth. Mass remains the same no matter where the mass is located or what form it takes, but the weight varies according to the variation in the force of gravity. One should not conclude from this that weightlessness (see section 8.8) experienced by astronauts results from an absence of gravity.

We need also be aware of the relationship of the mass of an object and the volume of space it occupies. For example, it is easy to determine that a kilogram of butter occupies approximately 1100 cubic centimetres and a kilogram of lead occupies about 90 cubic centimetres. Both materials have the same mass but occupy different volumes of space. It can therefore be said that the two materials vary in *density,* which is a measure of the amount of mass for a given volume. Density may be expressed in a variety of units as, for example, grams per cubic centimetre (g/cm^3) or kilograms per cubic metre (kg/m^3). The density of water is used as a standard, and is by definition one gram per cubic centimetre. The ratio of the density of a substance to the density of water is the *specific gravity* of that substance. Specific gravity has the same numerical value as density but no units are used. Thus lead, with a density of 11.34 g/cm^3, has a specific gravity of 11.34.

Newton's Law of Motion

Newton's first law states that:

Every object persists in a state of uniform motion in a straight line or at rest if it is not acted on by an external force.

The property of a body to resist being set in motion or to resist change in motion or direction is called its *inertia*. Inertia will vary directly with mass. For example, a large rocket with a large mass and therefore great inertia will require much more force to launch it against gravity than a small rocket with a small mass and therefore less inertia. The inertia of an object is measured in terms of its *momentum*, which Newton defined as being proportional to the object's velocity. Therefore the momentum of an object may be expressed as a product of the mass of the object and its velocity.

$$\text{Momentum} = \text{mass} \times \text{velocity}$$

From this we can conclude that without any external influence to vary the velocity an object will maintain its momentum or motion. If the velocity is zero the object will remain at rest.

The second law states:

The acceleration produced by an unbalancing force acting on a body is directly proportional to the magnitude of the force and inversely proportional to the mass of the body.

Acceleration is the rate of change of speed or direction with respect to time. In the case of an object moving in a linear direction, acceleration is a change of speed with time. In the case of uniform circular motion, there may be no change in the magnitude of the speed but there is a constant change in the direction and therefore a constant acceleration. The force, then, is the effort necessary to produce an acceleration.

What is meant by the terms directly and inversely proportional? When x is said to be *directly proportional* to y, we are saying that as x increases in value y will increase in proportion and vice versa. However, if x is *inversely proportional* to z, we mean that as x increases z is decreasing by a proportionate amount. Thus if x increases tenfold, z is reduced to $1/10$. If x is directly proportional to y and inversely proportional to z, then $x = y/z$. With this in mind, we can consider the second law in the following manner. With a to represent acceleration, F the magnitude of the force, and m the mass, we can represent Newton's second law as

$$a = \frac{F}{m}$$

or as more commonly presented, $F = ma$.

Let us consider a situation where $F = 0$, in which case $ma = 0$. However, we should understand that no material object will have a mass equal to zero. Therefore the only means by which the force could be zero is if the acceleration is zero. If acceleration is zero, the object must be traveling at a constant rate or be at rest. This statement, then, is an expression of the first law, from which it follows that the second law of motion includes the first as a special case.

Newton's third law introduces a new concept—the generation of two forces that are equal and opposite to each other. The most frequently stated version of the law is: For every action there is an equal and opposite reaction. This may lead to some misunderstanding, as the statement implies there is first an action followed by a responding reaction. This is not exactly true. The forces exist simultaneously and are of equal importance. Therefore the third law can be more accurately stated as follows:

If one body exerts a force on another, the second body exerts a
force equal in magnitude and opposite in direction on the first
body.

A good example of the application of Newton's third law is the manner in
which the thrust of gases in a rocket causes the rocket to move. Many
people have the mistaken belief that a rocket is propelled through space
by the push of the exhaust gases against the exterior air. If this belief were
true, it would be impossible for rockets to travel in outer space, since there
is no air to push against. In fact, rockets travel better in the vacuum of
space which provides less resistance to the rocket's flight. In actuality, the
gases generated by the rocket fuel apply a thrust against the rocket inside
the combustion chamber as the gases escape from the exit nozzle. This
thrust sets the rocket in motion in the opposite direction of that taken by
the escaping gases.

Universal Law of Gravitation

After the work accomplished by Galileo and Kepler and others, it was
inevitable that the law of gravitation would be formulated. All the factors
for its explication were available, but it required the genius of Isaac
Newton to combine the various ideas set forth by his predecessors into a
cohesive principle that explained the motions of the planets and the forces
that influenced them. The great plague in 1667 caused Newton to leave
Cambridge and move to the country where the enforced idleness gave him
the opportunity to reflect on the problem. Newton was not alone in this
endeavor. Two other men considered the problem independently,
Giovanni Borelli of Florence and Robert Hooke of England. Borelli
concluded that the planets constantly strove to move in two directions,
toward the sun and straight into space. By means of a weight on a piece of
string, Hooke demonstrated that if the weight were pushed it turned in a
circle. According to Galileo's law of inertia, the weight should travel in a
straight line but the string, like the influence of the sun on the planets,
forced the weight to follow a circular path.

Newton, unaware of the work by Borelli and Hooke, set himself the
task of identifying the nature of these forces with mathematical precision.
He recognized that the moon was attracted by the Earth in the same
manner as a stone on the Earth's surface. He reasoned (from his third law)
that the force of attraction of the Earth on the moon or the stone was as
large as the pull of the moon or the stone on the Earth. Since the force was
directly proportional to the mass of the object, he concluded that the force
between two objects would be proportional to the product of their masses,
or $F \propto Mm$. However, the moon was considered to be about 60 earth radii
distant from the Earth and would therefore be attracted with a force less
than that affecting the stone. From Kepler's third law Newton was able to
deduce that the attractive force between the two bodies decreased
inversely as the square of the distance between them, expressed as
$F \propto \frac{1}{r^2}$. Therefore the attraction of the Earth for the moon would be
$1/60^2 = 1/3600$ of the attraction of the Earth for the stone since the
distance from the moon to the Earth is 60 times the radius of the Earth.

Newton recalled that Galileo had established that an object falls to the
earth with an acceleration of 9.8 metres per second^2, and that the object
would fall 4.9 metres in the first second. If the force acting on the moon
was only 1/3600 of the force acting on an object on the Earth's surface,

then the moon must fall about 490 cm/3600 sec or 0.14 centimetres per second. Newton was able to calculate this rate of fall because he knew the velocity of the moon in its orbit and the radius of the orbit. Having accomplished this, Newton found the rate of fall smaller than it theoretically should have been and, being of precise mind, he was troubled by the results. Later it was found that the value of the Earth's radius (upon which the moon's orbital radius was based) was larger than previously thought, and when Newton recalculated his values he found the results to be exactly as he had previously predicted. From this work Newton was able to show that celestial bodies attract each other with a force proportional to the product of their masses and inversely proportional to the square of the distance between them (Figure 3.10).

Stated in mathematical form the law becomes

$$F \propto \frac{Mm}{r^2}$$

where M represents the mass of one body and m the mass of the second, r is the distance between them, and F is the force one body exerts on the other. This expression can be put in the form of an equation by replacing the proportionality sign with an equal sign and the universal constant of gravitation, G. Thus the law becomes

$$F = G \frac{Mm}{r^2}$$

The value for the universal constant of gravitation G was experimentally determined by Henry Cavendish in the late eighteenth century.

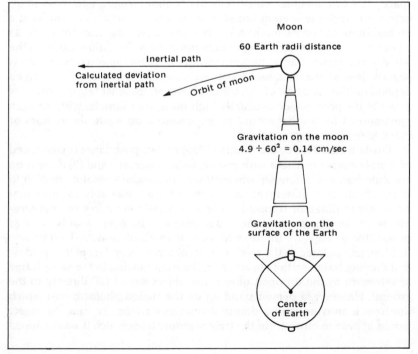

Figure 3.10 Newton's calculation of the Earth's gravitational effect on the moon.

Formulation of the law of gravity was a significant step toward solving the mysteries of motion in space. It was a basis for determining the motion and mass of planets and stars, and served as an explanation for the Earth's precession (see chapter 7) and the tidal action of the Earth's oceans. It was later useful in predicting the presence of unseen planets and stars on the basis of their gravitational influence on the associated visible planets and stars.

Newton, like others of his time, believed in the immutability of the universe and that the "luminiferous ether," a substance formerly thought to be necessary for the passage of light through space, being at rest, could serve as a basis for measuring the absolute motion of the bodies in space. A method that was later suggested to establish absolute motion of the Earth and the existence of the ether was to measure the motion of light in the ether. When the speed of light was determined with a fair degree of accuracy in the nineteenth century, such an experiment became possible, and was conducted in 1877 by A. A. Michaelson and E. W. Morley. Despite the extreme care taken and the accuracy of the instruments used, the results were negative. The speed of light could not be measured with respect to the ether, nor was the apparent motion of the Earth discernible. The experiment showed that the speed of light was independent of the motion of the observer and the motion of the source of the light. Ultimately it was realized that the only universal absolute is not something at rest, like the ether, but actually the fastest moving detectable thing in the universe—light.

3.6 ALBERT EINSTEIN (1879–1955)

Albert Einstein was able to utilize the ideas described above and form them into a new comprehensive theory. Einstein declared that there was no absolute frame of reference for measuring the motion of objects in space. The fundamental measurements of time, length, and mass of an object are dependent upon the state of motion of the object and have a mutual influence on each other. For example, time, or at least the clock, as a time-measuring device, is dependent upon mass. Pendulum clocks at the equator run faster than at higher latitudes because the force of gravity is slightly less at the equator. More accurate clocks are available which depend on the vibration of electrons. However, these vibrations will slow down in the presence of unusually high mass, for example, with the high gravitational forces that would be experienced on white dwarf stars or black holes (see chapter 14).

Einstein based his *Special Theory* (1905) on two postulates: (1) the speed of light is always constant with respect to all observers; and (2) there is no absolute frame of reference whereby we can measure absolute motion in space. With these statements as a basis Einstein was able to show that observers in different frames of reference would not agree on measurements of length, mass, time, and velocity. In other words, it was impossible to measure uniform motion in an absolute way. To illustrate this Einstein pointed to the motion of an object dropped from the window of a moving train. To the observer on the train moving in the same frame of reference as the dropped object, the object would fall directly to the ground. However, a person standing on the station platform past which the train is moving is in a different reference frame. To him the object would appear to move from the train window from which it was dropped toward the ground in a parabolic curve. From this example we can see that there is no such thing as an independent, absolute frame of reference against which the motion of the object can be described. To do this, the frame of reference must first be defined.

Einstein expanded these ideas to include the factor of acceleration and its influence on observers in different frames of reference, and from this predicted new laws of motion and a new law of gravitation. The General Theory provides equations with which it is possible to describe all natural phenomena regardless of whether the observer is at rest, moving uniformly, or accelerating with respect to some other object.

Einstein made several predictions based on the *General Theory of Relativity* (1916) which he said could be astronomically tested. One of these predictions had to do with the precession of Mercury's elliptical orbit which was observed to equal 574 seconds of arc per century. The precession was accounted for by Newtonian gravitational equations but the predicted rate was 44″ slower than was actually observed. This amount was accounted for by Einstein's General Theory. Mercury was chosen for this test because its orbital eccentricity is greater than that of other planets and the rotation of the orbit more easily measured. The Earth and Venus have orbits that are more circular, and therefore such measurements are extremely difficult to make. However, in recent years measurements of the orbital rotation of these planets have been found to be in essential agreement with the General Theory.

Einstein next suggested that light passing close to an object with a strong gravitational field, such as the sun or other star, would be slightly deflected. Such an effect cannot be tested in the laboratory, but can be observed by comparing the placement of stars very close to the sun's limb (outer edge) during an eclipse with the position of the stars when the sun is not in the vicinity (Figure 3.11). This test was made in 1919 during a total solar eclipse, and the results obtained were generally as predicted by Einstein. However, a number of subsequent tests yielded variable results, and it was concluded that the difficulties of making precise measurements of star positions during an eclipse were too great for the tests to provide good data.

Einstein also predicted that the wavelength of light from an object with a strong gravitational field such as a star, particularly a very dense star, would lengthen. According to the General Theory, strong gravitational fields have a slowing effect on time mechanisms. The rhythmical ticking of a clock or vibrations of atoms are slower on bodies with strong gravitational fields than on bodies with weak gravitational fields. Slower time would cause a longer wavelength of light and shift the spectrum of light toward the red end of the visible range (see chapter 4). Such shifts of the spectrum have been observed, but other interpretations of this phenomenon tend to weaken it as evidence in support of the General Theory.

One can conclude that experiments generally support the Special and General theories, and most scientists accept them as an accurate reflection of nature. Einstein's theories do not negate Newtonian physics. Only when velocities approach the speed of light would results differ; otherwise the predicted results would be identical for both theories. Einstein's theories describe natural phenomena more precisely than Newton's laws, possibly because more precise measurements are available. It would also be inaccurate to say that Einstein's theories are the last word on all phenomena relating to mass, motion, and time. For example, the existence

Figure 3.11 The relativistic bending of starlight is accomplished by the sun's gravitational field. As a result stars barely hidden by the sun appear to be displaced outward from their true position and are visible during an eclipse.

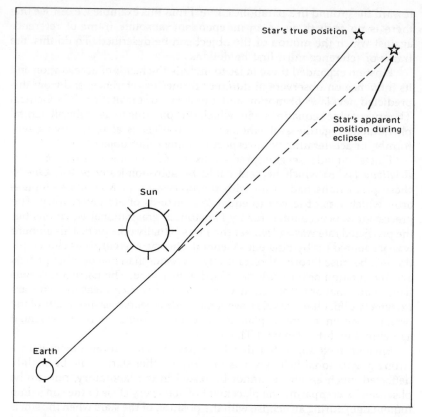

of antimatter on a large scale cannot be reconciled with the General Theory. Antimatter is a concept wherein matter is composed of atoms having protons, neutrons, and electrons with opposite charges than normal matter. These particles, called antiprotons, antineutrons, and positrons, have been detected. They are assumed to interact in mutual annihilation with their counterparts in normal matter, and as a consequence return to energy. The discovery of large masses of antimatter such as whole galaxies would require a reevaluation of the theories of relativity.

3.7 SUMMARY

Modern astronomy deviated from the concepts of classical astronomy by gradually rejecting those ideas which could not be supported by observational data. In the sixteenth century, Copernicus challenged the classical geocentric concept by suggesting that the observed motions of the planets and stars could be better and more simply described on the basis of a sun-centered universe, with the planets, including the Earth, orbiting the sun. This idea, while not immediately and universally adopted, did indicate a major shift in astronomical thought and marked the introduction of the era of modern astronomy.

Many scientists in the sixteenth century were opposed to the Copernican model of the universe, among them Tycho Brahe. Brahe collected great quantities of data on planetary motion, particularly of Mars, for the purpose of refuting the sun-centered or heliocentric theory and to promote his own model, the geostatic system. This system retained a stationary Earth at the center of the universe and had the planets orbiting the sun and the sun revolving around the Earth.

Johannes Kepler became interested in Tycho Brahe's work, not because of the geostatic system which ultimately fell into oblivion, but because of the great amount of data Brahe had collected on planetary motion. Kepler, primarily a mathematician, believed in the Copernican system and felt that the data would be valuable in proving the correctness of Copernicus's ideas. From the data Kepler developed, in the early part of the seventeenth century, three laws of motion, which represented a very important contribution to the science of astronomy.

At the same time that Kepler labored on his laws of motion, Galileo designed and built his telescope. With this instrument he was able to show that the moon had many earthlike features, that the sun had dark spots on its surface and rotated on its axis and that Jupiter had four satellites orbiting that planet, also making a number of other astronomical discoveries not possible without the telescope. As a result of the development of the telescope new discoveries were made at an ever-increasing rate, adding tremendously to the store of astronomical knowledge.

Basing his ideas on the work of Kepler, Galileo, and others, Isaac Newton made several highly significant contributions to science. First, he defined the nature of motion and mass through his three laws of motion. Second, he defined gravity in precise mathematical terms in his law of universal gravitation. With these laws it was possible to understand motions of objects in the universe and why every object maintained its position relative to every other object in the universe.

Newton believed in the immutability of the universe and that it was possible to achieve absolute points of reference from which absolute values for motion might be obtained. Einstein, in his Special Theory, showed that no such frames of reference existed and that it was only possible to describe the motion of an object relative to some other object also in motion. Later Einstein expanded upon these ideas in his General Theory to include the influence of acceleration on different frames of reference.

Einstein's theories do not replace those of Newton, but describe natural phenomena in more precise terms, especially when objects are approaching the speed of light. Newton's theories are still valid when considering speeds that we normally deal with. Einstein's theories may also not be the final word on matters relating to mass, motion, and time. Further refinements of his ideas are possible, which will increase the precision with which we describe nature.

QUESTIONS

1. What was Copernicus's motive for reviewing the arrangement of the geocentric system and what was his reason for suggesting the heliocentric system?

2. What was the structure of the Copernican system and what vestiges of the geocentric system did Copernicus incorporate into his new system?

3. What objections were raised against the system as described by Copernicus?

4. What contributions to astronomy were made by Tycho Brahe?

5. What ideas were stated by Kepler's first and second laws of planetary motion and what vestiges of the geocentric system did these laws eliminate?

6. What relationship between distance from the sun and period of revolution of a planet is expressed by Kepler's third law?

7. List at least five discoveries made by Galileo as a consequence of his use of the telescope.

8. Describe the difference between mass and weight.

9. Give Newton's three laws of motion. State an example to illustrate his third law.

10. What was the purpose of the Michaelson-Morley experiment and what were its results?

11. What were the tests suggested by Einstein to test his general theory?

12. Do Einstein's theories invalidate those of Newton? Under what circumstances are Einstein's theories more valid than Newton's?

FOR FURTHER READING

ANTHONY, H. D., *Sir Isaac Newton.* New York: Collier Books, 1960.

CASPAR, M., *Kepler 1571–1630.* New York: Collier Books, 1959.

KUHN, T. S., *The Copernican Revolution.* New York: Random House, 1959.

MASON, S. F., *A History of the Sciences.* New York: Collier Books, 1962.

TAYLOR, F. S., *Galileo and the Freedom of Thought.* London: Watt and Co., 1938.

THIEL, R., *And There Was Light.* New York: Mentor Books, 1957.

YOUNG, L. B., ed., *Exploring the Universe.* New York: McGraw-Hill, 1963.

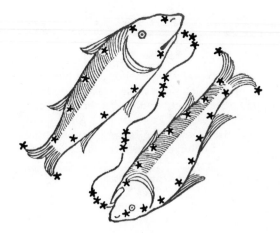

Pisces

CHAPTER 4

PROPERTIES OF LIGHT

> And God said: Be light made. And light was made.
> And God saw the light that it was good; and He divided the light from darkness.
> And He called the light Day, and the darkness Night; and there was evening and morning one day.
>
> —Gen. 1:3–5

Light is perhaps the first external stimulus of which one becomes conscious. It is a necessity in our world and the prolonged lack of it becomes an extreme deprivation. A person's ability to "see" was assumed by some ancient philosophers to be the result of some action on the part of one's eyes—in effect, a reaching out with a kind of tentacle that enabled one to reach and explore an object at a distance. The expression, "to cast one's gaze" upon objects is thought to come from this belief, but today we accept the idea that the eye is a receiver of some form of energy from the object toward which the eye is turned.

4.1 WHAT IS LIGHT?

For convenience, light is described as a wave phenomenon resembling the waves rippling on the surface of a quiet pond. The wave motion is characterized by a wavelength, λ, that describes the distance between successive wave crests. The number of wave crests passing a given point per second is the frequency f. From this it can be readily seen that the shorter the wavelength the greater will be the number of crests passing per second and therefore the greater the frequency (Figure 4.1). If waves are traveling at speed c, the relationship may be expressed as

$$c = \lambda f$$

where c is the speed of light.

Visible light represents only a small part of the total array of radiant energy. This array, called the *electromagnetic spectrum*, ranges from gamma radiation with a wavelength of 10^{-13} centimetres to longwave radio waves 100 kilometres or more in length (Figure 4.2).

Figure 4.1 Characteristics of a wave.

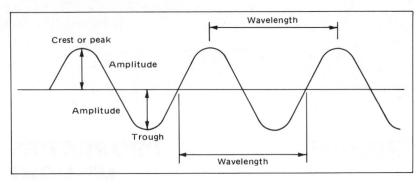

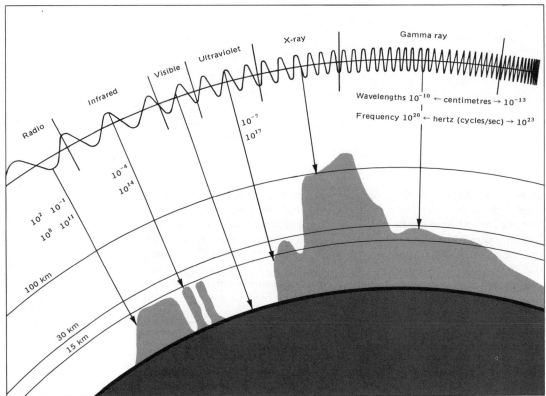

Figure 4.2 The components of the electromagnetic spectrum as they penetrate through the atmosphere are shown. The darkened areas represent radiation blocked at different elevations above the Earth's surface. (From a NASA diagram)

Gamma rays with a wavelength of less than 0.5 Å* originate in the interior of stars and are often emitted by radioactive elements during nuclear reactions. Longer wavelengths from 0.5 Å to 500 Å are X rays and can be detected photographically. The X rays are useful for studying bone structure and can be used in industry for checking the internal stresses of materials. Ultraviolet light, 500 Å to 3500 Å, is not visible to the eye and is sometimes called black light. The sun is a principal source of this form of radiation and is responsible for our sunburn. Radiation with wavelengths in the range of 3500 Å to 7500 Å comprises visible light. The range of visible light is divided into an array of colors (colors of the rainbow) which represents radiation at different wavelengths. Thus violet light occurs at

*The angstrom Å is a unit of length used to measure wavelengths of electromagnetic radiation. It is equal to 10^{-8} centimetres.

wavelengths of approximately 3500 Å to 4000 Å, whereas red light occurs at the other end of the visible range of 6500 Å to 7500 Å. Between 4000 Å and 6500 Å we find blue, green, yellow, and orange in ascending order of wavelength and descending order of frequency. At wavelengths of 7500 Å to 0.1 millimetre we experience infrared radiation, which again is in the invisible portion of the spectrum but can be detected as heat. The longer wavelengths, 0.1 millimetre to several metres, include the microwave region where one finds radiation useful in radio astronomy, radar, and television. At wavelengths of several metres or more the shortwave radio and regular radio bands are found, and although no limit in the longwave region of the spectrum has been reached, uses for the very long wavelengths have not been found and are not intensively studied. All these forms of radiation have one thing in common—they travel at the speed of light which is approximately 3×10^5 kilometres per second.

Light is useful in a variety of ways and particularly to the astronomer who gathers, focuses, and analyzes light in a continuing study of objects at tremendous distances from Earth. The study of light from distant objects is the major source of information about the universe, and without a clear understanding of light's optical properties it would not be possible to delve into these cosmological mysteries.

4.2 THEORIES OF LIGHT

Light has fascinated scientists and philosophers for centuries, but the exact nature of light is still not completely understood. The ancient Greek atomists believed that all matter could be subdivided again and again into ultimate particles incapable of being further divided. Light was considered to be matter, composed of tiny particles emitted from the light source. Aristotle rejected this concept and held that light resulted from some interference in the medium between the light source and the observer. From this we can see the beginnings of the centuries-old controversy about the intrinsic nature of light.

In the latter half of the seventeenth century, Christian Huygens compared the light phenomenon with sound. Sound was well recognized at that time as a wave disturbance that traveled at a known speed in the air but not in a vacuum. When Olaus Roemer discovered that light had a finite speed (see section 4.3), Huygens attempted to explain light as a wave phenomenon. He was able to present a mathematical analysis showing that reflection and refraction could be explained on the theory that light was composed of longitudinal waves (compressional waves) similar to sound waves. Since some medium was required to carry sound waves, Huygens assumed that the "luminiferous ether," a substance thought to fill all space, served to carry light waves through space.

Huygens' basis for the wave theory was the fact that light waves could pass through each other without apparent interference. He was of the opinion that if light were composed of particles they would collide and be deflected. Huygens also felt that the refraction or bending of a light beam entering water from the air was due to the slowing down of light as it traveled through water.

During this same period a rival theory was proposed by Isaac Newton, who strongly held the view that light consisted of a stream of particles or *corpuscles* rather than waves. He objected to the wave theory principally on the grounds that it did not appear to him to explain why light traveled in straight lines. Newton said that if light were a wave motion in the ether there would be a continual spreading of light equally in all directions. If this were the case there would be no sharp shadows, since the light would spread around corners in the same fashion as sound and water waves do.

Newton examined other properties of light. For example, he discovered that when a prism was employed to separate the sun's light into the colors of the spectrum, another prism in the reverse position made it possible to recombine the colors into white light again (Figure 5.25). From this he concluded that each color was composed of particles unique to that color, and by combining all the particles the eye would experience the sensation of whiteness. Newton also believed, contrary to Huygens, that when light particles from air entered a denser medium such as water the mutual gravitational forces between water and light particles increased the speed of the light particles, thereby causing them to be refracted.

Newton's corpuscular theory was accepted as the true theory of light for about one hundred and fifty years, although not without some criticism. This was not the result of any error in the wave theory or of Newton's ideas proving true. Rather they seemed to be accepted primarily because of Newton's tremendous reputation.

During the nineteenth and twentieth century, events occurred which helped resolve the controversy surrounding the wave theory versus the particle theory of light as described above. One of the more important concepts relating to light which developed was the idea of *field*. The gravitational pull of the Earth exists as a field with a magnitude and a direction at any point in the universe. This gravitational field influences objects in the universe even through the emptiness of space. In the latter half of the nineteenth century, James Clerk Maxwell was able to show a link between electricity and magnetism, and that these forces generated fields which could exist in space in the same manner as a gravitational field without any medium such as ether being present. The field concept did not explain what the electric and magnetic fields were, other than that they were closely related so that a change in one influenced the other. The field appeared to be some property of space.

In 1887, Heinrich Hertz, a German physicist, was able to demonstrate the existence of electromagnetic waves and the fact that, like light, they could be reflected and refracted and traveled with the same speed as light. The difference between electromagnetic waves and light appeared to be simply a difference in wavelength. It soon became apparent that light was actually a small part of a whole range of electromagnetic waves which moved at the same speed and differed only in their wavelengths (Figure 4.2).

Maxwell's electromagnetic theory explained many aspects of the light phenomenon but failed to explain the energy distribution of radiation emitted by a black body radiator. A *black body* is a theoretical object that will absorb all radiant energy incident upon it. At the same time they are also excellent radiators of energy. When heated a black body will emit radiation in all wavelengths in much the same manner as heated bodies such as the sun or other stars. Study of black body radiation furnishes much information about stellar emission and absorption of energy. For example, studies have shown that at higher temperatures the energy maximum occurs at the shorter wavelengths of the spectrum. It was felt that if the energy were distributed uniformly among all the wavelengths, then theoretically the energy curve should continue to rise steadily as it approaches the ultraviolet, but it does not. Instead energy emitted in the visible range of the spectrum at a given temperature increases to a maximum as the wavelength becomes shorter (Figure 4.3) and then decreases again in the ultraviolet range. The source of this energy is the motion of the atoms and molecules in the emitting body. As the temperature of absolute zero (–273.2°C) is approached the motion of the atoms gradually slows down, but as the temperature rises the motion increases. From Figure 4.3 it can be seen that as the temperature increases so does the amount of radiation, with the point of maximum emission occurring at progressively shorter wavelengths. A color change in the emitting body cannot be seen when energy maximums occur in the infrared portion of the spectrum where wavelengths are greater than 7500 Å, but we are sensitive to this radiation and can feel it as heat. As the temperature rises, energy maximum occurs in the visible range of the spectrum, and the color of the emitting body becomes red to orange to white to blue-white, indicating increasing temperature. The color change indicates a shifting of the wavelength of greatest radiation emission.

In 1900, Max Planck made an attempt to resolve this problem by assuming that bodies can radiate energy only in "packets," which he considered to be indivisible. The energy of a packet or *quantum* varied

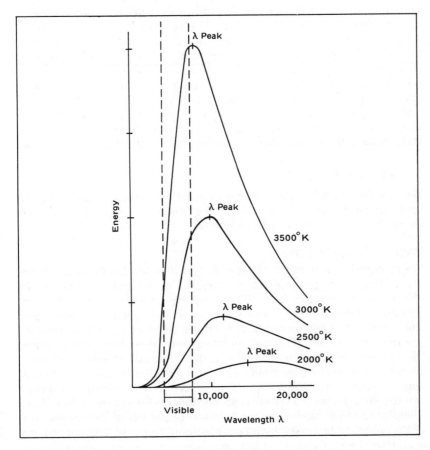

Figure 4.3 As the temperature of an object increases, the point of maximum emission of radiation occurs at progressively shorter wavelengths. Objects at 3500°K will have maximum emission just within the visible portion of the spectrum.

according to the frequency of the electromagnetic waves. Mathematically this is expressed as

$$E = hf$$

where E is the quanta of energy, h is Planck's constant of 6.6×10^{-27} erg · sec, and f is the frequency. The quantum of red light is equal to the energy of an electron accelerated by an electric potential of 1.8 volts or 1.8 electron volts. The quantum of blue light is 3 electron volts. From the equation it follows that ultraviolet light has much greater energy than visible light and is capable of removing electrons from atoms and leaving them electrically charged or ionized. The quanta of X ray may be thousands or even millions of electron volts, while in the longer wave radio frequencies the quanta may be very small. The amount of energy in the quantum seems to be related to the behavior of the different wavelengths.

A few years later, in 1905, Albert Einstein took the idea a step further by theorizing that the packets were integral parts of electromagnetic radiation and could not be subdivided. He called these packets of light energy *photons*, each photon having a certain amount of energy depending on its wavelength. This would determine the color of the radiating body. From this it would appear that there was a return of the particle versus wave controversy of the past, but photons do not behave as one might normally expect. It has been possible to study the behavior of photons by causing them to strike an electron. The resulting collision causes the photon to impart energy to the electron and recoil from the impact with the speed of light. How does the photon lose energy during such an encounter without a reduction in speed? It does so by an increase in wavelength, and we therefore have a discrete entity with measurable energy that moves in a manner indicating wave motion. How light behaves depends upon the nature of its interaction with matter. If the interaction is macroscopic, relatively speaking, then light is observed to behave as a wave. If the interaction is atomic or subatomic, light behaves as a photon.

4.3 PROPERTIES OF LIGHT

We now know a great deal about the optical and geometric properties of light, which knowledge is vital in the study of astronomy. A brief discussion of their characteristics follows.

Inverse Square Law

We recognize that light travels in a straight line. Shadows cast are sharp and distinct in bright sunlight and follow the geometric outline of the object obstructing the light. This property of light is important in astronomy, since it is the basis for the design and operation of telescopes and other optical equipment, which will be described later. The straight-line transmission of light is also an important prerequisite for measuring the intensity of light coming from distant celestial objects. From any source such as the sun, light radiates outward in all directions equally and approximates a constantly expanding sphere. From geometry we know that the surface area of a sphere is proportional to the square of the radius. Therefore if the distance from the light source is doubled, the surface area of the sphere is increased by a factor of four. The same amount of light energy will be available at the total expanded surface area as at the source, but the intensity for each unit of area will be less. The relationship is such that the brightness is inversely proportional to the square of the distance from the source (Figure 4.4). This means that at double the distance the light will be one-fourth as bright or by tripling the distance the brightness is reduced to one-ninth. This concept is important in determining the distance to some stars. The technique for accomplishing this will be discussed in chapter 13.

Reflection

Light transmitted through a medium of uniform density will, as stated above, travel in a straight line until the light strikes an opaque surface. Here the light not absorbed by the surface will be redirected or *reflected*. The manner in which the light approaches the surface will determine the nature of the reflection. A light ray approaching perpendicular to the reflecting surface will be reflected back along the same path. A line perpendicular to a surface or to another line is a *normal*. Therefore a ray approaching in this manner is said to be normal to the reflecting surface. An incoming ray, the *incident ray*, approaching obliquely to the surface forms an angle with the normal, the *angle of incidence* (i). This angle is equal to the *angle of reflection* (r) formed between the normal and the reflected ray (Figure 4.5). If it were possible for the reflecting surface to be absolutely

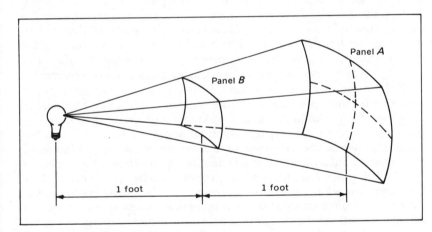

Figure 4.4 The diagram demonstrates the inverse squares law. Panel *A* is twice as far from the light source as panel *B* and therefore receives only one-fourth the light energy per unit of area.

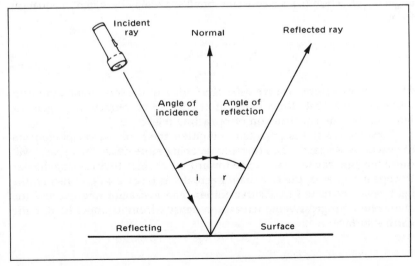

Figure 4.5 When a light ray is reflected from a surface, the angle of incidence is equal to the angle of reflection.

smooth, the light would be *regularly reflected*. However, the smoothest-appearing surface has some imperfections, causing the individual rays in a light beam to be reflected at different angles, and therefore the light is said to be *diffusely reflected*. Even apparently smooth surfaces will, upon microscopic examination, reveal an irregular surface, so that any light will be somewhat diffusely reflected, although the degree of diffusion will be slight.

Refraction

Another means whereby light can be made to deviate from straight line travel is to permit the light ray to pass from one transparent substance, such as air, through another substance of different density, such as glass or water. An increase in density will cause a reduction in the speed of light through the medium and will result in the light ray being bent or *refracted*. This phenomenon can readily be seen when we view, from an angle, a stick partially submerged in water. The part of the stick beneath the surface appears bent in relation to the part of the stick above the surface (Figure 4.6).

To describe what takes place let us consider a light ray entering a pane of glass. Light falling along the normal to the glass surface will continue through the glass, its direction unchanged. A ray of light approaching the glass obliquely forms the *angle of incidence* (i) with the normal (Figure 4.5). As the light enters the glass it will be bent toward the normal forming the *angle of refraction* (r'). By the same process, light leaving the glass and entering the less dense air would be refracted away from the normal.

Prior to the seventeenth century, physicists considered the angle of refraction (r') to be proportional to the angle of incidence (i), and that the doubling of one would result in the doubling of the other. While this relationship is close to being true when the angles are small, it fails as the angles increase in size. However, in the seventeenth century, Willebrord Snell worked out a quantitative law describing the relationship between i and r', a law which was independently confirmed a few years later by René Descartes. The relationship, now known as *Snell's law*, states that whenever light passes from a transparent medium of one density to that of another density, the ratio of the sine of the angle of incidence (i) to the sine of the angle of refraction (r') is a constant. Snell's law may be stated in equation form as

$$\frac{\text{Sin } i}{\text{Sin } r'} = n$$

The constant n is the relative *index of refraction* of the second medium with respect to the first. Knowledge of the index of refraction of glass is important in the construction of refracting telescopes.

Light from a star is refracted as it enters the Earth's atmosphere from the vacuum of space. The atmosphere, being more dense than space, will cause the light ray to be refracted toward the normal, thus causing the star to appear higher in the sky than it actually is (Figure 4.7). In reality, the light ray is refracted in a curve rather than a straight line because the atmosphere progressively increases in density from its upper limit to the Earth's surface.

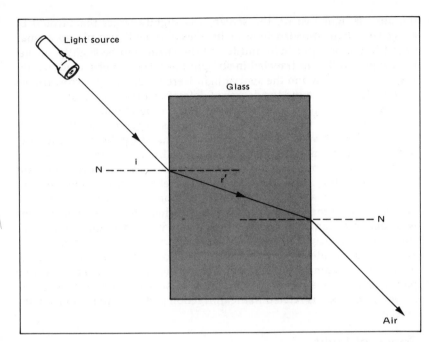

Figure 4.6 Light is bent or refracted when passing from one medium into another of different density, as, for example, from air into glass.

Figure 4.7 Atmospheric refraction causes starlight to be bent as it enters Earth's atmosphere. The star appears at a higher altitude than it actually is. Only stars directly overhead (seen along the normal) will appear in their correct position.

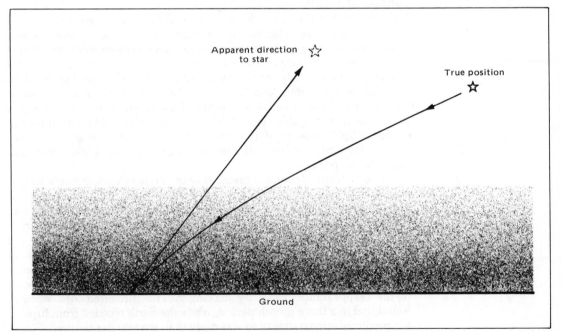

Diffraction

Another instance wherein light does not adhere strictly to the principle of straight-line travel occurs when light is restricted by very small apertures. Under these circumstances, Francesco Grimaldi in the seventeenth century discovered that light was bent or *diffracted* as it passed through the tiny opening. When allowing light to pass through a pinhole, he observed that the area of light cast on the screen was slightly larger than would be expected if light traveled in straight lines. Grimaldi also placed a small object in the beam of light and found the shadow to be slightly smaller than expected if light traveled in straight lines. He also observed that the edge of the shadow and the spot of light were bordered by narrow fringes of color. Grimaldi found the degree of defraction to be very small and only detectable if a tiny aperture was used in the experiment.

A satisfactory theory for diffraction was suggested in 1814 by A. Fresnal, who showed that diffraction was caused by the propagation of light in a wave form through a small aperture. Each point on the wave front serves as a source of new wave action in all directions. However, along the front, wave action from one point will cancel out wave action from the neighboring point so that only the forward motion remains. The exception to this occurs at the ends of the wave front where there is no neighboring wave action to cancel the propagation of the wave in all directions and a slight bit of "leakage" of the wave form occurs. This would be analogous to water waves entering a harbor through an opening in a breakwater and the waves spreading into the region behind the breakwater. The resolving power of telescopes, that is, the ability to separate two closely spaced objects in the sky, is related to the diffraction pattern of the light from the objects.

Speed of Light

Until the time of Galileo, light was assumed to be an instantaneous event in all parts of the universe, that is, an event in one part of the universe was instantly seen in all parts of the universe. Even Galileo made use of this assumption to measure the speed of sound, but later in reevaluating this concept he attempted to measure the speed of light. He placed an assistant on a hill with a lantern and himself on another hill about a mile away. Galileo flashed a light with his lantern and the assistant returned the signal as soon as he saw Galileo's light. The experiment failed to determine the speed of light because of human inability to react with the necessary speed. However, Galileo's attempt did suggest the possibility that light had a finite speed.

In 1676, Olaus Roemer, a member of the French Academy of Sciences, was able to determine the speed of light by measuring the period of revolution of Io, one of the Galileon satellites of Jupiter. The satellite revolves around Jupiter in 42 hours, 27 minutes, and 33 seconds, a period that could be determined precisely by timing each emergence of the satellite from behind the planet. During the Earth's annual orbit around the sun, the Earth's distance from Jupiter varies considerably. Roemer reasoned that if light had a finite speed, there would be a measurable delay in the reappearance of Io. By making careful measurements, Roemer found that in a three month period, while the Earth receded from Jupiter, the predicted appearance of Io was more than ten minutes delayed (Figure 4.8). Six months later when the Earth moved toward Jupiter again, the delay was diminished. From this information, Roemer calculated the speed of light at a little over 1.4×10^5 miles per second.

Many measurements of the speed of light have subsequently been made. In 1849, Hippolyte Fizeau made use of Galileo's principle but with considerable refinement. Fizeau allowed light from a source to be reflected by a mirror after passing between the teeth of a turning wheel. At a slow speed of rotation, the light went from the source to the distant mirror and

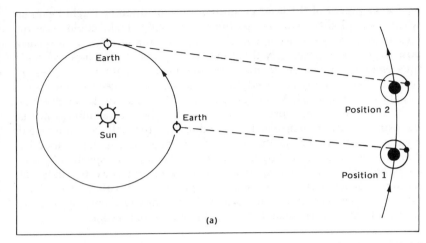

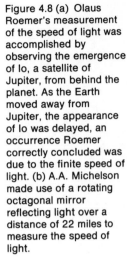

Figure 4.8 (a) Olaus Roemer's measurement of the speed of light was accomplished by observing the emergence of Io, a satellite of Jupiter, from behind the planet. As the Earth moved away from Jupiter, the appearance of Io was delayed, an occurrence Roemer correctly concluded was due to the finite speed of light. (b) A.A. Michelson made use of a rotating octagonal mirror reflecting light over a distance of 22 miles to measure the speed of light.

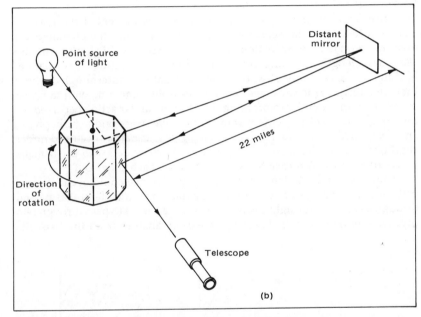

returned through the same space by which it had left. As the wheel rotated faster the light would be interrupted by a tooth, and at a faster rate still, light would pass outbound through one space and return through the next. Knowing the rate of rotation of the wheel, the distance of the gaps between the teeth on the wheel, and the distance the light beam had to travel enabled Fizeau to calculate a fairly accurate speed for light.

Albert Michelson repeated the experiment in 1923, measuring light speed between two mountains in California that were 22 miles apart. He had surveyed the distance to the nearest inch and used a rotating eight-sided mirror to relay the beam back and forth. Several years later he conducted the experiment using a long evacuated tube, which permitted him to measure the speed of light in a vacuum. From this experiment he was able to obtain a figure of 186,271 miles per second for the speed of light.

Development of the laser permitted the generation of a light beam in which every light wave was of the same length. Light could now be measured with great accuracy. By using an atomic clock it was also possible to count accurately the number of waves produced per second, thus enabling physicists to measure the speed of light to a degree of accuracy never before achieved. In 1972, it was announced that the speed of light was 186,282.3939 miles per second or 299,792.4562 kilometres per second. These figures represent the speed of light in a vacuum, which is the usual condition in space. The speed of light through any transparent medium is influenced by the index of refraction and is determined by dividing the speed of light by the index of refraction. The speed of light in air is only slightly less than that in a vacuum, but since the index of refraction of water is 1.33, the speed of light in water is about 2.25×10^5 kilometres per second, and in glass with an index refraction of 1.5, the speed of light is approximately 2×10^5 kilometres per second.

Interference

Evidence favoring the wave theory of light was presented by Thomas Young in 1802 when he demonstrated that light exhibits the phenomenon of *interference*. He drew an analogy in terms of two series of ripples with the same wavelength traveling over the water and meeting at a channel leading to a larger body of water. The wave patterns intermingled where they met, resulting in the crest of one wave train coinciding with the crest of the other and thereby reinforcing the crest. A similar action occurred for the troughs. Under these circumstances the waves were said to be in phase, an effect called *constructive interference*. *Destructive interference* occurred under conditions where the crests of one wave train coincided with the troughs of the other, in which case the waves canceled each other out.

Young was able to demonstrate that interference could occur in light patterns. He used a relatively simple apparatus in which light was passed through a slit, and this light beam in turn was allowed to pass through two parallel slits (Figure 4.9). The light waves emanating from the two slits

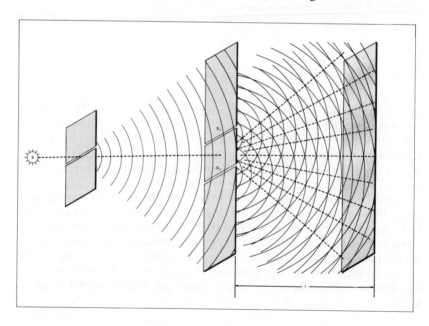

Figure 4.9 Schematic diagram of Thomas Young's double slit experiment showing constructive and destructive interference and the resulting light and dark bands seen on a screen.

overlapped due to diffraction, resulting in constructive and destructive interference and a pattern of alternating light and dark bands of light on the screen. Such patterns could not be reconciled with Newton's particulate theory of light, but the wave theory provided a good explanation for this behavior of light.

Polarization

Huygens considered light waves as longitudinal, but modern theory finds that light consists of transverse waves in which the vibrations are at right angles to the direction of propagation of wave motion. Light from the sun or a star or, for that matter, any ordinary source is considered to be *unpolarized*, in that there are many vibrations at all angles perpendicular to the direction of light wave motion. Light waves may be *polarized*—a condition where the vibrations of the waves are at least partially aligned—by use of certain mineral crystals such as tourmaline. If light is allowed to pass through a thin disk of the crystal and the crystal rotated, no change in the character of the light will be apparent. However, if one crystal is placed behind another, then the transmitted light will vary from maximum intensity to almost no light as one of the disks is rotated through 90° (Figure 4.10).

The structure of tourmaline is such that vibration of unpolarized light in one plane only would be allowed to pass through the disk. The light

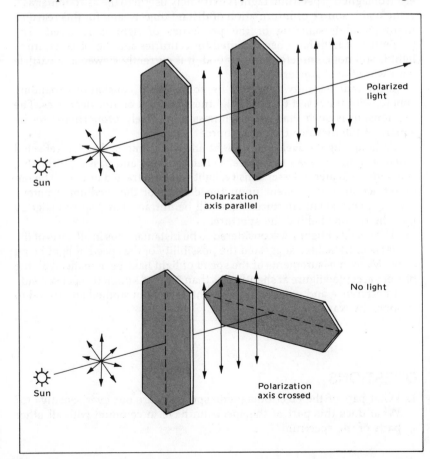

Figure 4.10 Light passing through one tourmaline crystal is polarized and will pass through the second crystal if it is oriented in the same direction as the first. If turned 90°, the second crystal will block the passage of polarized light rays.

Figure 4.11 A mechanical
model shows that a
transverse wave
controlled by slit *A* in a
vertical position will pass
through slit *B* but not slit
C.

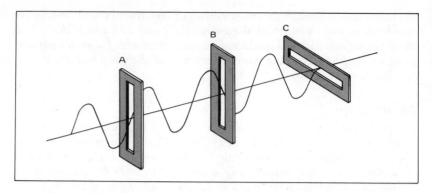

Figure 4.11 A mechanical model shows that a transverse wave controlled by slit *A* in a vertical position will pass through slit *B* but not slit *C*.

thus transmitted is said to be polarized. This light will pass through a second disk if its polarizing axis is parallel to that of the first. If the polarizing axis of one of the disks is turned 90° so that the two axes are at right angles to each other, light will not pass through the second crystal. Figure 4.11 illustrates this principle with a mechanical model.

4.4 SUMMARY

Light, a vital necessity in our world, is a small segment of the total electromagnetic spectrum. Light is extremely useful to the astronomer as it is the major source of information of distant objects, and for this reason a thorough understanding of the properties of light is required. The properties of light have been studied for centuries and the exact nature of light is still not completely understood. It is currently viewed as a particle with wavelike characteristics.

Light radiates outward from its source as a constantly expanding sphere, with the result that the brightness per unit of area decreases. The relationship is such that the brightness is inversely proportional to the square of the distance from the source.

Although light travels in a straight line when unimpeded, it is reflected when striking an opaque surface. When entering and passing through a medium of greater or lesser density, light will be refracted or bent. The rate of refraction is dependent upon the density of the medium it enters. Through very small apertures light will be diffracted and spread slightly into the regions behind the aperture.

The speed of light was considered to be instantaneous in all parts of the universe until Galileo suggested the possibility of the speed of light being finite. Many measurements of the speed of light have been made, with the present accepted figure at slightly less than 3×10^5 kilometres per second.

Interference and polarization of light have been studied and found to support the wave concept of light.

QUESTIONS

1. What part of the electromagnetic spectrum are our eyes sensitive to? What does this part of the spectrum have in common with all other parts of the spectrum?

2. Compare briefly the theories of light as presented by Christian Huygens and Isaac Newton.

3. What is the relationship between temperature changes and the wavelength at which an object emits maximum radiation?

4. Assume that objects A and B are emitting the same amount of light. If we see object A as sixteen times brighter than object B, how much further away from us is object B than A?

5. Assuming that Galileo and his assistant were 15 kilometres apart when attempting to measure the speed of light, what length of time did it take the light from Galileo's lantern to reach his assistant?

6. Make a diagram showing reflection and refraction of light. Identify all angles formed.

FOR FURTHER READING

Feynman, R. P., R. B. Leighton, and M. Sands, *The Feynman Lectures on Physics.* Reading, Mass.: Addison-Wesley, 1964.

Holton, G., and D. H. D. Roller, *Foundations of Modern Physical Science.* Reading, Mass.: Addison-Wesley, 1958.

Minnaert, M., *The Nature of Light and Colour in the Open Air.* New York: Dover, 1954.

Ripley, Jr., Julien A., *The Elements and Structure of the Physical Sciences.* New York: John Wiley & Sons, 1964.

Sears, F. W., and M. W. Zemansky, *University Physics.* Reading, Mass.: Addison-Wesley, 1964.

Hercules

CHAPTER 5

TOOLS THE ASTRONOMER USES

Nothing could be more noble than to contemplate the manifold wisdom of the Creator, but not with the gaze of vulgar admiration but with a desire to know the causes, and to feed upon this beauty by a more careful examination of their mechanism.

—*Jeremiah Horrocks*

We tend to think of observatories as sites where astronomers concentrate their observational activities around the use of the telescope. This has not always been the case, since the telescope was not invented until the beginning of the seventeenth century. However, records indicate that observatories of various types existed in antiquity (see chapter 2). Sometimes the entire structure itself served as the means of making measurements, as at Stonehenge, and at other times, as in Babylonia and Egypt, temples and towers were used as sites for study of heavenly bodies.

5.1 ANCIENT TOOLS

One of the first instruments used by the Greeks and later by the Arabs was the *astrolabe* (Figure 5.1), a device used to measure the altitude of celestial objects and to determine their position and movements. In medieval times the astrolabe was reintroduced into Europe and later was useful to the great explorers as an instrument to guide them in their voyages of discovery. In its simplest form the astrolabe carried two pointers attached to a circle marked in degrees in radial position. One pointer was fixed and the other was free to rotate. Both pointers of the astrolabe were provided with a sight.

The *armillary sphere* (Figure 5.2) was somewhat more complex than the astrolabe, being a device that represented the great circles of the sky including the equator, the horizon, the meridian and polar circles, and the ecliptic circle. Its exact origin is unknown, but it was used by the ancient astronomers including Erastosthenes, Hipparchus, and Ptolemy, and subsequently by the Arabs, before it was introduced into Europe.

The oldest of the modern European observatories was built by the astronomer Regiomontanus in 1472 in Nuremberg, where he constructed

Figure 5.1 The astrolabe
is a device used for
centuries to tell time and
measure angular dis-
placement of stars, and
to find equinoxes and
solstices of the sun. (The
Granger Collection)

Figure 5.2 Armillary
sphere built by Santucci
of Pommerenia com-
pleted in 1598. The
armillary sphere repre-
sented the great circles
in the sky including the
ecliptic, equator, merid-
ian, and polar circles.
(Bettmann Archive)

a variety of instruments useful in measuring position and motion of the
planets and stars. Observatories were also constructed in Samarkand
about 1420, in India (Figure 5.3), and in China (Figure 5.4).

The astronomers working at Samarkand were noted for their accurate
observational data. However, credit for the techniques and precision of
modern astronomy belong to Tycho Brahe, who in the sixteenth century
was able, by patient attention to details in the construction of his
equipment, to provide the most accurate data available to astronomers
until fairly recent times (Figure 5.5). He accomplished this by improving
upon existing equipment, by developing new sights, and by enlarging

instruments so the scales could be more finely divided. In addition, he refined the measurement of the passage of time by referring observations to the position of the sun. He also increased the number of planetary observations from a few isolated points in the orbit to frequent observations all along the orbit. Despite all these improvements observational astronomy was still limited by the human eye. It wasn't until 1608, when Hans Lippershey, a Dutch lens grinder, developed the first telescope, that the first major advance in centuries was possible in the science of astronomy.

Figure 5.3 Ancient royal observatory of Delhi, India. (Bettmann Archive)

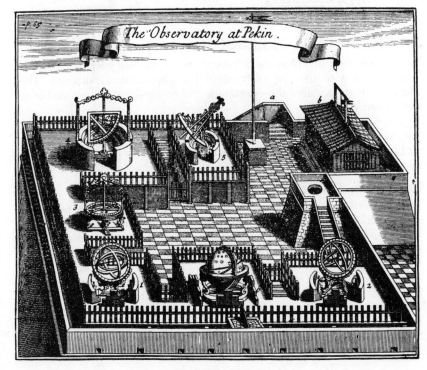

Figure 5.4 An observatory of China in the seventeenth century. (Bettmann Archive)

Figure 5.5 Tycho Brahe in his observatory on the island of Hven, 1602. (Bettmann Archive)

5.2 THE OPTICAL TELESCOPE

Hans Lippershey invented the telescope, but to Galileo goes the credit of being the first to use the instrument for astronomical purposes. The telescope made possible the examination of celestial objects not visible to the human eye.

There are two fundamentally different types of optical telescopes, but they serve the same primary function, that of gathering light within the visible range of the spectrum and focusing the light to produce a brighter image.

Refracting Telescope

One of the basic types of optical telescopes is the refracting telescope, which makes use of an objective lens to bend or refract light to a focus. The *objective* is the principal image-forming component of a telescope. In chapter 4 we discussed how refraction occurs when light passes from air into a more dense medium such as glass. The amount of refraction can be readily measured, as it is a physical property of the transparent material and will always be the same for a given substance under similar conditions. A ray of light entering the curved surface of the objective lens (Figure 5.6) at L_1 will be refracted toward the normal N_1 forming angle r_1. When leaving the lens at L_2 the ray will be refracted away from the normal,

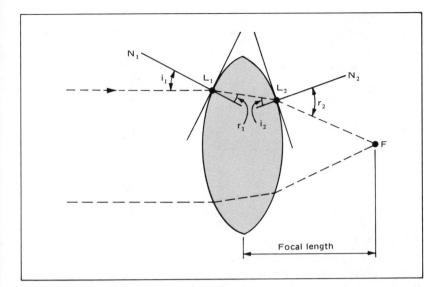

Figure 5.6 Refraction of light through a lens will focus the light to a focal point.

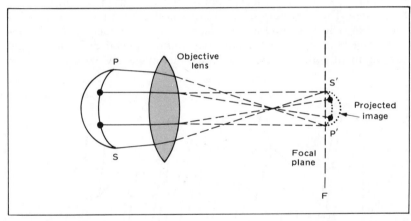

Figure 5.7 Light rays from an extended object like the moon will be focused on the focal plane.

and angle i_2 will be smaller than angle r_2. The net result is that light rays from a single distant object approach the lens in a parallel manner and are brought to a single focus at the *focal point F*. Light from several distant sources or from different parts of a single large object such as the moon would enter the lens from slightly different directions and therefore would focus, not on a single point, but on the same *focal plane* (Figure 5.7). The distance from the center of the lens to the focal plane or focal point is the *focal length*.

In photographing portions of the sky, the photographic plate is placed so that the active surface of the plate will coincide with the focal plane, and all images will be focused on the plate. While this arrangement is used for photographing distant objects, it is not suitable for visual observation. For visual observation a second lens or combination of lenses is required, which is the *eyepiece*. There are several arrangements possible for an eyepiece, but the primary function of all is to refract light rays so that they enter the eye in a parallel manner, although with an inverted image, and to magnify the image (Figure 5.8).

Figure 5.8 An eyepiece is used to redirect light rays so that they enter the eye in a parallel although inverted manner. The eyepiece will magnify the image.

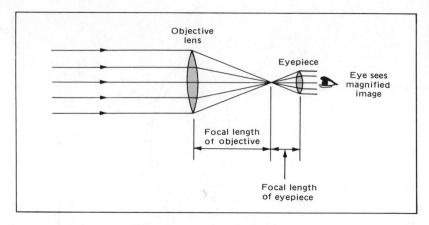

Reflecting Telescopes

The reflecting telescope was devised by Isaac Newton to overcome some of the deficiencies inherent in the refracting telescope. Instead of permitting light to pass through a lens, a concave mirror is used to bring light rays to a focus. The curvature of the mirror is that of a parabola, which is capable of reflecting and focusing light on the prime focus or focal plane. Reflecting telescope construction may follow several different designs, varying mainly in the manner in which the image is viewed (Figure 5.9).

Figure 5.9 Several designs are possible with the reflecting telescope: (a) direct focus as used in large telescopes such as the 200 inch; (b) Newtonian focus with flat secondary mirror to redirect the light through the side tube; (c) Cassegrain with a convex secondary mirror that spreads the beam and increases the focal length of the objective; (d) Coude focus with a convex secondary mirror and a third flat mirror to reflect light along the polar axis for use with spectroscopic equipment.

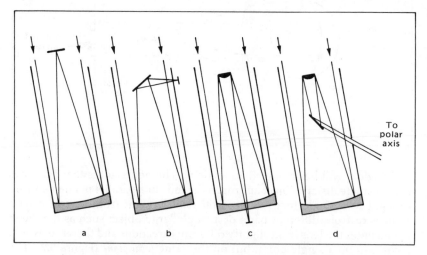

In the large telescopes such as the 200-inch* Hale telescope on Mt. Palomar (Figure 5.10), the *prime focus* design is used. The image is reflected from the objective mirror to the focal plane, where a mount for a photographic plate is positioned. This is located in an observer's cage at the upper end of the tube, where the astronomer sits and operates the telescope (Figure 5.11). Although seemingly large, the cage blocks only about 15 percent of the light entering the telescope.

In the smaller telescopes, popular with amateur astronomers, the *Newtonian focus* is used. This type is suitable for direct viewing and was the original design developed by Newton. It has a secondary mirror, which

---

*Telescopic measurements are commonly given in the English system and this convention will be continued here. All other measurements will be in the metric system.

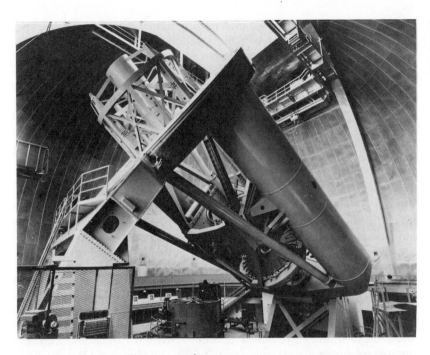

Figure 5.10 Palomar Observatory 200-inch Hale telescope pointing north. (Courtesy of Hale Observatories)

redirects light, focused by the objective mirror, out through a hole in the side of the tube. The image may then be magnified by an eyepiece placed at the exit hole in the side of the tube.

The *Cassegrain reflector* has a convex secondary mirror, which reflects the light back toward the objective mirror and out through a hole in the center of the objective mirror. This arrangement allows the observer to view through the lower end of the telescope and provides for a long focal length with a relatively short tube for more convenient portability of the instrument.

Figure 5.11 Observer in prime focus cage of 200-inch Hale telescope. The reflecting surface of the 200-inch mirror is visible. (Courtesy of Hale Observatories)

A modification of this design is used in the large reflecting telescopes for redirecting light to a convenient location for spectrographic analysis. This is the *Coude focus*, which employs a convex secondary mirror to reflect light back toward the objective mirror. A third mirror placed between the objective mirror and the convex secondary mirror is set to reflect the light out through the side of the tube at the lower end of the telescope. This third mirror is positioned to reflect the light along the polar axis, about which the telescope is automatically turned to compensate for the Earth's rotation. In this way the spectrographic equipment may be permanently installed in a room to permit control of the temperature.

The Schmidt Telescope

The Schmidt telescope is, in a manner of speaking, a compromise between the refracting and the reflecting telescopes, designed to reduce some limitations of both types. The Schmidt telescope utilizes a spherical mirror as the objective to reduce chromatic aberration and coma (Figure 5.12). A correcting lens, sufficiently thin so that it does not introduce any appreciable chromatic aberration, is utilized to correct for spherical aberration of the spherical mirror. (See section 5.3 for a discussion of chromatic and spherical aberration and coma.) The image of this system is projected onto a curved focal plane, which makes the telescope suitable for photographic work only. It does provide a unique system for obtaining excellent photographs over a wide-angle view of the sky (Figure 5.13).

The Solar Telescope

The primary function of a solar telescope is to provide a means of physically examining the surface of the sun and for spectroscopic studies of the sun. To accomplish this, the solar image is projected to a darkened area, in some instances below ground level. One of the problems in solar observation is thermal disturbance, whereby atmospheric conditions distort images due to radiation of heat from the Earth's surface. Some observatories have chosen to locate their sites overlooking a lake where the quiet waters provide more stable thermal conditions.

Figure 5.12 Diagram of the Schmidt telescope which utilizes a thin correcting lens to correct for spherical aberration on the spherical mirror. This system permits the entire surface of the mirror to be used.

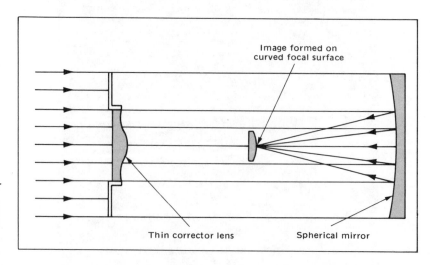

Image formed on curved focal surface

Thin corrector lens   Spherical mirror

Figure 5.13 The 48-inch Schmidt telescope with viewer at guiding eyepiece. (Courtesy of Hale Observatories)

5.3 LIMITATIONS OF OPTICAL TELESCOPES

Both the refracting and the reflecting telescopes have some limitations. The refracting telescope has a defect known as *chromatic aberration*, due to the fact that the various wavelengths of visible light are refracted at slightly different angles through a given lens. Thus wavelengths yielding blue and violet light, which are refracted the most, will focus at a point closer to the objective lens than the average for all the light, whereas red light is bent the least and focuses beyond the focal plane (Figure 5.14). This

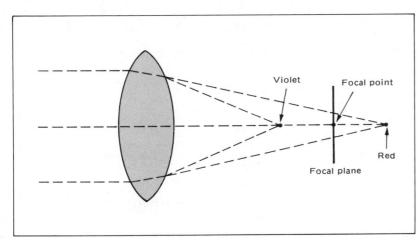

Figure 5.14 Chromatic aberration results from the fact that different wavelengths of visible light are reflected at slightly different angles. Violet light, refracted the most, focuses closer to the objective than red light, which is refracted the least.

Figure 5.15 Great tele-
scope built by Hevelius in
the seventeenth century.
Due to the long focal
length no tube was avail-
able and the lens was
suspended by wires.
(Bettmann Archive)

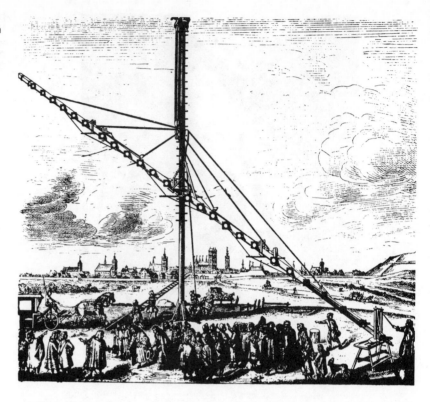

results in a color defect manifested by a fringe of color around the image of some distant object. An early attempt to overcome this problem was to build telescopes with long focal lengths of 50 metres or more (Figure 5.15). This permitted the use of objective lenses with a minimum of curvature and thus reduced the amount of bending of light with less dispersion. However, the telescopes were cumbersome and extremely difficult to operate and therefore quite unsatisfactory.

Another attempt to reduce chromatic aberration was made by John Dolland, an English optician in the mid-eighteenth century. He developed a compound *achromatic lens,* which consisted of a double convex lens made of crown glass and a plano-concave flint glass lens (Figure 5.16). Because of

Figure 5.16 A single con-
vex lens fails to bring
violet and red colors to a
focus, but a compound
lens made of glass of
slightly different densities
will bring these colors to
a common focus.

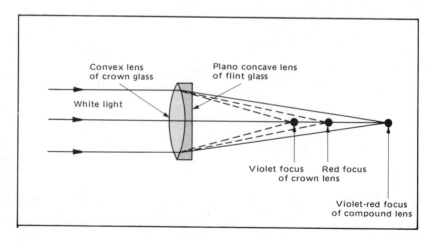

the shapes and the slightly different indices of refraction of the lenses, a correction in the refraction of the different wavelengths of light reduced chromatic aberration. By the careful selection of the wavelengths of light in which a celestial object was to be viewed, it became possible to make achromatic lenses that produced a satisfactory image with less color distortion. Refracting telescopes used for visual observation in the nineteenth century generally had lenses which brought yellow and green light to the same focus. It was felt that this served the best purpose, since the eye was not as sensitive to red and violet light as to the light in the center of the visual range, and therefore the fringes of color from red and violet light would be less troublesome. A disadvantage of the achromatic lens was that four surfaces had to be ground instead of two.

Another fault in both refracting and reflecting telescopes is *spherical aberration*. This occurs where the lens or mirror is shaped to follow the curvature of a circle (Figure 5.17), in which case the image projected near the center of the lens or mirror is focused further from the objective while the image projected from the periphery is focused closer to the objective. Spherical aberration may be corrected in a refracting telescope by using two lenses with different indices of refraction, and the elements may be constructed to reduce both spherical and chromatic aberration. In a reflecting telescope a mirror whose surface is ground in the shape of a parabola is used (Figure 5.18). Although more difficult to grind than a

Figure 5.17 Spherical aberration results when the shape of a lens (a) or mirror (b) follows the curvature of a circle. Images projected from the periphery of the lens or mirror are focused nearer to the objective than images projected from the center.

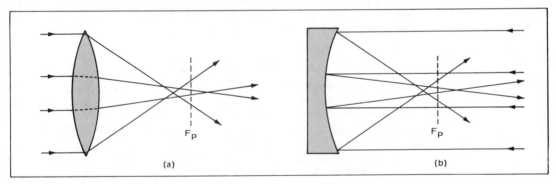

(a)

(b)

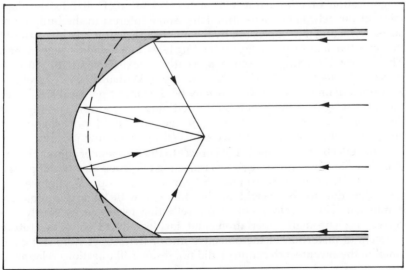

Figure 5.18 Spherical aberration in reflecting telescopes may be corrected by using a concave parabolic mirror.

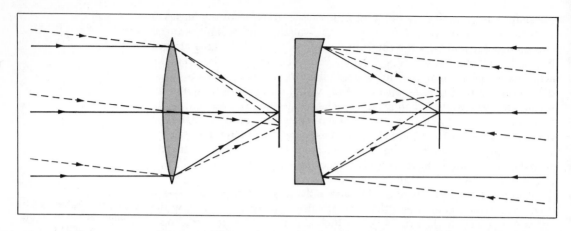

Figure 5.19 Images formed from light not entering the telescope parallel to the axis of the objective will appear as small elongated objects.

spherical mirror, the parabolic mirror will have the advantage of reflecting all parallel light rays from the same source to a single focus.

An additional form of distortion is *coma*, which is the distortion of the image formed from light that does not enter the telescope parallel to the axis of the objective (Figure 5.19). Although noticeable in both refracting and reflecting telescopes, it is a more serious problem in large telescopes with parabolic mirrors. Star images near the center of the field of view will be sharp pinpoints of light, whereas star images on the periphery of the field will appear as tiny cometlike objects; hence the name coma for this form of aberration.

5.4 TELESCOPE PERFORMANCE

One of the primary questions asked about the performance of a telescope is its magnifying ability. Because of small size, poor quality glass, and relatively short focal length, early telescopes were comparatively low powered. Galileo's best instrument magnified 33 diameters. To increase magnification (and solve the problem of chromatic aberration), telescopes were built with longer focal lengths, some going to the extremes depicted in Figure 5.15. This was not a practical solution to either increasing magnification or eliminating chromatic aberration. Dolland's development of the achromatic lens stimulated some interest in the building of larger refracting telescopes, but by the nineteenth century the diameter of the objective lens of the largest refracting telescope was only nine inches. The largest refracting telescope presently in existence is the 40-inch refractor located at the Yerkes Observatory in Williams Bay, Wisconsin. A 40-inch diameter objective lens may be the largest practical size, since the great weight of the lens is supported only by the rim. The slightest distortion due to sagging or bending of the lens would more than offset any gain in light-gathering ability achieved by increasing its size.

The reflecting telescope is not subject to the problem of support of the lens. The objective mirror used in a reflecting telescope may be supported on the back as well as around the rim and, by careful engineering, distortion due to the weight of the mirror can be almost completely eliminated. The development of the reflecting telescope by Newton to correct for chromatic aberration, and incidentally to solve the future problem of distortion of the objective mirror (telescopes were still quite small in the seventeenth century), did not resolve all questions relating to telescope construction. Difficulty was experienced in grinding suitable surfaces that were capable of accurately reflecting light to a common focus. In addition, until the twentieth century the objective mirrors were made of metal (speculum: 2 parts copper, 1 part tin with a small amount of arsenic for whiteness), which tarnished readily and required constant resurfacing to maintain good efficiency. Despite these drawbacks many fine instruments were built, and by the end of the eighteenth century William Herschel, an English astronomer, had constructed a 48-inch reflector made of a bronze alloy, which reflected about 60 percent of the light when in its most highly polished condition (Figure 5.20).

In 1845, William Parsons of Ireland, the third Earl of Rosse, constructed the largest metallic objective mirror ever built. The 72-inch disk required 17 years in building and weighed four tons when completed. The tube was almost 45 feet in length and had to be enclosed to prevent the wind from causing it to sway (Figure 5.21).

In the twentieth century, glass had replaced metal as the material for making objective lenses and mirrors. The largest reflecting telescope in service by mid-century was the 200-inch Hale telescope located on Mount Palomar (Figure 5.10). A still larger reflecting telescope was constructed in the Soviet Union located in Zelenchukskaya, northwest of Tiflis in the Caucasus mountains. The mirror is about 236 inches in diameter and housed in a tube 82 feet in length. The telescope was placed in operation in 1976.

An increase in the size of the telescope does not necessarily result in an increase in magnification. Rather, the increase in diameter will improve the performance of the telescope by increasing the amount of light from the distant object that the telescope can bring to a focus. This increase in light or *light-gathering ability* is dependent upon the area of the objective lens or mirror. If we compare one telescope with another or with the human eye, we find that the light-gathering power is proportional to the

Figure 5.20 Sir William Herschel's grand telescope. The 48-inch objective mirror was made of a bronze alloy that reflected 60 percent of the light. (Yerkes Observatory photograph)

Figure 5.21 William Parsons' 72-inch reflector was the largest metallic objective mirror ever built. (Yerkes Observatory photograph)

square of the diameter of the aperture. This relationship may be expressed as

$$\text{Light-gathering ability} = \frac{(\text{Diameter of objective 1})^2}{(\text{Diameter of objective 2})^2} \qquad (5.1)$$

On this basis the 200-inch telescope on Mount Palomar may be compared with the 100-inch telescope on Mount Wilson, and the Mount Palomar instrument will be shown to have four times the light-gathering ability of the Mount Wilson telescope. By gathering more light a much sharper image is available without an increase in magnification.

Magnification may be expressed as the amount by which the object's apparent distance from the observer has been reduced. Magnification is based on the relationship of the focal lengths of the objective and the eyepiece and may be expressed as

$$M = \frac{F}{f} \qquad (5.2)$$

where M is the magnification in diameters, F is the focal length of the objective, and f is the focal length of the eyepiece. As an example, if the focal length of the objective is 100 inches and f equals two inches, then $M = 100/2 = 50$ diameters. An object viewed through a telescope with these characteristics would appear 50 times larger than when viewed by eye, or would appear to be 1/50 of the real distance from the observer.

The light from individual stars is not magnified, but in examining stars the light-gathering ability of a telescope is important. In viewing objects within the solar system (planets, satellites), both magnification and

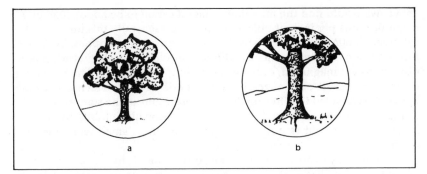

Figure 5.22 Increasing magnification decreases the field of view. Low magnification (a) permits seeing a large field. Increased magnification (b) reduces the field of view.

light-gathering ability need to be considered. If we have two telescopes with the same aperture (same light-gathering ability), but one has an eyepiece permitting twice the magnification of the other, it will form an image twice the diameter or about four times the area. However, each unit of area of the more highly magnified image will be only one-fourth as bright, making details more difficult to see. This kind of relationship is of great importance when objects are being photographed.

Another factor in telescope performance that must be considered is the fact that magnification cannot be increased indefinitely. Any increase in magnification will also decrease the field of view (Figure 5.22). The *field of view* of any telescope can be determined from the telescope's magnification and the field of view of the eyepiece given in degrees by the manufacturer. This may be expressed as follows:

$$\text{Field of view} = \frac{\text{Apparent field of eyepiece}}{\text{magnification}} \quad (5.3)$$

If the eyepiece has a field of 25° and the telescope with that eyepiece magnifies 50 diameters, then the actual field of view is 0.5°, which is about the same angular diameter of the moon as seen in the sky.

Practical limits to the amount of magnification have been established and are pretty much dependent upon the diameter of the pupil of the human eye. To decrease the amount of magnification requires that the diameter of the beam of light exiting from the eyepiece increase. If this increase becomes greater than the diameter of the pupil of the eye, then all of the image will not enter the eye but will be blocked by the iris. The night-adapted eye has an aperture of about 0.75 centimetres (0.3 inch), and by using the following relationship it is possible to determine the lower limits of magnification.

$$M = \frac{1}{d}$$

where M is magnification in diameters per inch of objective aperture, and d is diameter of the beam of light exiting from the eyepiece. Hence if the pupil of the eye can ideally accept light from the eyepiece 0.3 inches in diameter, the minimum magnification would be about $3.3M$. This means the minimum magnification for a 12-inch telescope would be 36 diameters. The minimum diameter to which the average human eye can contract is about 0.06 centimetres (0.025 inch). Using the same relationship as above, we would find the maximum magnification to be about 40M or 40 magnifications per inch of objective aperture. Generally this limit may be extended to 60M on a good viewing night with excellent optics. Increased magnification creates some additional problems in that atmospheric distortions are magnified, causing "twinkling" of the stars. The twinkling effect is the result of changes in atmospheric density, which causes the light to be refracted to differing degrees and results in the stars apparently changing position and brightness. This occasionally inhibits the use of large telescopes, but is not a great problem where small telescopes are concerned.

Another way to measure telescope performance is by the *resolving power* of the telescope, that is, the ability to aid the eye in separating two objects that are close together. For example, the human eye is capable of distinguishing letters on an eye chart at a standard distance of about 6.1 metres (20 feet). At greater distances the letters cannot be distinguished because, as experiments have shown, objects closer than 6 minutes (0.1°) of arc cannot be seen separately by the eye. Several lights close together can be seen as separate objects at 10 metres, but would appear as one light at 300 metres. The same is true for distant stars. Mizar, the second star in the handle of the Big Dipper (Ursa major), appears as a single star to the eye, yet even a small telescope will resolve it into the two components of a double star system.

Light waves tend to interfere with each other when focused by the objective, causing the formation of interference rings or a diffraction pattern around the image of a star and resulting in a slightly hazy image. If the interference pattern of two stars overlap, the stars will appear as one object. Increased magnification will not eliminate the problem, but increasing the aperture will. A relationship known as *Dawes rule* measures the ability of a telescope to resolve the images of distant objects in seconds of arc as follows:

$$L = \frac{4.56}{a} \text{ seconds of arc} \tag{5.4}$$

where L is the limit of resolution of the telescope and a is the diameter of the objective in inches. Thus a 10-inch telescope can resolve the image of two stars into separate components if they are at least 0.456 seconds of arc apart.

5.5 SPECTROSCOPY

Only a small percentage of the telescopic work done by modern astronomers represents actual viewing. Most of the work is accomplished by photographic methods or by analyzing the light through a spectroscope. A *spectroscope* utilizes a prism or a diffraction grid to spread light from a source such as the sun or a distant star into the various colors of the spectrum. In Figure 5.23 a schematic diagram of the spectroscope is shown. Light from a source passes through a slit and goes through a collimating lens, which causes the rays to enter the prism in a parallel manner. The rays from the spectrum are then directed into a camera and onto a photographic plate. The prism is used in a spectroscope if the source of light is faint, as from a distant star or galaxy. In the event the light

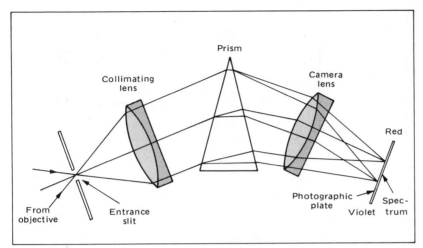

Figure 5.23 Schematic diagram of a spectroscope.

source is bright, for example, from the sun or a bright planet or star, then it is possible to use a diffraction grid, which has the advantage of providing a larger spread to the spectrum and allows for more detailed analysis.

The Spectrum

Work on spectroscopy can be traced back to Isaac Newton, who was one of the first men to examine the spectrum in some detail as an outgrowth of his work on lenses. He was able to show that the order of the array of colors never varied, although he tested the idea with various prisms and under a variety of conditions (Figure 5.24). Newton did find that by passing light through one prism and then through a second reversed prism, the light recombined into white light (Figure 5.25). He also discovered that by passing a single color through a second prism, no further color change resulted, indicating that the color was essentially pure (Figure 5.26).

Further work on the prism in 1802 by William Wollaston, an English scientist, revealed the existence of certain dark lines in the spectrum when the sun was used as the light source. Wollaston was unable to discover any reason for the lines, so he ignored them. At a later date (1814), Joseph von Fraunhofer observed the same phenomenon and considered the pos-

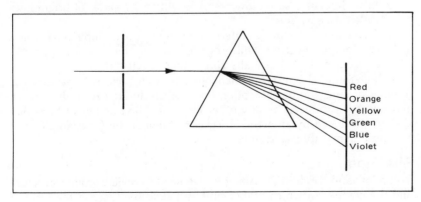

Figure 5.24 Newton found that the array of colors formed by passing a beam of light through a prism never varied in order.

Figure 5.25 Newton was
able to show that the
colors of the spectrum
resulting from the pas-
sage of light through a
prism recombined into
white light when passed
through a second re-
versed prism.

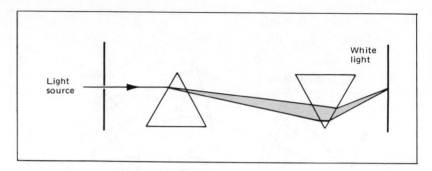

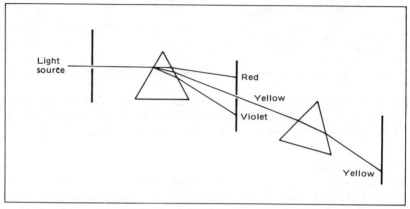

Figure 5.25 Newton was able to show that the colors of the spectrum resulting from the passage of light through a prism recombined into white light when passed through a second reversed prism.

sibility that the dark lines represented the dividing lines between the various colors. He discarded this idea when, with better equipment, he was able to observe hundreds of dark lines in the solar spectrum. He assigned letters to the more prominent lines, which are still identified as the *Fraunhofer lines* and are useful in designating approximate positions in the visible spectrum.

In the mid-nineteenth century Gustav Kirchhoff was able to show that spectral lines were produced by certain elements under prescribed conditions, regardless of the source of light. These conditions, sometimes referred to as Kirchhoff's principles of spectral analysis, are (Figure 5.27):

1. If a solid, liquid, or gas is in an incandescent state at high pressure, a *continuous spectrum* is emitted.
2. Incandescent gases at low pressure will emit a series of bright lines called *emission lines.*
3. A cool low-pressure gas intercepting light from a source producing a continuous spectrum will produce *absorption lines,* which appear as dark lines on the normally continuous spectrum.

Kirchhoff found that each element yielded a characteristic pattern and number of emission or absorption lines which differed from the pattern and number produced by any other element. The number of lines that may be produced by an element varies tremendously; for example, iron has more than a thousand lines.

The Atom

To understand Kirchhoff's principles we need to review atomic structure and the manner in which atoms emit and absorb energy in the form of light to form spectral lines.

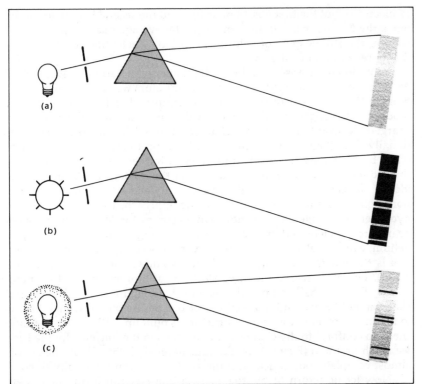

Figure 5.27 Kirchhoff's principles of spectral analysis. An incandescent source (a) emits light in a continuous spectrum. An incandescent gas at low pressure (b) will emit bright lines of the emission spectrum, and a cool gas at low pressure (c) will produce absorption lines in the continuous spectrum.

Early in the twentieth century, through the work of E. Rutherford and others, the atom was found to consist mainly of empty space. At the center is a positively charged nucleus, composed of protons with positive charges and neutrons with zero charge. Surrounding the nucleus are a number of electrons corresponding in number to the protons, but with a negative charge. The proton and neutron are about equal in mass, each having a mass 1836 times greater than that of the electron. Each element contains a number of protons and electrons characteristic of that element. Thus hydrogen has one proton and one electron, helium has two protons and two electrons, and so forth to the last natural element, uranium, which has 92 protons and electrons. The atomic number of the element corresponds to the number of protons that occur in the element. Atoms of an element with the same number of protons may have different atomic masses because the number of neutrons varies. These atoms are called *isotopes* of that element. For example, there are three known isotopes of hydrogen, each with one proton and one electron. *Protium,* the most common form, does not have any neutrons and has an atomic mass of one. *Deuterium* has one neutron in the nucleus, giving it an atomic mass of two, and *tritium* has two neutrons in the nucleus, giving it an atomic mass of three. It should be noted in the use of the term isotope that all three forms of hydrogen described are isotopes. It is incorrect to consider the most common form as the atom and the other two as isotopes.

The atomic mass is generally the average mass of the isotopes of any particular element. The atomic mass is based on the *atomic mass unit,* which is defined as one-twelfth the mass of *carbon 12.* Thus, for example, there are two isotopes of chlorine: ^{35}Cl and ^{37}Cl. Both forms have 17 protons and

17 electrons, but the first has 18 neutrons and the second has 20 neutrons. Since the atomic mass of an element is the average relative mass of the isotope, the atomic mass of chlorine is 35.453, which indicates that ^{35}Cl is the more common of the two.

Why do atoms absorb and emit light energy in various wavelengths? An empirical approach to the structure of the atom was suggested by Niels Bohr in the early part of the twentieth century. He suggested an atomic model for hydrogen in which electrical forces bind a single negatively charged electron to a positively charged proton. The electron is conventionally described as revolving around the proton in specific orbits or *shells*, but in reality the orbits do not exist. Rather, the electron forms a three-dimensional vibrating cloud around the nucleus of the atom. The idea of shells is used for the sake of simplicity and schematic representation. Following this somewhat simplistic approach, we may say that in hydrogen the single electron normally occupies the lowest energy shell nearest the nucleus and the atom is in a low energy state (Figure 5.28a). In helium, with two electrons, both occupy the lowest energy or first shell but this is the maximum number of electrons allowed in this shell (Figure 5.28b). Therefore two of the three electrons of lithium, the next element, occupy the first shell and one the second shell (Figure 5.28c). The second shell can hold a maximum of eight electrons before the third shell begins to fill. In this manner the atomic structure is built up, with electrons filling shells to conform to the charge of the protons in the nucleus. By the time we reach uranium (Figure 5.29) the four inner shells are filled, the fifth is almost completely occupied, and the sixth and seventh partially filled.

Now let us return to hydrogen, with one electron in the first shell. Under these circumstances the atom is said to be in the *ground state* or low energy state, which implies that should the electron be in an outer shell a high energy state exists. This means that an outside source of energy has been imparted to the atom, permitting the electron to break the bond holding it in the shell and causing it to move to an outer shell. Upon return to the low energy state or ground state, the electron reoccupies the first shell and energy is emitted in the form of a photon of light. The amount of energy is equivalent to the energy required to place the electron in the high

Figure 5.28 (a) Hydrogen atom with one proton in the nucleus and one electron in the first shell. (b) Helium atom with two protons and two neutrons in the nucleus and two electrons in the first shell. (c) Lithium atom with three protons and four neutrons in the nucleus and two electrons in the first shell and one in the second shell.

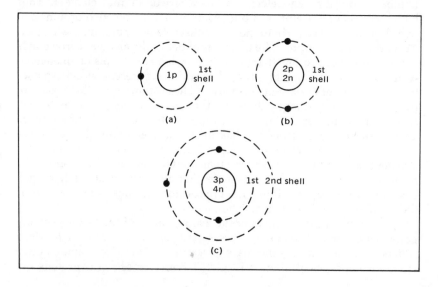

energy state. The wavelength of the light will be dependent upon the shell from which the electron leaves to return to the low energy state and the shell to which it returns (Figure 5.30). For example, an electron leaving the fifth shell to return to the ground state would emit a photon of light with a wavelength of 949 Å, which is in the ultraviolet portion of the spectrum and not visible to the human eye. An electron going from the fourth shell to the second shell would emit light in the wavelength of 4861Å, which is in the blue region of the visible spectrum.

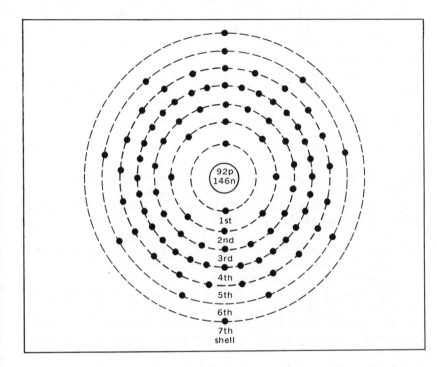

Figure 5.29 The uranium atom showing distribution of 92 electrons in seven shells.

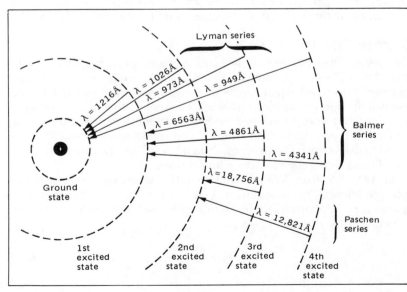

Figure 5.30 In the hydrogen atom energy in the form of a photon of light is emitted when the electron returns from a high energy state to a low energy state. The wavelength of the emitted photon is dependent upon the shell from which the electron leaves to return to a lower energy state. The diagram shows the Lyman, Balmer, and Paschen series of spectral lines in the hydrogen atom.

In the case of the sun or a star, huge gaseous bodies at very high temperatures (up to tens of thousands of degrees), numerous violent collisions occur between the atoms, raising the electrons to a higher energy state. The tendency is for the electron to return to the lowest energy state, which it does immediately in about 10^{-8} seconds. Since there are many possible high energy states an atom can achieve, there will be a variety of wavelengths of light emitted by the atom. The high temperature and density of the gas in the star will result in many collisions, and an electron may be caused to return to a high energy state repeatedly. The frequency and rapidity with which these events occur in a high-temperature, high-pressure gas cause the normally sharp emission lines of the spectrum to overlap and blend, resulting in the *continuous spectrum*, which includes all wavelengths of light.

The sun and stars have atmospheres that surround these bodies as cooler gases at low pressure. The stellar atmospheres contain the same elements as do the stars, and the photons of light from the star (continuous spectrum) can be absorbed by the atoms of gas in the atmosphere. Thus the photon represents another form of energy whereby the electron may be raised to a higher energy state. When this occurs the electron immediately returns to ground state, emitting a photon of light in the same wavelength as was absorbed. However, the possibility that the photon is emitted in the same direction traveled by the original photon is remote. Therefore it is unlikely that the emitted photon will enter the spectroscope and record that specific wavelength. The spectrum as seen through the spectroscope will then appear with dark lines at the positions of the wavelengths characteristic of the elements in the stellar atmosphere. This is the *absorption spectrum*, which is found to be extremely useful in determining the composition of the gas in a star.

Energy may also be imparted to a gas in the form of heat or by electrical means, and thus excite the atoms and cause electrons to achieve a high energy state. The return of the electron to ground state is accomplished with the release of a photon of light at the appropriate wavelength. This emission of light at the wavelengths characteristic of the atom emitting the light may be recorded by a spectroscope as an *emission spectrum*, and in this way reveals the composition of the object under study.

Zeeman Effect

The spectral lines may be modified by the presence of a magnetic field. In 1896, Pieter Zeeman discovered that the sodium lines of an emission spectrum broadened when a tube of glowing sodium vapor was placed between the poles of an electromagnet. He went on to find that normally single spectral lines broadened or became double or triple under the influence of the magnetic field. This became known as the *Zeeman effect* and served to detect magnetic fields on distant bodies, another useful application of the spectroscope in astronomical work.

In 1895, William Wein found that the wavelength of maximum radiation intensity in the spectrum, expressed as λ_{max} (lambda maximum) in angstroms, was inversely proportional to the temperature in degrees Kelvin. This may be expressed as

$$T°K = \frac{constant}{\lambda_{max}\text{Å}} \tag{5.5}$$

where Wein's constant is equal to 2.898×10^7 Å–°K. From this relationship it can be seen that as wavelength, at which a body such as a star glows with the greatest intensity, decreases the temperature increases. This coincides with the relationship between the color of a star and its temperature. A red star is cooler than a yellow star which in turn is cooler than a white star, since the color is related to the wavelength of light being radiated at greatest intensity.

Doppler Effect

Spectroscopic analysis is also useful in detecting motion toward or away from an observer by making use of the *Doppler effect*, first described in 1842 by Christian Doppler. His principle can be most aptly described in terms of the sound emitted by a train whistle as it approaches or leaves an observer. Sound from the train whistle will approach an observer in waves, which are manifested by the alternate compression and rarefaction of the molecules of the medium (generally air) through which the sound is traveling. If the train is moving toward the observer the sound waves are compressed; they have a shorter wavelength and are detected as a high-pitched sound. If the train is moving away from the observer the sound waves emitted are stretched out into a longer wavelength, resulting in a low-pitched sound.

Light also travels in waves, although light is itself a wave phenomenon and does not require a medium through which it must travel. The same phenomenon that occurrs in sound is also detected in light. A distant star moving away from an observer, for example, would have the spectral lines shifted toward the red end of the spectrum as compared to the spectral lines of a stationary system. The wavelengths of the light emitted from the object moving away from the observer are stretched out, taking on the characteristics of longer wavelengths found at the red end of the spectrum. On the other hand, the wavelength of light of objects moving toward the observer would be compressed and the spectral lines shifted toward the blue end of the spectrum. The amount of shift is related to the velocity of the object. This has been a most useful tool in establishing radial motion of stars and many distant galaxies. Almost all galaxies appear to be moving away from an observer on Earth, so the phenomenon has become known as the *red shift*.

To determine the direction and speed of a star or galaxy, a comparison photograph is made with the same spectroscope of some element such as iron. The comparison spectrum is made just before and after the stellar spectrum, and the displacement of the lines of the stellar spectrum indicates the radial motion being taken by the star. Using the Doppler principle, it is possible to calculate the velocity V of the star from the following equation:

$$V = c\,\frac{\lambda - \lambda_\circ}{\lambda_\circ}\ \text{kilometres per second} \qquad (5.6)$$

where c is the speed of light (3×10^5 km/sec), λ (lambda) is the observed wavelength, and $\lambda_\circ$ is the true wavelength as seen in the comparison spectrum.

The above equation will yield satisfactory results where the velocities of stars and nearby galaxies is in the vicinity of 30,000 kilometres per second or less. However, for larger red shifts the theory of relativity is invoked, and a more complex relationship is described by the following:

$$\frac{\lambda - \lambda_{\circ}}{\lambda_{\circ}} = \left(\frac{1 + \dfrac{v}{c}}{1 - \dfrac{v}{c}} \right)^{\frac{1}{2}} - 1 \tag{5.7}$$

The implications of the calculated values will be dealt with later in the discussion on Hubble's law (chapter 16) and quasars (chapter 17). Velocities in either case are designated positive (+) if the object is receding from the observer, resulting in a red shift in the spectrum, or negative (-) if moving toward the observer, manifested by a shift toward the violet end of the spectrum.

5.6 AUXILIARY EQUIPMENT

Although the spectroscope represents one of the most versatile instruments used in conjunction with the telescope, there are a number of other pieces of equipment that also add greatly to the usefulness of the telescope.

Camera

Most observational work by the professional astronomer is accomplished through the use of a camera. In fact, it may be said that the large telescopes are in effect giant cameras with light from the objective focused onto the photographic plate. A photograph taken of a celestial object or segment of the sky has several distinct advantages over visual observations. One advantage is that the photograph provides the astronomer with a permanent record of an observation made with a degree of accuracy not possible with the human eye. The photograph may be studied at leisure and compared with other photographs taken over a period of years of the same celestial region. In this way relative motion of the stars may be determined. Long time-exposures make it possible to resolve many faint and distant objects. In fact, objects a hundred times too faint to be visible to the eye may be successfully photographed. Painstaking study of such photographs have resulted in the discovery of such objects as the planet Pluto and the invisible components of multiple star systems (see chapter 15). The photographs may be inspected by many individuals and are available for future inspection and explanation of celestial phenomena, as, for example, they are used in this book.

Photometry

The amount of light coming from a star, or the stellar magnitude (see chapter 13), provides information on relative brightness. This may be measured by means of a photomultiplier tube (Figure 5.31). Light passing through the telescope is brought to a focus on a photocathode. When this occurs electrons are emitted which are attracted from one positively charged dynode to another. As an electron strikes a dynode more electrons are emitted, resulting in the generation of an electric current many times greater than initially received. The electric current generated is a measure of the light coming from the star and can be equated with its brightness.

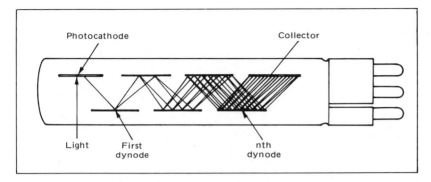

Figure 5.31 The photo-
multiplier tube.

It is also possible to use the same technique in measuring temperature of stars by means of such radiation-sensitive devices as thermocouples or thermopiles. A thermocouple consists of two wires of different composition joined together. When the point of attachment is subjected to heat, a measurable electric current is produced. By placing a thermocouple at the focal point of a telescope, infrared radiation (heat) from a star can be detected and the induced current, measured as a function of that radiation, will provide a gauge for the temperature of the star.

Image Tube

The image tube, a device which emits electrons when exposed to light, is used to enhance the image received by the telescope. The photo-sensitive surface of the tube is placed at the telescopic focus. Light from the celestial object generates a stream of electrons, which are focused directly onto a photographic plate or screen similar to a television screen for viewing. The enhanced image provided by the image tube permits much faster photographic work and reduces the time required at the telescope compared with the older procedures.

5.7 RADIO ASTRONOMY

We have thus far discussed the detection from the Earth of radiation at or near the visible portion of the spectrum, which represents only a narrow segment of the total spectrum. We have also mentioned those parts of the spectrum filtered out by the atmosphere, but there remains a large segment of the spectrum that was essentially ignored until the 1930s when Karl Jansky, a Bell Telephone scientist, observed radio signals coming from a source beyond the Earth. This segment includes radiation ranging in wavelength from 0.25 centimetres to 100 metres. Radio waves with a longer wavelength are reflected into space by the atmosphere and may be viewed only from some vantage point above the atmosphere.

Radiation in the radio portion of the spectrum is invisible to us: there is no way we are aware of whereby our bodies can detect radio signals in the same way that we see light. Jansky's discovery of radio waves from space prompted the development of the radio telescope after World War II and ushered in radio astronomy by providing a window on space in the radio portion of the spectrum, from which a great deal of information has already been obtained.

Radio telescopes function in the same manner as optical telescopes, in that the radio telescope is capable of reflecting radio waves to a focus in the

Figure 5.32 Schematic
diagram of a radio tele-
scope.

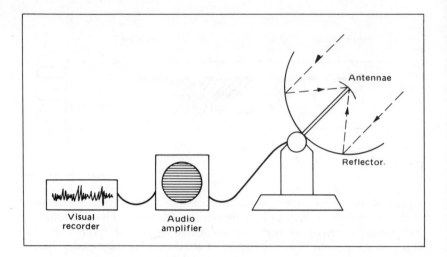

Figure 5.32 Schematic diagram of a radio telescope.

same way that light is reflected to a focus by optical telescopes. Radio
waves are reflected to an antenna, which serves to convert the incoming
radiation into electrical energy. A radio receiver amplifies the electrical
energy, permitting the recording of radio waves coming from space on
tape as an audio or visual signal (Figure 5.32).

By comparison with the mirror or lens of optical telescopes the radio
"dish" is huge. Movable parabolic dishes of approximately 90 metres (300
feet) in diameter are becoming common (Figure 5.33). These radio
telescopes are constructed so they can be pointed in any direction above
the horizon. Fixed radio telescopes can be built even larger, and depend
upon the motion of the Earth for scanning the sky. The 300-metre (1000-
foot) dish at Arecibo, Puerto Rico, is an example (Figure 5.34) of such a
device. The dish was constructed by placing a wire mesh over a natural
depression in the hills. Although fixed, some directional flexibility is
obtained by mounting the antenna on a movable trolley suspended from a
cable over the dish.

Figure 5.33 The 65-metre
(210-foot) tracking and
communications antenna
of the world-wide NASA/
JPL deep space network
near Goldstone, Califor-
nia. (JPL/NASA)

The surface of the dish is not as critical as is the surface of an optical mirror or lens. The 200-inch mirror, for example, was ground to an accuracy of a fraction of the wavelength of light, requiring highly skilled artisans and resulting in very high cost. The radio dish may be made of wire mesh if the holes are no larger in diameter than a small fraction of the wavelength of the radio waves. For the shorter radio waves a solid surface is more suitable, but it need not necessarily be smooth. A rough surface, which can generally be built for less expense, is satisfactory so long as the irregularities are no greater than a fraction of the wavelength of the radio waves to be reflected. This represents a distinct cost advantage for the radio telescope.

The radio telescope has several other advantages. Viewing with a radio telescope is not limited to nighttime; rather, it may be used day or night even during overcast weather, since radio waves are not appreciably influenced by these factors. In addition, radio waves can penetrate the gas and dust clouds that interfere with optical viewing in the central portion of the Milky Way and, because of the large size of the radio telescopes, very faint signals far beyond the capability of the optical telescope may be detected.

A disadvantage of the radio telescope is its tendency to pick up unwanted signals from electrical equipment or such natural phenomena as lightning. Another problem is its inability to resolve small details. This ability is vital to separate a source of radiation from among several closely related objects. The ability to resolve close objects is dependent on the size of the dish and the wavelength of the radio waves being observed. Because of the relatively long wavelengths involved, a very large dish is required.

Figure 5.34 The 300-metre (1000-foot) radio telescope (left photo) at Arecibo, Puerta Rico. The reflector is fixed in a depression in the hills. The antenna (right photo) is mounted on a movable trolley for directional flexibility. (The Arecibo Observatory 1000-foot telescope is part of the National Astronomy and Ionosphere Center, operated by Cornell University under contract with the National Science Foundation.)

The problem has been overcome by using two radio telescopes a known distance apart as an *interferometer*. The signal from a radio source will arrive at each antenna via very slightly different paths, and will therefore be slightly out of phase or interfere with each other. The angle formed by the line from the source and the baseline is a function of the degree of interference. If the length of the baseline is known, then it is possible to accurately determine the direction of the source. Very good results have been obtained by using radio telescopes that are as much as 10,000 kilometres apart, for example, in the United States and Sweden.

5.8 OBSERVATIONS FROM SPACE

Up to now we have discussed observations from the Earth's surface in that portion of the electromagnetic spectrum known as the visible spectrum, which represents only a very small portion of the total spectrum. Gamma radiation, X ray, and most of the ultraviolet radiation are screened out by the atmosphere and can only be examined by instruments at high elevation (above 35 kilometres) or in orbit around the Earth. Infrared radiation from space can be detected from the Earth's surface, but water vapor in the Earth's atmosphere reduces the efficiency of viewing. Infrared radiation can be more readily measured from high elevations. For example, at 12 kilometres above the Earth's surface, water vapor in the atmosphere is only about 50 percent of that found in the atmosphere at the surface. At 30 kilometres water vapor is reduced to about one percent. With the advent of the airplane, balloons, and manned and unmanned satellites, astronomy has taken a giant step toward viewing the sky in regions of the spectrum heretofore not visible.

Infrared astronomy was greatly enhanced by the development of an infrared sensing device developed in 1963. The infrared detector was cooled to a temperature a few degrees above absolute zero by immersing it in liquid helium. This made the device remarkably sensitive, and when placed at the focal plane of a telescope it was able to detect extremely faint infrared emissions from celestial objects. Its effectiveness was further increased by raising the infrared telescope sufficiently high in the atmosphere to reduce the obscuring influence of atmospheric water vapor to a minimum.

Ultraviolet astronomy is mainly dependent upon observations made above the larger part of the atmosphere by satellite. Ultraviolet rays are effectively screened by the atmosphere's ozone layer, except for a small segment of the spectrum just beyond the violet region. The Orbiting Astronomical Observatory (OAO) has proved to be particularly valuable in ultraviolet astronomy. Launched in 1968, it is a satellite composed of eleven telescopes, the largest of which is 16 inches. Solar energy cells, receiving energy from the sun, provide electric power to operate the receiver, transmitter, and other powered components. The equipment has been successful in "observing" objects at wavelengths down to about 1000 Å, a feat not possible from the Earth's surface.

X-ray astronomy deals with the region of the spectrum from about 100 Å to a little less than 1 Å. These radiations, fortunately for life on Earth, are effectively screened out by the atmosphere, and can only be observed at elevations of greater than 100 kilometres above the Earth's surface. Rockets and orbiting satellites are required to detect this form of radiation, and since the advent of the space age a number of X-ray sources have been discovered. The first X-ray detecting satellite (Uhuru) was launched in 1970 from a floating platform off the east coast of Africa. The satellite rotated once every twelve minutes as it orbited the Earth, scanning the sky in a comprehensive search for X-ray sources.

Radar astronomy has become a useful tool, but is useful only in the solar system. A radar beacon, sent from the Earth toward the moon or nearby planets or other objects in the solar system, is reflected back to Earth and the return beacon carefully studied for information on the characteristics of the object's surface features, velocity, and rotation. Objects beyond the solar system are too distant for such study, since the time required for the radar signal to reach a star and return is very great. The time required for a signal to reach even the nearest star and return would be in excess of eight years, and the returning signal would in all probability be too weak to be detected.

5.9 SUMMARY

Astronomical observations made in ancient observatories yielded much valuable data, despite the fact that telescopes did not exist until the seventeenth century. Most of the observations made during this pretelescope period were used to determine position and motion, and were made with a degree of accuracy that permits use of the information for comparative purposes even today.

The telescope was a major advance in astronomy. The refracting telescope makes use of a lens to bring light to a focus and the reflecting telescope utilizes a mirror to accomplish the same purpose. Each has advantages and disadvantages and attempts have been made to overcome these or minimize their effects. Chromatic aberration, spherical aberration, and coma are major limitations to the use of the optical telescope.

Telescope performance is determined by several factors. One of the primary measures is magnifying power. Magnification is accomplished by the eyepiece, and the amount of magnification is a function of the relative focal lengths of the objective and the eyepiece. Light-gathering ability is a measure of the amount of light from a distant object that a telescope can bring to a focus. It is dependent upon the area of the objective lens or mirror. The field of view, representing the area of the sky being viewed, and the resolving power or ability of the telescope to separate two closely positioned objects are also means by which the telescope performance is measured.

The spectroscope, when used in conjunction with a telescope, is another powerful tool used by the astronomer. The spectroscope permits detailed examination of light coming from a distant object, and yields information on composition, temperature, magnetic fields, and radial motion.

Although the spectroscope is recognized as a major addition to equipment used in the study of the skies, there are a number of other devices available as well. The camera serves to provide a permanent record and can record light from objects too dim to be seen with the human eye. A photomultiplier measures the light from stars and permits making comparisons of relative brightness. The image tube is useful in enhancing an image, making it easier and quicker to photograph.

The advent of the space age has opened up a number of additional possibilities for viewing. Forms of electromagnetic radiation screened out by the atmosphere in the infrared, ultraviolet, and X-ray regions of the spectrum can be studied. Even radar has been found useful in studying nearby objects in the solar system.

Radio astronomy has opened yet another window on space. The discovery of radio waves in the longwave region of the spectrum has resulted in the construction of large radio telescopes for viewing these radiations. Information obtained in this manner has resulted in some major advances in the overall view of the universe.

QUESTIONS

1. Briefly outline the development of instruments used in astronomical work up to the invention of the telescope.
2. What is the primary function of a telescope? Briefly describe the two basic types of optical telescopes.
3. Make a sketch of the various designs used in reflecting telescopes.
4. What are the principal limitations found in optical telescopes and how can these be remedied?
5. How much greater than the eye is the light-gathering ability of the 200-inch telescope if the eye has an aperture of 1/8 inch?
6. What properties of a telescope determines its (a) light-gathering ability, (b) magnifying ability, (c) field or view, (d) resolving power?
7. How many times greater would the diameter of the telescope objective need to be to increase its light-gathering ability by 25 times? How would this increase in light-gathering ability affect its magnification and its resolving ability?
8. How does the Schmidt telescope differ in construction from other optical telescopes? What is the principal advantage of the Schmidt telescope over other telescopes?
9. What are Kirchhoff's principles of spectral analysis?
10. An electron can move from the second shell to ground state in only one way and from the third shell to ground state in two ways: 3 to 1 and 3 to 2, 2 to 1. How many possible ways are there from the fourth shell to ground state; from the fifth shell to ground state?
11. Describe the Zeeman effect.
12. What is the Doppler effect as it applies to light?
13. What are the advantages of photographing celestial objects over that of viewing them visually?
14. How does the radio telescope differ from the optical telescope?
15. What are the advantages of radio telescopes over optical telescopes?
16. What portions of the spectrum are best viewed above the Earth's atmosphere? Why?

FOR FURTHER READING

Hey, J. S., *The Evolution of Radio Astronomy.* New York: Science History Publications, 1973.

Howard, N. E., *The Telescope Handbook and Star Atlas.* New York: Thomas Y. Crowell Co., 1967.

Miczaika, G., and W. Sinton, *Tools of the Astronomer.* Cambridge, Mass.: Harvard University Press, 1961.

Page, T., and L. W. Page, *Telescopes; How to Make Them and Use Them.* New York: The Macmillan Company, 1966.

Cassiopeia

CHAPTER 6   **THE SOLAR SYSTEM**

I find that by arranging these objects in a certain successive regular order, they may be viewed in a new light, and, if I am not mistaken, an examination of them will lead to consequences which cannot be indifferent to an inquiring mind.

—William Herschel

As a general rule, textbooks on astronomy follow a pattern in which one section is devoted to the solar system and another to stellar characteristics and the universe. It should be recognized that the solar system, although occupying a major portion of a text, represents only an infinitesimal portion of the universe. The more extensive coverage may be due to the fact that people have been much more intimately acquainted with the solar system since ancient times. However, this imbalance in knowledge of the solar system versus the rest of the universe is rapidly dwindling. New techniques have broadened our understanding of the universe beyond the solar system. At the same time exploration of the solar system has raised new questions that may be resolved only by direct study of the planets. The solar system, as the human abode, is possibly the limit of direct physical exploration of the universe for some time to come (Figure 6.1).

6.1 THE PLANETS

Our view of the solar system has evolved in the past 2500 years from an earth-centered to a sun-centered system. The heliocentric concept, developed by Copernicus, replaced the geocentric ideas held by the ancient Greeks and led to a more complete understanding of the true organization of the solar system. The sun, the most prominent known astronomical object in the sky, was recognized as the dominant member of the solar system, orbited by the planets Mercury, Venus, Earth, Mars, Jupiter, and Saturn, arrayed in that order from it. With the invention of the telescope three additional planets were discovered: Uranus, found beyond Saturn, in 1781; Neptune, more distant than Uranus, in 1846; and Pluto, the most distant from the sun, in 1930. This completes the list of nine known principal objects that we currently recognize as planets occupying the solar system.

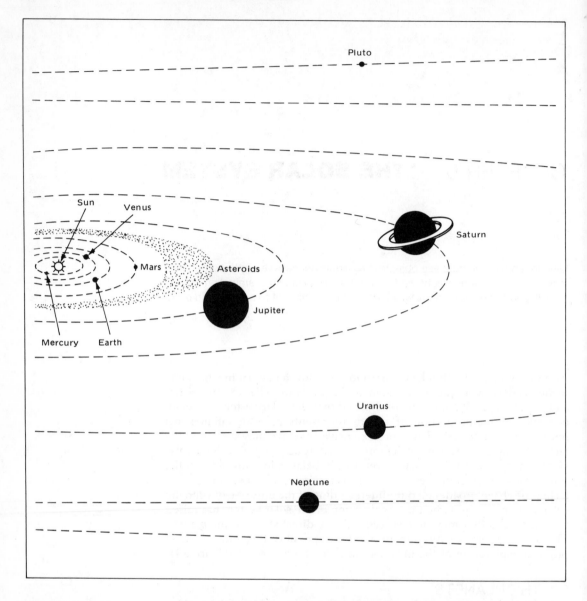

Figure 6.1 The solar system, composed of the planets, their respective satellites, and asteroids revolving around the sun.

If we viewed the solar system from the vantage point of the nearest star, it would be impossible to see the planets with our present technology. This is true, first, because the brilliant sun envelops the planets in its light and obscures the relatively dim reflected light from the planetary surfaces, and second, because the planets represent a very small percentage of the total mass of the solar system and are quite tiny by comparison with the sun. The nine planets comprise only 0.14 percent of the mass compared with the sun which contains 99.85 percent. The small balance remaining is incorporated in the satellites, a ring system associated with Saturn, comets, asteroids, meteoroids, and miscellaneous dust and gas.

Mass and Density

The mass of the individual planets or other major objects in the solar system may be determined by use of Newton's modification of Kepler's third law. This states that the product of the sum of the masses of two bodies revolving around a common center of gravity and the square of their mutual period of revolution equals the cube of the semimajor axis of their orbit, or

$$(m_1 + m_2)P^2 = a^3$$

or

$$(m_1 + m_2) = \frac{a^3}{P^2} \tag{6.1}$$

For planets orbiting the sun or for binary stars (see chapter 15), all masses m are expressed in terms of solar mass, P is expressed in years, and a in astronomical units. For the planet-sun system, it is convenient to express mass of the planets in terms of the Earth's mass equal to 1.00 (Table 6.1). By applying the equation given above we can determine the mass of any planet in terms of the Earth's mass if we know the period of revolution and the distance of one of the satellites from that planet. For example, it is possible to observe that Miranda, M_m, a satellite of Uranus, M_u (seventh planet from the sun), orbits that planet in 1.4 days at a mean distance of 1.28×10^5 kilometres, values that may be designated P_m and a_m respectively. We also know that the moon's sidereal period, $P_{\mathbb{C}}$, or the length of time it takes the moon to complete one revolution around the Earth, is 27.3 days (656 hours). The distance $a_{\mathbb{C}}$ between the Earth $M_\oplus$ and the moon $M_{\mathbb{C}}$ is equal to 3.84×10^5 kilometres. From this information we can form a ratio in which the earth-moon system is related to the other planet-satellite system, and we have

$$\frac{M_u + M_m}{M_\oplus + M_{\mathbb{C}}} \cdot \frac{P_m^2}{P_{\mathbb{C}}^2} = \frac{a_m^3}{a_{\mathbb{C}}^3}$$

Since the masses of the two satellites are relatively small compared to their respective planets, we may neglect them, and we have

$$M_u = \frac{P_{\mathbb{C}}^2}{P_m^2} \cdot \frac{a_m^3}{a_{\mathbb{C}}^3} M_\oplus \tag{6.2}$$

Substituting the relevant data and considering the mass of the Earth as 1.00, we obtain the mass of Uranus, M_u, in terms of the Earth's mass.

$$M_u = \left(\frac{27.3}{1.4}\right)^2 \cdot \left(\frac{1.28 \times 10^5}{3.84 \times 10^5}\right)^3 \cdot M_\oplus = 14\,M_\oplus$$

The mass for each planet relative to the Earth may be calculated in this manner, as can the mass of each planet relative to the sun. Since the mass of the Earth in tons has been experimentally determined (see chapter 7), the mass of all members of the solar system may be so calculated. Accurate mass determinations for Mercury and Venus, which have no satellites, have been obtained recently by measuring the influence of these planets on artificial satellites flying nearby. An accurate mass for Pluto is not yet available.

Each planet will be described in more detail in later chapters, but there are some general characteristics of the planets as a group that can be discussed to some advantage at this point. For example, it is possible to classify the planets in two ways: on the basis of common physical characteristics such as mass and density and on the basis of position with respect to the Earth.

In considering the physical characteristics of planets, we find when examining the data in Table 6.1 that the relative masses (Earth = 1.00) and diameters permit dividing the planets into two distinct groups. One group, the *terrestrial planets*, would include Mercury, Venus, Earth, and Mars. These are small in size, of low mass and relatively high density, and are grouped close to the sun. The second group, the *gas giants*, include Jupiter, Saturn, Uranus, and Neptune. These planets are large, of greater mass and low density, and are grouped further from the sun than the terrestrial planets. In making a comparison of the planetary densities, we see that the terrestrial planets as a group have relatively high densities, averaging approximately 5.0 grams per cubic centimetre. The gas giants, on the other hand, have densities which are only slightly more than that of water. In fact, Saturn has a density less than that of water, and if a large enough body of water could be found Saturn would float. Pluto, the planet most distant from the sun, is an enigma. It does not fit the pattern of the gas giants in spite of its proximity to them. It appears more like the terrestrial planets which are close to the sun, but Pluto's mass and diameter are not known for certain, and therefore its density is in doubt (Table 6.1).

It will be useful to become familiar with the nomenclature used to describe planetary positions. The nomenclature, already known to the ancient Greeks, may be applied equally well to lunar and planetary positions, as well as the positions of comets, asteroids, and other objects in the solar system.

The planets are divided into two groups: the *inferior planets*, those planets between the Earth and the sun including Mercury and Venus and the *superior planets*, the planets beyond the Earth with respect to the sun (Figure 6.2). Although the terms were originally used with the geocentric system, they apply equally well to the heliocentric arrangement because the motions of Mercury and Venus, as viewed from the Earth, are quite different from that of the other planets.

Elongation is the angle formed by imaginary lines drawn from the Earth to the sun and the Earth to a planet and defines the angular distance of the planet from the sun as seen from the Earth. The angle is described as *eastern elongation* if the planet is east of the sun (following the sun across the sky) and *western elongation* if the planet is west of the sun (leading the sun across the sky). The greatest angular distance that an inferior planet can attain is called its greatest (eastern or western) elongation. For Mercury this angle is 28° and for Venus, 48°.

Special terms are applied to elongations of significance. For example, when elongation is 0° the planet is said to be in *conjunction*. When Mercury and Venus are between Earth and the sun, they are in *inferior conjunction*, and when opposite the sun from the Earth all planets are in *superior conjuntion*. Any planet whose position forms a 90° angle with the Earth and the sun is said to be at (eastern or western) *quadrature*. If the elongation is 180°, that is, with the Earth between the planet and the sun, the position is called *opposition* (Figure 6.3).

TABLE 6.1 PHYSICAL PROPERTIES OF THE MAJOR PLANETS IN THE SOLAR SYSTEM

Planet		Discovery	Equatorial Radius (km)	Mass (Earth = 1.00)	Density (g/cm³)	Surface Gravity (Earth = 1.0)	Obliquity	Number of Satellites	Greatest Magnitude (see section 13.2)
Mercury	☿	Antiquity	2,432	0.055	5.4	0.38	7°	0	-1.9
Venus	♀	Antiquity	6,050	0.815	5.2	0.91	-179°	0	-4.4
Earth	⊕	—	6,378	1.000	5.52	1.00	23.5°	1	—
Mars	♂	Antiquity	3,394	0.107	3.9	0.39	25°	2	-2.8
Jupiter	♃	Antiquity	68,700	318.0	1.4	2.74	3.1°	14	-2.5
Saturn	♄	Antiquity	57,550	95.0	0.71	1.17	26.7°	10	-0.4
Uranus	♅	Herschel 1781	25,050	14.5	1.32	0.94	-98°	5	5.7
Neptune	♆	Adams Leverrier 1846	24,700	17.2	1.63	1.15	29°	2	7.6
Pluto	♇	Tombaugh 1930	2,900	0.1	4.85?	0.5?	?	0	

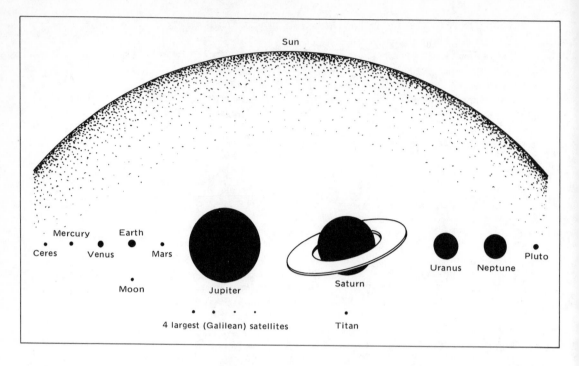

Figure 6.2 The sun and planets drawn to scale.

A planet in opposition or conjunction may not be at exactly 0° or 180°, because the orbits of the planets are not all on the same plane. The *ecliptic*, or the apparent annual path of the sun across the sky, results from the motion of the Earth around the sun. The imaginary plane thus formed, the *ecliptic plane*, serves as a reference plane from which the inclination of the orbital planes of the other planets are measured. The angle of inclination is generally very small, for most planets a few degrees, but for Mercury it is 7° and for Pluto, the outermost planet, it is 17°. Thus the angle at conjunction or opposition for these planets is not exactly 0° or 180° but the terms are useful nevertheless.

Motion

The motions of the planets are controlled by the sun, and all exhibit a similarity in direction of motion around the sun. As viewed from Polaris, the polar star, the planets as well as all other objects in the solar system revolve in a counterclockwise direction around the sun, or if viewed from the Earth the planets move across the sky from west to east against the background of seemingly stationary stars.

With the exception of Venus and Uranus the counterclockwise direction is the favored direction of rotation of the planets on their axes. Rotation can most easily be determined by close scrutiny of the movement of some distinguishing feature on the surface of the planet. After numerous observations of the feature passing the line of sight, it is possible to establish, with reasonable accuracy, the rate of rotation. In instances where surface features are indistinct, it is possible to make use of the Doppler shift of the light spectrum reflected from the surface of the planet. As the planet rotates one edge turns toward the Earth. The result is that light reflects in a shorter wavelength than light reflected from the center of the planetary disk. By the same token, light from the edge of the planet turning away from the Earth will have a slightly longer wavelength. This technique has been found useful in measuring the rotation of Uranus and Neptune, where distinct surface features are not visible. Radar has also been found useful in measuring rotation, as radar beams sent from Earth and reflected from the surface of a planet are Doppler-shifted as a result of the planet's rotation. The technique is limited to nearby planets, because the signal becomes too weak to be accurately detected from distant planets. The rotation of Mercury and Venus have been determined in this manner. Distinguishing features on these planets are difficult to see, because Mercury is very close to the bright sun and Venus is completely enshrouded in clouds.

If we look at the relationship of the sun and the planets with respect to the ecliptic, we find some interesting points that should be noted at this time. For example, the ecliptic does not coincide with the sun's equatorial plane, as one might expect from the manner in which the solar system is thought to have formed (see section 6.4). The sun's equator is inclined 7° from the ecliptic, and the orbital angle of inclination of only one planet, Mercury, corresponds to that. We also find that the *obliquity*, or the inclination of the equatorial planes of the planets to their orbital planes, is quite marked for most of the planets. Jupiter is an exception, as its equatorial plane is within a few degrees of coinciding with its orbital plane. By contrast, the Earth's equatorial plane is inclined 23.5° from its orbital plane, which by definition is also the ecliptic. The equatorial planes of Venus and Uranus are inclined by amounts greater than 90°, which implies retrograde (clockwise) rotation for these planets if viewed from a position north of the solar system. We have noted earlier that planetary motion (revolution and rotation) is described as viewed from a position north of the solar system, and obliquity is measured from this prospect. Venus' obliquity of 179° places its equator almost on the orbital plane but with the planet upside down in the sense that counterclockwise rotation is now observed from a south polar view. Uranus' obliquity of 98° means that the equator is now almost perpendicular to the orbital plane, and the counterclockwise rotation of the axis is seen a few degrees below the

Figure 6.3 Configuration of the inferior planets (a) and superior planets (b).

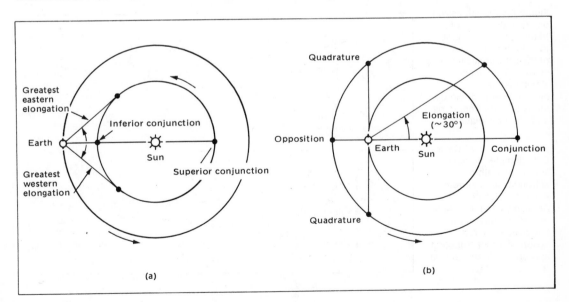

(a)   (b)

orbital plane. This gives Uranus the appearance of rolling along during part of its travels in orbit.

Distances

The mean distances of the planets from the sun range from 58 million kilometres for Mercury to 5910 million kilometres for Pluto. Each planet is approximately one and a half to two times the distance of its nearest internal neighbor from the sun. Distances within the solar system are, for convenience, expressed in terms of *astronomical units* (AU). Toward the end of the seventeenth century, G. D. Cassini defined the AU as the mean distance of the Earth from the sun. The modern definition is the mean length of the semimajor axis of the orbit of a hypothetical object with the mass and period of revolution around the sun that K. F. Gauss assumed for the Earth. The semimajor axis of Earth's orbit is 1.000000230 AU, but this is rounded off to 1.00 AU for convenience. The first attempt to measure this distance in miles was made by two Frenchmen in 1672. During that year Mars was in a favorable position of opposition and its parallactic displacement, as seen from Paris and French Guiana, was measured by triangulation (see section 13.1) (Figure 6.4). The astronomical unit was calculated to be approximately 87 million miles (140 million kilometres).

In 1877, David Gill made another attempt to accurately measure the astronomical unit by using the Earth's rotation to provide him with an adequate baseline (Figure 6.5). On Ascension Island in the Atlantic Ocean, Gill made paired observations, twelve hours apart, of Mars during a five-month period. The results were accurate, yielding a distance of about 150 million kilometres from the Earth to the sun. This distance has been confirmed many times by observations obtained by improved techniques.

An easy method for remembering the distances in terms of astronomical units is provided by a relationship known as *Bode's law*, although this is not a physical law in the strict sense. Bode's law was first suggested by Titius of Wittenberg in the eighteenth century and publicized by Bode in an astronomical journal. To each number in a series—0, 3, 6, 12 (found by

Figure 6.4 French astronomers making use of triangulation attempted to determine the length of the astronomical unit in 1672. They made simultaneous measurements of the distance to Mars from Paris and French Guiana.

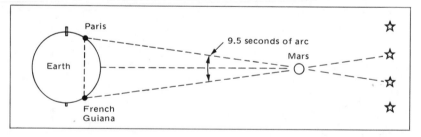

Figure 6.5 In 1877 David Gill attempted to measure the length of the astronomical unit from Ascension Island in the Atlantic Ocean. He made use of the Earth's rotation making paired observations of Mars first at *A* and 12 hours later at *B*.

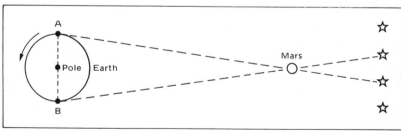

doubling the previous one)—4 is added and the sum is divided by 10. The resulting sequence is the approximate mean distance of the planets to the sun expressed in astronomical units. The calculated distances according to Bode's law corresponds quite closely to the measured distances of the planets with the exception of Neptune and Pluto (Table 6.2).

It is difficult to envision the magnitude of the solar system from a simple listing of the distances in a table. By constructing a diagram based on a recognizable scale, it may be possible to judge the extent of the distances to the Earth's neighbors, and we can visualize the immensity of the space that scientists plan to explore. If we consider a scale where the Earth's distance to the sun (one AU) is equal to 10 centimetres, then Mercury would be 4 centimetres and Venus 7 centimetres from the sun. Mars would be 16 centimetres and the asteroids would be distributed between 10 and 40 centimetres, with most in orbit around 28 centimetres from the sun. A distance of 52 centimetres would reach Jupiter and it would be 96 centimetres to Saturn. One would need to go 191 centimetres from the sun to reach Uranus, 301 centimetres to Neptune, and Pluto would be 395 centimetres from the sun. We have thus far traveled to the moon (¼ millimetre on this scale) and sent unmanned probes to Mercury, Venus, Mars, Jupiter, and Saturn. The task of reaching Uranus, Neptune, and Pluto with unmanned probes is great but can be accomplished with existing knowledge and technology.

Continuing in this line of thought leads to consideration of the prospect of reaching the stars. The real distance to the nearest star is so great that billions of kilometres would be too small a measure. On the scale used above (10 cm equals 1 AU) the nearest star would be 27 kilometres away. Even the astronomical unit would not suffice for expressing these immense distances, so instead the light year is used. The *light year* represents the distance light travels in one year moving at a velocity of 3×10^5 kilometres per second. This makes one light year equal to 62,400 astronomical units or 9×10^{12} kilometres. Proxima Centauri, the nearest star to the sun, is part of a triple star system that is 4⅓ light years from the sun. The possibility of reaching even this nearby group is quite remote with the knowledge and technology available to us at present.

Once the distance in kilometres of a planet from the sun is known, it is possible to determine the velocity of a planet in its orbit. Although planetary orbits are elliptical, the degree of ellipticity is relatively small for most planets, so we can assume a circular orbit and still maintain a reasonable degree of accuracy. The circumference of the orbit can be calculated, once we know the radius of the orbit (distance from planet to the sun), from the following relationship:

$$circumference = 2\pi r$$

where r is the orbital radius in kilometres and π equals 3.1416. By careful observation the period of revolution of each planet has been determined, and from this it is now possible to calculate the velocity, as follows:

$$v = \frac{2\pi r}{p} \tag{6.3}$$

where v is the velocity, $2\pi r$ the circumference, and p the period of

TABLE 6.2   DISTANCES, MOTIONS, AND RELATED ORBITAL DATA OF THE MAJOR PLANETS IN THE SOLAR SYSTEM

Planet	Distances 10⁶km	AU	Sidereal Period (Revolution)	Synodic Period (Days)	Rotation	Escape Velocity (km/sec)	Orbital Velocity (km/sec)	Inclination of Plane of Orbit to Ecliptic
Mercury	57.91	.3871	88 days	116	58.6 days	4.2	47.9	7°
Venus	108.21	.7233	225 days	584	243 days	10.3	35.0	3.33°
Earth	149.60	1.0000	365¼ days	—	23h 56m 4.1s	11.2	29.8	0°
Mars	227.9	1.5237	687 days	780	24h 37m 22.6s	5.1	24.1	1.85°
Jupiter	778.3	5.2037	11.8 yrs	399	9h 50m	61	13.1	1.31°
Saturn	1428.0	9.5803	29.6 yrs	378	10h 14m	36	9.6	2.49°
Uranus	2872.0	19.1410	84 yrs	370	10h 49m	22	6.8	0.77°
Neptune	4498.0	30.5982	165 yrs	368	16h	23	5.4	1.78°
Pluto	5910.0	39.4387	248 yrs	367	6.4 days?	?	4.7	17.17°

revolution. Since the velocity of the planets is expressed as kilometres per second, the period of revolution would need to be stated in seconds instead of days or years. From Table 6.2 it is possible to see that the more distant a planet is from the sun, the slower is its velocity in orbit.

Once the distance from the Earth to a planet is known, it is also possible to determine the size of a planet by measuring its angular diameter. The angular diameters of planets or other objects in space are quite small (with the exception of the sun and moon) and are usually measured in seconds of arc. The diameter of a planet or other object, subtending angle alpha (α) in seconds of arc at distance r, can be determined from

$$d = \frac{2\pi r \alpha}{1,296,000''} = \frac{r\alpha}{206,265''} \qquad (6.4)$$

where d is the diameter of the object in kilometres, r is the distance of the object in kilometres and alpha (α) is the angular diameter in seconds of arc. Diameters of various bodies in the solar system have also been determined by measuring the light reflected from the surface and comparing this amount of light with light reflected from a similar surface where the size of the object is known. The assumption is made that the surface reflectivity per unit of area for both objects is the same, and any variation in the amount of light will be a function of their relative sizes. This method is used for small objects where measurement of the angular diameter may be extremely difficult.

6.2 THE SATELLITES

All but three of the planets have satellites. These range in size from Titan, associated with Saturn and about 5700 kilometres in diameter, to tiny Deimos, one of the Martian satellites. Deimos is irregularly shaped and about 9-by-13 kilometres in diameter. Most but not all of these satellites revolve around their respective primaries in a counterclockwise direction. Mercury, Venus, and Pluto have no natural satellites, so the 34 known satellites are associated with the remaining six planets. Only Titan is known to have an atmosphere, but Pioneer 10 and 11 have yielded data indicating that the four Galileon moons of Jupiter may have atmospheres. A few others may be covered by water, ice, or frozen gases, but this has yet to be established.

We found by grouping the planets that certain characteristics and comparisons could be more readily made and the same is true for the satellites. The simple listing of 34 satellites does not yield their true nature. For example, we can separate the satellites into several groups, as shown in Table 6.3.

TRUE AND CAPTURED SATELLITES IN THE SOLAR SYSTEM   TABLE 6.3

Planet	True Satellite	Captured Satellites	Other
Earth	—	—	1
Mars	2	—	—
Jupiter	5	9	—
Saturn	8	2	1
Uranus	5	—	—
Neptune	1	1	—

The *true satellites* are those located on the equatorial plane of the planet with which they are associated, and it is assumed that this relationship existed from the time the solar system was formed. The second group, the *captured satellites*, have orbits that do not coincide with the equatorial plane of the related planet, and it is thought that these objects became satellites at some time since the formation of the solar system.

One must note that the moon does not fit the characteristics of a true satellite, but neither can it be said to be captured. The moon does not orbit the Earth's equatorial plane; rather, its orbit is quite close to the ecliptic, which is what one would expect of a planet. If we studied the combined orbits of the earth-moon system (see chapter 8), we would find that the moon constantly "falls" toward the sun, and the moon's orbit is always concave with respect to the sun. This is not true for any of the other satellites, which in some portions of their orbits "fall" away from the sun.

Saturn too has a satellite anomaly in the form of a ring system. The ring system, which will be described more fully in chapter 10, is composed of many fragments that have not coalesced into a single body. This may be due to the fact that the rings exist within a limiting distance from the center of the planet, called *Roche's limit*. This limiting distance was defined by E. Roche in 1849, and represents the distance from the center of a planet within which a solid satellite will not form. The distance is equal to 2.44 times the planet's radius.

There is another relationship between the planets and their satellites that we may explore. This relationship is the force responsible for maintaining the satellites in their respective orbits. We recognize immediately that the force is gravity and, in addition, that both the planet and the sun contribute to this force. The sun's huge mass places it at a distinct advantage in attracting another object compared with a planet attracting the same object. However, in measuring the force of gravity we must also take into account the distance between the sun and the satellite as compared to the distance between the planet and the satellite, and in this case the planet has a distinct advantage. The relationship between these forces may be determined by the following equation.

$$\frac{F_p}{F_s} = \frac{M_p}{M_s} \cdot \left(\frac{D_s}{D_p}\right)^2 \tag{6.5}$$

where F, M, and D refer to force, mass, and distance, and subscripts p and s to planet and sun.

From this relationship it is possible to calculate the ratio of the forces applied by the sun and the planet on a particular satellite. It would be found, for example, that Jupiter applies a force on its satellite Ganymede almost 300 times greater than the sun. If this procedure is followed for all satellites, a rather interesting relationship is revealed. All the true satellites would indicate a force ratio in favor of the planets by a factor of 35 times or more. Deimos, a Martian satellite, falls slightly below this but is nevertheless still included as a true satellite for reasons stated previously. All the captured satellites are attracted by their respective planets with a force less than 35 times that exerted by the sun, with the exception of the Neptunian satellite Nereid, which has a value slightly more than 35. The five outer satellites of Jupiter have a force favoring the planet of slightly more than one. This indicates that the planet is just barely able to retain these satellites and prevent them from being removed from their orbits by the gravitational force of the sun. The most unusual circumstance is revealed with respect to the moon. Here we find that the sun attracts the moon with a force twice that of the force exerted by the Earth, thereby reinforcing the theory that the moon is closer to being a planet orbiting the sun than a satellite of the Earth.

6.3 OTHER SOLAR SYSTEM INHABITANTS

In 1801, Ceres, the largest (750 kilometres in diameter) of the asteroids or minor planets, was discovered in orbit between Mars and Jupiter. Many thousands of asteroids have been detected in this part of the solar system, but most are less than a few kilometres in diameter. Only a few are more than 100 kilometres in diameter.

Although not of significant mass, comets are significant in numbers and revolve around the sun in highly elliptical orbits. By 1705, Edmund Halley had calculated the orbits of several dozen and noted that some very bright comets were observed at intervals of 75 to 76 years. He also noted that the elements of the orbits were very similar and concluded that this represented the appearance of a single comet with a period of 76 years traveling an elongated orbit around the sun. The comet became known as Halley's comet.

Comets warm up as they approach the sun, and the particles vaporize to form a gaseous cloud around the body of the comet. When close to the sun solar radiation propels material away from the comet to form a tail. Generally such tails are associated only with large comets. Comets deteriorate with each revolution around the sun and ultimately break up and disappear. New comets take their place, coming, it is theorized, from a vast comet cloud thought to exist at distances up to 150,000 astronomical units from the sun. The influence of nearby stars and the sun's gravitational force may cause some of the comets to periodically move toward the center of the solar system.

Within the interplanetary spaces are countless numbers of small particles orbiting the sun in the same manner as the planets. These particles, called *meteoroids*, are too small to be seen even with the best optical aids, but when they encounter the Earth and enter its atmosphere they are seen as "shooting stars" or, more correctly, *meteors*. Meteoroids seldom survive the flight through the Earth's atmosphere, but larger particles, seen as *fireballs*, may last to strike Earth's surface, and when they do they are recognized as *meteorites*.

6.4 ORIGIN OF THE SOLAR SYSTEM

Now that we have a view of the structure and components of the solar system we can ask, How did all this start? A successful theory of the origin of the solar system will be expected to account for the properties of the sun and the planets, including their masses, chemical composition, and motion. The theory must take into consideration the spacing of the planets and their satellites, the relationship of the satellites and their respective primaries, and existence of the asteroids, comets, and other components of the solar system. Questions on the observed relationships among the various components, as well as any irregularities in the system, must be answered. From the theory we should also be able to make inferences about the formation and existence of other planetary systems in the universe.

A theory which answers all of the known facts and questions is not yet available, but there have been a number of attempts to develop such a theory. In general two basic approaches have been used. One approach assumes an existing sun affected by some cataclysmic event that resulted in the formation of the planets. Such a theory was first postulated in 1749 by G. Buffon, a French naturalist who suggested that the planets were formed as a result of a collision between the sun and a comet. The collision presumably scattered material from the sun to form planets. The theory covering this type of event became known as the *close encounter hypothesis*, and in 1900 was modified by F. R. Moulton and T. C. Chamberlain of the University of Chicago. Their version suggested that a visiting star passed so close to the sun that material was drawn from the sun, forming the planets. If this had occurred then the solar system would be unique in the universe, since celestial collisions are now thought to be extremely rare if they occur at all. Another cataclysmic-event theory on the formation of the planets was that the sun was once a component of a binary (double star) star system, a circumstance not unusual in the universe (see chapter 15). A visiting star or some other phenomenon led to the breakup of the sun's companion and part of the debris remained to form the planets. There is also the *random capture theory*, which states that the normal sun captured interstellar material from which the planets and other objects in the solar system were formed.

Since learning more about the formation of stars, none of the above models for solar system formation are considered seriously any longer, so we may turn to the second concept wherein the sun and the planets are thought to have originated from interstellar dust and gas at approximately the same time. This idea is not new. The basic concept was first suggested by Immanuel Kant and Pierre Simon LaPlace in the eighteenth century. Since then, the idea has been modified considerably and has evolved into the modern *solar nebular theory*. The theory has not yet been perfected and does not account for all the various characteristics of the solar system. However, several bits of observational evidence are available that support its basic premise, namely, that stars are formed from interstellar dust and gas (see chapter 14). One is the fact that young stars are almost always found in dense clouds of interstellar dust and gas, suggesting that the stars are formed from these clouds. Second, most stars tend to occur in small groups or clusters of up to several thousand, which leads to the assumption that they were formed from the same body of interstellar dust and gas.

Star formation appears to result when a massive interstellar cloud becomes unstable. At that point the internal gravitational force of the cloud overcomes its internal pressure, and the resulting collapse causes the density of the dust and gas to increase. The density will not be uniform throughout the cloud, but there will be concentrations of higher density, which appear to lead to the initial stage of star formation. The material forming the star, now rotating, continues to condense with the bulk of the mass tending to concentrate toward the center. Such a rotating entity is considered to be highly unstable and apt to divide into two and occasionally more fragments revolving around a common center of gravity. This is thought to be the manner in which double-star systems are formed, and since at least 50 percent of the star systems are of this type, the theory is consistent with observation. Planetary systems are considered variations of multiple-star systems. A star cloud with a small angular momentum (the rotational equivalent of momentum) would result in a single, central, newly forming star and a disk composed of less than one percent of the dust and gas mass (Figure 6.6).

In the dust and gas cloud forming the new sun, we would find that in addition to hydrogen, the most abundant element in the universe, there is some helium and carbon, nitrogen, and oxygen present. The last three elements should be readily recognized as those which, along with hydrogen, are essential in the formation of life. In addition, there are other elements present such as silicon, magnesium, and iron, all metals that are critical in the formation of planetary rock. As we shall see later in chapter 14, these elements are formed in previously existing stars, which explode and add the elements to interstellar material to be incorporated into new stars. Therefore it is reasonable to conclude that the sun is really a second generation star, since it includes material from preexisting stars.

We can assume that the composition of the dust and gas that formed the sun and the planets is representative of the material comprising the entire interstellar cloud. If this is correct, then the elements present on the Earth and other solar system planets would also be found on planetary systems associated with other stars formed from the same interstellar cloud. As a result there is the distinct possibility that life could have formed on other planets suitably placed for life to develop. These conditions will be discussed at greater length in chapter 19.

The question of how the planets and other components of the solar system were formed has not yet been completely resolved. There are at

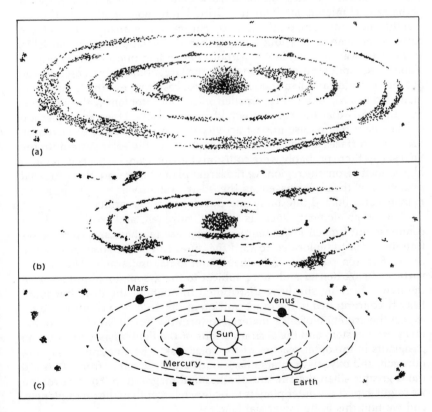

Figure 6.6 The solar system originated from a cloud of dust and gas (a). The cloud contracted and rotated forming a disk which condensed to the elements important in the formation of the planets (b). By accretion the elements built up to become larger and larger particles until the planets were formed (c).

present two major views on the formation of the planets, but each has some inherent difficulties.

One model sees the tiny fragments of the solar nebula formed into *proto-planets*. In each proto-planet the dust gravitated toward the center to form a rocky core surrounded by a gaseous envelope or atmosphere. The inner planets may possibly have lost most of their atmosphere during this early stage due to the emission of matter (solar winds) from the sun, and in this way became the terrestrial planets. The outer planets, somewhat further from the newly forming sun and therefore less influenced by it, retained extensive atmospheres. An objection to this model is that little or no impact of one object into another is envisioned during the formative stage of the planets, yet observation of the surface features on the moon and several planets indicates extensive impact during this early period.

Another view accepted by many astronomers is that the planets and other objects in the solar system were formed by the accretion process. The dust and gas cloud collapsed to form the sun, and in the process the cloud began to rotate and formed a disk. Mathematical models have shown that such a disk is unstable, and that most of the material would collect at the center to form the sun. Angular momentum, however, maintained a small percentage of the material in the disk at several thousand degrees Kelvin. These high temperatures resulted from the initial collapse of the gas cloud, but there was no means for maintaining the temperatures in the disk, so heat was lost by radiation. When temperatures cooled to approximately $1200°K$, certain elements such as silica, iron, and magnesium condensed out to form molecules of solid iron and magnesium silicates, the basic rock materials. These particles, each perhaps one micron in diameter, drifted into one another and stuck together to form conglomerations of particles. Eventually a few of the particles became large enough to gravitationally capture other particles, and in this way became larger without depending upon random impact alone. Also, under these circumstances, the larger objects increased in size at a more rapid rate than the smaller ones, resulting eventually in a few major bodies surrounded by smaller ones. One problem with the accretion theory is that the process would be very slow. However, accretion does explain the intense impact activity that seems to have occurred on planetary surfaces during the formative stages of the solar system. In the final stages there probably were collisions between large objects moving at high velocity from the regions of the larger planets in toward the terrestrial planets. But these decreased as the solar system was swept clear and planets consolidated internally and developed stable orbits.

The other elements such as carbon, nitrogen, and oxygen did not condense until much lower temperatures were reached, and were therefore more apt to be located toward the outer rim of the solar disk in the form of frozen particles of methane, ammonia, and water. The silicate particles being denser and more subject to the gravitational influence of the newly forming sun were distributed throughout the disk, but more heavily concentrated in the inner portion of the disk. As a result we could expect the accretion of rock material to form the terrestrial planets in the inner solar system and the accumulations of the lighter gases to form the gas giants in the outer reaches of the solar disk. The amounts of carbon, nitrogen, and oxygen in interstellar clouds are far greater—from 10 to 100 times greater—than the amount of silica and magnesium. For this reason we might expect the outer planets to be much larger than the inner planets, and we find this to be the actual fact.

There is one additional factor that requires explanation, and that is the reason for Jupiter's huge size. If the frozen gases tended to form in the outer reaches of the solar disk, we could expect the outermost of the gas giants to be the largest. This, as one can see from Figure 6.2, is not the case. Jupiter and Saturn are composed not only of methane, ammonia, and water but also great amounts of hydrogen. Hydrogen was a major component of the original gas cloud, and is the lightest element and therefore most easily escapes the influence of a gravitational field. However, Jupiter and Saturn's masses became sufficiently great to overcome this tendency of hydrogen and large amounts of the gas were accumulated, accounting for the huge size of these two planets. It is thought that, during their formative stages, the terrestrial planets also had some hydrogen in their atmospheres. However, some astronomers feel that the hydrogen was dissipated into the outer solar system by the action of the solar winds, which emanated from the sun at high velocities during that early period.

Several additional difficulties exist which are common to all models of planetary formation. One of these is the presence of the asteroids. Possibly the large mass of Jupiter prevented these objects from forming a planet, but this is not known for certain. It is hoped that future space probes will obtain information that would provide an explanation for the presence of the asteroids. Another difficulty relates to the origin of the comets. Their rather wide-ranging orbits leads to the suspicion that comets may represent materials left over from the original cloud forced into the outer regions of the solar system by the solar winds. Detailed analysis of comets would aid in solving some of the mysteries relating to the early formative stages of the solar system.

6.5 AGE OF THE SOLAR SYSTEM

Having reviewed some of the theories presented to explain the formation of the solar system, we can next turn to the problem of determining how long ago this event may have occurred. The solar system is complex, but there are characteristics which would indicate the system was formed at one time rather than in piecemeal fashion. Measuring the rate of radioactive decay of certain elements, for example uranium U^{238} to lead Pb^{206}, the half-life of which is 4.6 billion years, has provided a method for determining the age of rock. The radioactive dating of some rocks on Earth indicate that they solidified about 3.8 billion years ago. From this we may assume that the Earth is at least that old. Measurements made on stony meteorites, similar in composition to the mantle of the Earth, reveal an age of 4 to 4.5 billion years, and samples brought back from the moon indicate a similar age. Age determination by means of radioactive dating technique of meteorites and lunar samples probably provide us with a more accurate indication of the age of the solar system than samples from the Earth. This is true because these objects are not subject to the intensive geologic processes that have significantly altered the Earth's surface since it was first formed. From this sort of evidence we can assume that the age of the solar system is at least 4.5 billion years.

Theory has made it possible to predict the ratio of the abundance of various radioactive isotopes of given elements at the time they were erupted by an exploding star (see chapter 14). Such matter, mixed with hydrogen, constituted the raw materials for a new star containing heavy elements such as our sun. From a measurement of the abundance ratios of some of the radioactive elements now in the solar system, it is possible to determine when these elements were formed in the interior of a former star. This turns out to be approximately 6 billion years ago, which means the solar system could be no older than this. From such information we can assume that the sun is 5 to 5.5 billion years old, with the planets and other solar system inhabitants from 0.5 to one billion years younger.

6.6 SUMMARY

After 2500 years of viewing the solar system as Earth-centered, Copernicus developed the heliocentric or sun-centered concept, with the planets, including the Earth, orbiting around the sun. The sun includes most of the mass of the solar system, and the planets and all other objects comprise only 0.14 percent of the mass. By means of Newton's modification of Kepler's third law, the mass of the major units in the solar system may be determined. The planets revolve around the sun in a counterclockwise direction (as viewed from the north) and, with the exception of Venus and Uranus, rotate counterclockwise on their axes. The planets orbit the sun approximately on the ecliptic—a plane arbitrarily defined by the Earth's orbital plane. The distances of the planets from the sun follow a pattern described by Bode's law: each planet is 1.5 to 2 times the distance of its nearest inner neighbor from the sun. Once the distance of the planet from the sun is known, it is possible to determine the planet's diameter and calculate its orbital velocity.

Not a great deal is known about the individual planetary satellites, but some characteristics indicate that there are two types. One, the true satellite, is found orbiting the equatorial plane of its primary and, it is assumed, was formed there at the same time the planet was formed. The other type is captured, that is, it came under the gravitational influence of a planet after the solar system was formed and continues to orbit that planet. The moon fits neither of these conditions and may be another planet.

Other inhabitants of the solar system are the asteroids, comets, and meteoroids, but all these together are rather insignificant from the point of view of total mass.

As yet no completely satisfactory theory on the formation of the solar system has been proposed that answers all the questions raised on structure, composition, and motion. One group of theories assumes a cataclysmic origin, wherein some external object collided with or made a close approach to the already existing sun. The tidal forces of such an event, it is suggested, removed material from the sun, which formed into the planets. Theories of this type are not seriously considered at this time.

The theory of solar system formation currently advocated is based on the idea that the sun, planets, and all other objects in the solar system were formed from an interstellar cloud of dust and gas. The cloud initially collapsed into a rotating, disk-shaped structure, with the largest proportion of the mass concentrated in the center to form the sun. The small amount of mass remaining in the disk formed into the planets and other objects through accretion.

Radioactive dating has been used as a means of determining the age of the solar system. Available evidence from this technique leads to the conclusion that the solar system is from 4.5 to 5.5 billion years old.

QUESTIONS

1. Why is it not possible to see the planets of the solar system from a planet orbiting the nearest star?
2. Determine the mass of Saturn based on data of Titan in Appendix 3.
3. Using positional nomenclature, give the position of the following planets:
   (a) Venus seen in transit across the sun
   (b) Jupiter seen on the meridian (directly overhead) at midnight
   (c) Mars on the meridian at sunrise
   (d) Saturn opposite the sun as seen from Earth
4. How may the period of rotation of a planet be determined?
5. What is Bode's law and how does it work?
6. What is the velocity of Jupiter in its orbit? Make use of equation 6.3 to show how you arrived at the answer.
7. What are the differences between the true satellites and the captured satellites?
8. Into which category of satellite does the moon fit? Why?
9. What aspects of the solar system does a successful theory on the formation of the solar system account for?
10. In outline form describe the theory currently considered acceptable on the formation of the solar system.

EXERCISES AND PROJECTS

1. Locate Venus as soon after sunset as possible. At this time it may easily be seen with a good pair of field glasses. Every few days for the next several weeks, or months, if possible, plot Venus' position with respect to the stars in its background. When Venus is close to the horizon right after sunset stars may be difficult to see because of the light from the sun below the horizon. Positions of Venus will have to be estimated. Use the appropriate star chart in Appendix 6, or, when practical, make your own star chart of the western sky. Fill it in as more stars become visible. If there is a good western horizon, the same observations may be made for Mercury.

FOR FURTHER READING

Cook, A. H., *Physics of the Earth and Planets.* New York: John Wiley & Sons, 1973.

Hartmann, W. K., *Moons and Planets.* Belmont, Calif.: Wadsworth Publishing Co., 1973.

Kopal, Z., *The Solar System.* London: Oxford University Press, 1973.

Wood, J. A., *Meteorites and the Origin of the Planets.* New York: McGraw Hill, 1968.

Aquarius

CHAPTER 7   **THE EARTH**

What is the Earth but a lump of clay surrounded by water?

—Bhartrihari

What is Heaven? A globe of dew.

—Shelley

The Earth is the third planet from the sun, the largest of the terrestrial planets, and the fifth largest of all the planets in the solar system (Table 6.1). It is the only planet in the solar system thus far known to support life. Studying the Earth as a planet has the advantage of permitting data obtained from other planets to be interpreted in earthlike terms. For this reason the characteristics of the Earth are discussed as a preliminary to studying the other planets. Studying the Earth in its entirety presents some problems, because only a small portion is visible at any one time to a single observer on its surface. Certain aspects, such as the physical shape and some astronomical motions, could be viewed better by an observer on the moon.

In this chapter the major physical features of the Earth and its nearby environment will be discussed first. Then we will examine the various motions of the Earth, followed by a consideration of such aspects as timekeeping, seasons, measurements in space by means of celestial coordinates, and life on Earth.

7.1 SHAPE, SIZE, MASS, DENSITY

Shape

We think of the Earth as round—a sphere, to be more exact—as did Aristotle 2500 years ago. His reasons for claiming that the Earth was a sphere were quite sound, in that he observed objects on the Earth's surface disappear below the horizon he was traveling away from and new objects appear above the horizon he was approaching. This was particularly noticeable with respect to the positions of constellations in the sky when he traveled north or south. Aristotle occasionally saw an eclipse of the moon when the Earth passed directly between the sun and the moon. He observed that the shadow of the Earth cast on the moon during an eclipse was an arc, representing a portion of a circle. Additional evidence is provided by ships sailing to sea and gradually sinking below the horizon to disappear from view. This could be interpreted in two ways: one as the result of the ship sailing on a spherical Earth, or another as the ship falling off the edge of a flat Earth. Magellan's voyage around the Earth in 1521 finally resolved this point in favor of the spherical Earth. More recently, additional proof in the form of photographs of the Earth from space have become available, which along with the orbiting of numerous artificial satellites around the Earth amply demonstrate the Earth's sphericity (Figure 7.1).

What has in reality been proven from satellite data is that the Earth is not an exact sphere, but rather an *oblate spheroid*. The Earth is slightly compressed at the poles and bulging at the equator as a result of the Earth's rotation. By means of very exacting measurements made with the aid of artificial satellites, the shape has been further refined to reveal that the Earth bulges slightly more in the southern hemisphere and has a slightly indented south polar region (Figure 7.2).

Size

Because of the oblateness of the Earth, the polar diameter of 12,693.45 kilometres is slightly less than the equatorial diameter of 12,726.41 kilometres. The degree of *oblateness* is defined as the difference between the equatorial diameter and the polar diameter divided by the equatorial diameter. This turns out to be 1/298. The equatorial circumference is 40,055.1 kilometres.

Measuring the diameter and circumference of the Earth is not a recent

Figure 7.1 A clear view of the Earth in space taken from ATS III 36,000 kilometres in space.(NOAA)

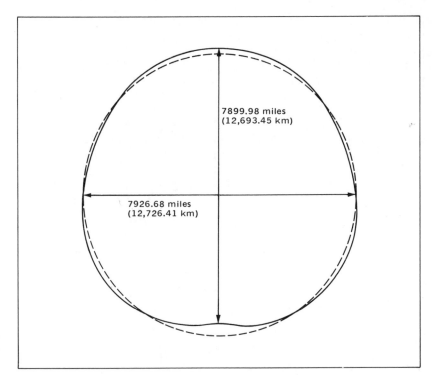

accomplishment. In the third century B.C. a Greek astronomer, Eratosthenes, was able to measure the circumference of the Earth. He noted that on the longest day of the year in Alexandria, a stake in the ground cast a shadow of slightly over 7° of arc at noon (Figure 7.3). At the same day in Syene (Aswan), the sun was directly overhead and cast no shadow. The angular difference between the location of the two cities was 1/50 of a complete circle, and the distance between the two cities was 5000 stadia (920 kilometres). From this information Eratosthenes calculated the circumference of the Earth and found it to be equivalent to about 2.5×10^5 stadia or about 4.6×10^4 kilometres. By dividing by 2π he was able to determine that the radius of the Earth was equivalent to approximately 7330 kilometres. These calculations were somewhat higher than present-day figures, but Eratosthenes did not have the advantage of modern precision equipment to measure angles or distance.

Mass

We have previously referred to the Earth's mass in relative terms where $M_\oplus$ equals 1.00, and now a means of quantifying the mass will be illustrated. Several experiments to "weigh the Earth" have been performed in the past. We will use P. von Jolly's method (1881) to illustrate the principle (Figure 7.4). A flask of mercury (M_{Hg}), accurately weighed, was carefully counterbalanced on a balance. Then a five-ton lead weight (M_{Pb}) was placed beneath the mercury. The gravitational attraction of the lead was sufficient to offset the balanced mercury, so a small mass (Ms) was added to counterbalance the mercury again. The distance (d) between

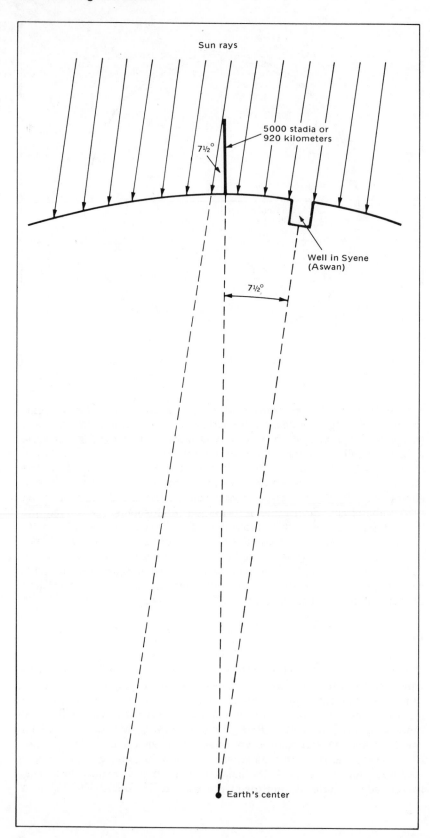

Figure 7.3 Eratosthenes measured the circumference of the Earth in the third century B.C. by noting the difference in the angle of the sun's shadow at Alexandria and Syene on the longest day of the year.

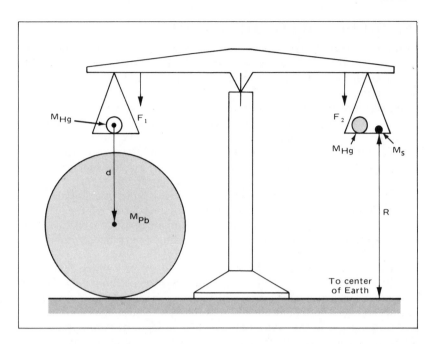

the centers of mass of the mercury and lead were carefully measured, and the masses of these objects and the small mass added were determined. The data provided the means of calculating the mass of the Earth ($M_\oplus$) from the following proportion:

$$\frac{Ms M_\oplus}{R^2} = \frac{M_{Pb} M_{Hg}}{d^2}$$

where R equals the radius of the Earth. Solving for $M_\oplus$ we obtain

$$M_\oplus = \frac{M_{pb} M_{Hg} R^2}{Ms\, d^2}$$

All factors on the right-hand side of the equation are known, and the mass of the Earth can be determined to be

$$M_\oplus = 5.98 \times 10^{24} \text{ kilograms}$$

Density

The density of the Earth, or for that matter any object, may be calculated once we know the mass and the volume of the object. The volume of a sphere is obtained by the equation $4/3\ \pi R^3$. It is assumed that the Earth and most other celestial objects are essentially spherical, so all that is needed for volume determination is the radius. From this, the density of the Earth is found to be 5.52 grams per cubic centimetre (Table 6.1). A comparison with other planets shows the Earth to have the highest density of any of the planets in the solar system.

7.2 STRUCTURE OF THE EARTH

Although scientists have not penetrated more than 10 to 12 kilometres into the crust of the Earth, a knowledge of the Earth's density has provided some insight into its internal structure. The Earth is not a homogeneous mass, but appears to be made up of lighter materials on the surface and progressively heavier materials toward the center. The evidence for this is the fact that the density of the surface material is about 2.65 grams per cubic centimetre, while the average density for the Earth, as stated above, is 5.52 grams per cubic centimetre. The conclusion may be drawn that the density of the rock toward the center of the Earth is somewhat greater than that on the surface in order to achieve an average of 5.52 g/cm³. Additional information comes from the varying speed with which earthquake waves travel through the Earth's interior and the manner in which the waves are refracted, which indicates that the Earth is made up of concentric spheres of progressively denser materials surrounding a very dense core (Figure 7.5).

The Earth's surface is made up of the land masses, which cover about 29 percent of the surface with an average elevation of about 0.8 kilometres above sea level, and the oceans, which cover the remaining 71 percent and are an average depth of about 4 kilometres. The continents extend beyond the tide line along the continental shelf to a depth of approximately 200 metres. Beyond this is the continental slope, which drops off more abruptly to the ocean basins (Figure 7.6). The continental slope is considered by some scientists to be the zone where ocean basins, composed primarily of basaltic rock, join the continental masses, composed mostly of less dense granitic rock.

Figure 7.5 Cross-section of the Earth, showing crust, mantle, and core.

The surface of the Earth represents the outer portion of the outer layer

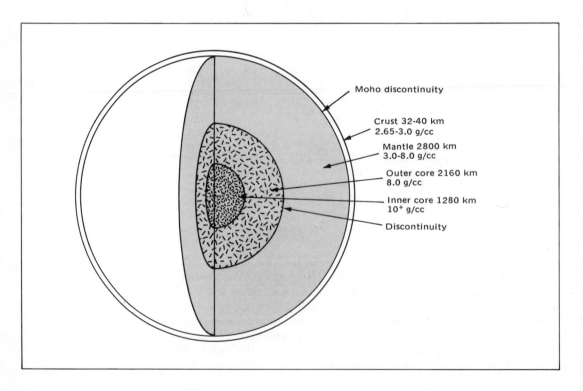

Moho discontinuity

Crust 32-40 km
2.65-3.0 g/cc

Mantle 2800 km
3.0-8.0 g/cc

Outer core 2160 km
8.0 g/cc

Inner core 1280 km
10⁺ g/cc

Discontinuity

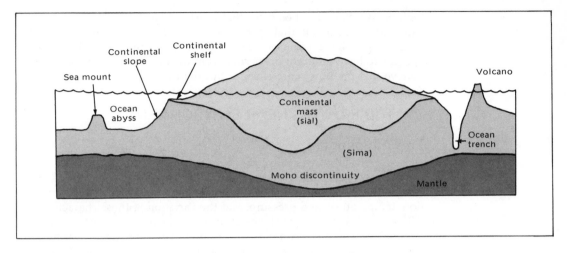

Figure 7.6 Hypothetical profile of the crust showing continental mass and some features of the ocean basin. Vertical scale is exaggerated.

or *crust* of the Earth. The crust is a thin shell ranging in thickness from 5 kilometres beneath the oceans to 40 kilometres or more in the continents. The crust is composed of relatively lightweight rock, with the composition of the continental rock varying somewhat from that of the rock comprising the ocean basins. The continents are mainly granitic rock, which is composed of minerals high in silicon and aluminum. An Austrian geologist, E. Suess, in the nineteenth century suggested the term *sial* for this material after the chemical symbols for silicon (Si) and aluminum (Al). This granitic material is relatively lightweight and appears to "float" on the more dense basaltic rock found on the ocean floor and beneath the continental masses. The basaltic rock contains silicon and greater amounts of magnesium than granites, causing Suess to suggest the term *sima.*

A more or less abrupt change occurs in the rock at the base of the crust marking the beginning of the mantle. This was first noted in 1909 by the Croatian geologist A. Mohorovicic, who discovered a change in the velocity of earthquake waves at what became known as the base of the crust. The zone of change is now known as the *Moho* (short for Mohorovicic) *discontinuity,* which refers to velocity discontinuity or change.

Beneath the Moho discontinuity, the mantle, made up of increasingly denser rock, extends to a depth of about 2900 kilometres, where another important discontinuity occurs that marks the boundary between the mantle and the *core.* The density of the mantle ranges from 3.3 g/cm^3 just below the Moho to an estimated 8.0 g/cm^3 at the boundary between the mantle and the core. Other discontinuities of less importance occur within the mantle.

The core of the Earth is a sphere about 6800 kilometres in diameter, presumably composed of very dense material in a fluid state, possibly of iron-nickel composition. The core is thought to have a density in excess of 10 g/cm^3, although neither iron nor nickel achieve this density on the surface. It is felt that the higher-than-normal density levels of these materials result from the extremely high pressures occurring at that depth.

Although the core has never been reached, indirect evidence indicates its fluid state. Secondary earthquake waves (transverse waves) will not pass through a fluid medium, and this type of earthquake wave does not pass through the Earth's core. This evidence, together with the extremely high pressure and temperatures that must exist at the Earth's center, lead geologists to conclude that at least the outer core is in a fluid state. The inner core, with a radius of 1350 kilometres from the Earth's center, appears to have different properties from the outer core. Some geologists believe that the inner core is solid, but the evidence is not conclusive.

7.3 THE EARTH'S NEAR ENVIRONMENT

The Atmosphere

Although the atmosphere appears to extend one-to-two Earth radii above the surface of the Earth, 90 percent of the gases exist within the first 30 kilometres. The atmosphere is described and studied in terms of composition, temperature, and pressure, and the variation of these characteristics with altitude. However, such descriptions are inadequate in describing the atmosphere, because complex air motions resulting from heating and chemical changes resulting from the ionization of certain gases produce phenomena in the atmosphere that are quite complex in nature. The fact that the atmosphere is very dynamic, and that it is continually changing characteristics with time of day, season, year, and even with solar cycles, is now beginning to be realized.

In the previous chapter the suggestion was made that action of the solar winds was responsible for removal of the Earth's primitive atmosphere, and that the primitive atmosphere differed from our present atmosphere. Some of the evidence for this is based upon a comparison of the abundances of elemental gases such as neon and nitrogen with nonvolatile elements of the Earth, and a comparison of these ratios with their cosmic abundances. One such comparison reveals that there is one ten-billionth the amount of neon with respect to silicon in the Earth than might be expected from cosmic abundances of these elements. Consequently, it is believed that most of the original gases were lost early in the Earth's evolution and a secondary atmosphere formed somewhat later. The most likely source of a secondary atmosphere is volcanic activity, and it was quite different from the present one. The secondary atmosphere contained such gases as ammonia, methane, sulfur oxides, hydrogen sulfide, nitrogen oxides, carbon dioxide, and copious amounts of water vapor, but no elemental oxygen. Some of these compounds, as we shall see shortly, were important in the development of life on the Earth. The initial formation of atmospheric oxygen came from the photodissociation of water and, to a lesser degree, of carbon dioxide. *Photodissociation* is the breakdown of the water molecule into hydrogen and oxygen by the action of sunlight. Subsequently, the major portion of oxygen came from the photosynthetic process of plant life that gradually evolved on the Earth. Nitrogen was formed from ammonia. The nitrogen was retained in the atmosphere while the hydrogen, being a lighter gas, escaped into space.

The present atmosphere is a mixture of gases, including nitrogen and oxygen as the major components and small amounts of argon, carbon dioxide, and other elemental gases (Table 7.1). Water vapor is also present in the atmosphere in amounts up to 4 percent by volume, but it is generally not included as one of the atmospheric components. The combined weight of these gases in the atmosphere is such that about one kilogram of air rests on each square centimetre of Earth's surface at sea level. This represents a total atmospheric weight of 6×10^{15} tons.

PERCENTAGES OF ATMOSPHERIC GASES BY VOLUME   TABLE 7.1

Gas	Symbol	Percent of Volume
Nitrogen	N_2	78.08
Oxygen	O_2	20.95
Argon	A_2	0.93
Carbon dioxide	CO_2	0.03
Other gases		0.01
(neon, helium, krypton, hydrogen, xenon, ozone, radon)		

A human contribution to the atmosphere is the pollutants coming mainly from the combustion of various fossil fuels. Toxic gases, vapor, and solids are discharged by this process, which when reacted on by sunlight, contaminate the atmosphere with substances collectively called *smog*. In some areas the smog concentration has reached levels so harmful to plants that the growth of certain crops is no longer feasible. The levels of concentration of smog are harmful to humans as well, and government at all levels is now requiring a decrease in smog-producing activities. These include standards for automobile exhaust emissions and a requirement that industries switch to fuels with less smog-producing properties. The success of these and other measures will determine whether or not life will survive on the planet Earth.

The atmospheric components listed in Table 7.1 are mainly found in the lower atmosphere or *troposphere*. The troposphere is one of the major distinct layers of the atmosphere and extends from the surface of the Earth to 8 kilometres above the poles and 15 kilometres above the equator. Within this layer the gases generally occur as a turbulent homogeneous mixture, comprising about 75 percent of the total mass of the atmosphere. Temperatures within this layer decrease with elevation, since most of the heat comes from infrared radiation (heat) from the Earth, thus heating the atmosphere from the bottom up. This results in temperatures in the troposphere decreasing from an average of 20°C for the total Earth's surface down to –80°C at the *tropopause* (upper limit of the troposphere) above the equator and –55°C above the polar regions. The differences in temperature at the tropopause are due to the variation in elevation above the Earth's surface.

In the *stratosphere*—the layer which occurs above the troposphere and extends to an altitude of approximately 50 kilometres—temperatures increase with altitude. The temperature seems to be influenced by the presence of a layer of ozone (O^3) in the *stratopause*, or upper limit of the stratosphere. Ozone has the capacity to absorb energy coming from the sun and reradiate this energy as heat, thus causing the stratosphere to show a higher temperature at its upper boundary and a decreasing temperature downward toward the tropopause. This increased temperature with altitude is thought to contribute to stratification of the gases with very little vertical mixing or general turbulence in the stratosphere.

Above the stratosphere is a layer that is quite turbulent. This is the *mesosphere*, a layer extending up to an altitude of approximately 90 kilometres, in which the temperature decreases with an increase in altitude. The decrease in temperature continues to the *mesopause*—the upper layer of the mesosphere—where the lowest measured atmospheric temperature of –140°C is recorded.

Up to the altitude so far discussed (90 kilometres), the atmospheric gases are a fairly homogeneous mixture; thus, this portion of the atmosphere is called the *homosphere*. Above this elevation, the gases tend to separate according to their atomic weights and to stratify, with the heavier gases such as oxygen at the bottom of the zone, helium above the oxygen, and hydrogen in the upper extremes of the atmosphere (Figure 7.7). This is the *heterosphere*, a zone which includes the *thermosphere* and the *ionosphere*.

Figure 7.7 Profile of the near Earth atmosphere and the upper extremes of the atmosphere.

The thermosphere lies above the mesosphere and is the region in which the temperatures increase with elevation. The increase in temperature results from the absorption of ultraviolet radiation from the sun by monoatomic oxygen, and causes temperatures in this zone to rise to as high as 1275°C at an elevation of 700 kilometres. The heat implied by a temperature of 1275°C cannot be felt, because the density of gas atoms and

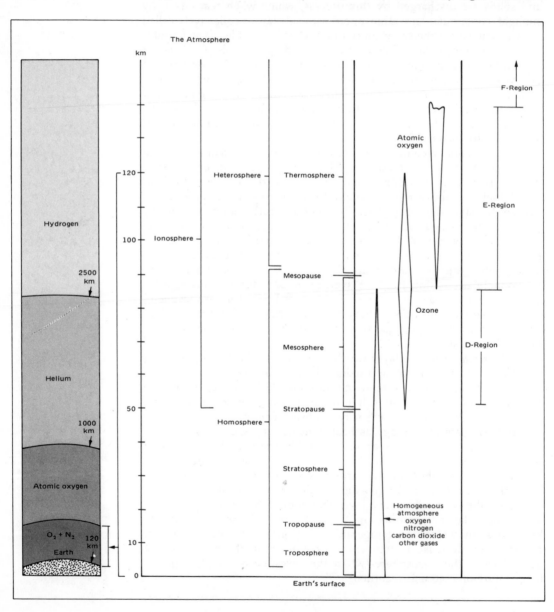

charged particles is so low that transmission of energy by collision is very small. However, the energy levels of the individual particles is sufficiently high to be characteristic of such temperatures.

The presence of monoatomic oxygen results from the effect of solar radiation, particularly X-ray and shortwave ultraviolet radiation. Such electromagnetic energy causes molecular oxygen to be split into atomic oxygen at elevations as low as 50 kilometres. This form of oxygen increases until it is more abundant than molecular oxygen at approximately 200 kilometres. Such activity, beginning at 50 kilometres, overlaps the thermosphere and extends up to the limit of the atmosphere. The presence of free electrons and positively charged ions has caused this region to be called the ionosphere. The ionosphere is subdivided into a number of subregions, such as D-region, E-region, F-region, identified basically according to electron density. These subregions are important because of the influence the free electrons and the positively charged ions have on radio waves and radio communication.

The Magnetosphere

The existence of magnetism has been known for more than 2000 years, but the fact that the Earth itself is a huge magnet was not demonstrated until 1600, when Sir William Gilbert explained how the magnetic compass may be used for navigation. The Earth's magnetic field was viewed much like that shown in Figure 7.8, in which the small circle represented the Earth and the lines represented the lines of force of the magnetic field.

Since the advent of satellite exploration in 1957, the view of the Earth's magnetic field has changed, and the term *magnetosphere* has been coined to identify the region of Earth's magnetic influence in space (Figure 7.9). The *magnetopause* is the boundary which separates the region close to the Earth, where there are strong and oriented magnetic fields derived from the

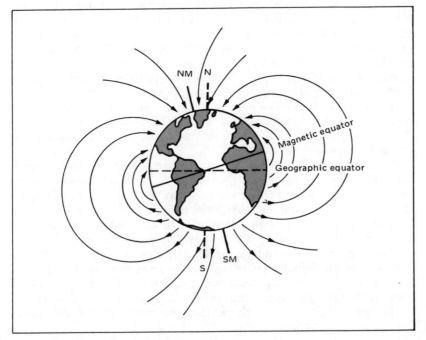

Figure 7.8 Pre-space age view of the Earth's magnetic field.

Figure 7.9 A diagram of
the magnetosphere es-
tablished by satellite
experiments.(NASA)

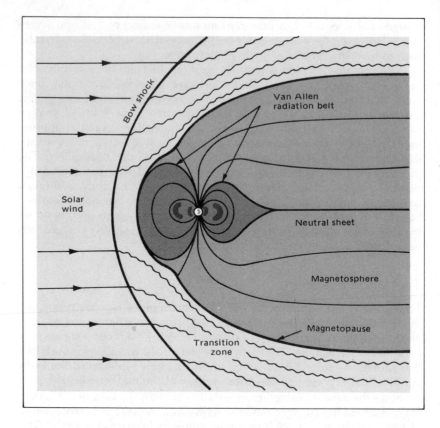

planet itself, from the region outside, where there are weak and fluctuating
fields derived from the sun. The solar winds impinging on the Earth's field
cause field lines from the polar regions to be swept back in a direction
opposite the sun to form the Earth's *magnetic tail.* Field lines in the tail
below the plane of the magnetic equator point predominantly away from
the sun, those above the plane point toward it. Between the two at the
equatorial plane is a *neutral sheet* in which field reversal is accomplished.

The flow of solar wind through interplanetary spaces is supersonic.
Therefore, a shock wave, the "bow shock," is created upstream of the
Earth's field in the same manner that a shock wave is created upstream of
an obstacle in a river or by a boat plowing through the water. Between the
bow shock and the magnetopause is a region of turbulence insofar as field
direction and particle motion are concerned. This region of turbulence is
called the *transition zone.*

Within the confines of the magnetosphere are the *Van Allen radiation
belts,* discovered by James A. Van Allen in 1958. He described radiation
belts as two concentric belts of charged particles, one approximately 1½
radii from the Earth and the other 3 to 4 radii. Some scientists now prefer
to speak in terms of a single radiation zone with a slot between where
radiation levels are lower, rather than two well-defined belts.

The radiation belt contains protons and electrons trapped by the
Earth's magnetic field. Proton energies in the outer portion of the belt are
too low to contribute to particle count, and this led early observers to
conclude that this zone was composed primarily of high-energy electrons.

Recent satellites have continued the exploration of the Earth's radiation belt, and mapped it in greater detail. The more the belt is studied the more complex it seems. First, the belt is strongly affected by the magnetic field of the Earth, and second, both the belt and the magnetic field are influenced by the activities of the sun. These interrelationships continue to be studied to obtain a greater knowledge about our near-earth environment and to provide information on magnetic field activities around the other planets.

Why is there such interest in the Earth's atmosphere and magnetosphere? An understanding of the Earth's atmosphere provides some insight into activities around the seven other planets now known to have atmospheres. Atmospheric conditions on Earth play a large role in the kind of weather we experience on the surface. Large-scale air mass movements are the principal means by which heat energy is transferred from one part of the globe to another. At the same time the atmosphere and magnetosphere protect us from harmful radiation and particles emitted from the sun.

More specifically, the astronomer is interested in atmospheric conditions because they have a decided influence on his ability to "see." The effects of clouds and haze or smog are evident. However, even on clear nights air movement and the continually changing densities of the atmosphere can cause the stars to "twinkle," resulting in less than ideal seeing conditions. At the same time there is the overall refraction of light caused by the fact that light is entering a more dense medium (see chapter 4) as it enters the atmosphere. This causes the observed altitude of a star to be somewhat greater than its true altitude (Figure 4.7). Most astronomical work now done in the visible range of the spectrum is accomplished by means of photography. The problem of atmospheric turbulence, aggravated by increased magnification, causes lengthy time exposures to be blurred. Photographs taken from manned and unmanned satellites have shown details of unsurpassed clarity not possible from the bottom of our atmospheric ocean, even under the most ideal conditions.

7.4 EARTH'S MOTIONS

Since the sixteenth century, astronomers have recognized that the Earth moves in a variety of ways within the confines of the solar system. We limit the discussion in this chapter to those motions that can be observed relative to the sun. There are additional motions in which the Earth participates in conjunction with the entire solar system, which will be discussed in a later chapter. Some of the motions readily recognized include *rotation*, or the turning of the Earth on its axis, and *revolution*, or the movement of the Earth in orbit around the sun. These, along with *precession*, a motion not so readily apparent, will be discussed here.

Rotation

The rotation of the Earth is evidenced by the daily movement of celestial objects (sun, stars) from east to west across the sky. This motion was incorrectly attributed by the ancient Greeks to the movement of these objects in space, but ample evidence now exists that the Earth moves. One of the first proofs was provided in the nineteenth century by the French physicist J. B. L. Foucault, who demonstrated the rotation of the Earth in the Pantheon in Paris by means of a huge, swinging pendulum (Figure 7.10). A 25-kilogram weight attached to a metal thread about 65 metres in

Figure 7.10 Foucault
pendulum at Griffith Ob-
servatory, Los Angeles,
California. A pendulum
experiment by J. B. L.
Foucault proved the axial
rotation of the Earth.
(Griffith Observatory)

Figure 7.10 Foucault pendulum at Griffith Observatory, Los Angeles, California. A pendulum experiment by J. B. L. Foucault proved the axial rotation of the Earth. (Griffith Observatory)

length was suspended over a calibrated disk to indicate the direction of swing. When the pendulum was set to swinging, the direction of swing slowly began to veer. Since gravity, the only force operating on the weight, was able to attract the weight only toward the center of the Earth, it was concluded that the drift of the swinging weight was due to the Earth's rotation.

In discussing the Earth's shape, we described it as an oblate spheroid wherein the Earth bulged at the equator. A malleable object suspended in space and subject only to the force of gravity would take on the shape of a perfect sphere. An object rotating in space would bulge at the equator, since a portion of the gravitational force is dissipated in maintaining the rotation. Since the greatest rotational force is experienced at the equator and less at the poles, gravity will have a greater influence at the poles and less at the equator. This would result in a bulging equator, a phenomenon that can only be attributed to a rotating Earth.

Additional proof of the Earth's rotation may be provided by observing the trajectory taken by a missile fired from the equator northward. Such a projectile would be expected to follow a course along a meridian of longitude during its entire flight if the Earth were nonrotating. However, the missile seems to veer to the right. This is called the *Coriolis effect*, discovered by G. G. de Coriolis in 1835. If the projectile were sent south from the equator, it would appear to veer to the left. These apparent motions can only be explained on the basis of a rotating Earth.

The Earth's rotation serves as a convenient means of telling time, each rotation being equivalent to one day of 24 hours. There are, however, two ways in which this period may be measured. One is by selecting a distant star and noting when it crosses the celestial meridian at noon or midnight.* The interval of time from one transit of the star across the meridian to the next transit is one rotation. This interval, called the *sidereal* (star) *day,* represents the true rotation of the Earth through 360°.

Although the sidereal day may be measured as the period between two successive transits of any bright star across the meridian, the starting point is usually the transit of the vernal equinox across the meridian on March 21. The *vernal equinox,* discussed later in this section, is the point in the sky occupied by the sun as it crosses the celestial equator apparently moving north. If we set our watches according to this arrangement and consider the sidereal day as a 24-hour period, we find that the sun will cross the meridian four minutes later the next day, and in a month's time the sun will cross the meridian about two hours later. In six months we would eat our noonday meal at midnight, and in one year's time we would again have the sun at the meridian at noon. The sidereal day is therefore not a satisfactory means for telling time.

The second method of measuring the Earth's rotation makes use of the interval between one transit of the sun across the meridian at noon until the next transit. This interval is the *solar day,* which we use for telling time. Since the Earth moves in orbit (Figure 7.11) each day, the solar day is approximately four minutes longer than the sidereal day. Thus

$$1 \text{ sidereal day } = 24\text{h sidereal time}$$
$$\text{or } 23\text{h } 56\text{m } 4.091\text{s of mean solar time}$$

$$1 \text{ solar day } = 24\text{h mean solar time}$$
$$\text{or } 24\text{h } 3\text{m } 56.55\text{s sidereal time}$$

We might at this point make a note of the fact that the days are divided into 24 equal hours. This was adopted from a scheme used by the ancient Greeks and Egyptians, who possibly derived it from the Babylonians. Originally daylight and darkness were divided into 12 hours each. Since the periods of daylight and darkness are not uniform throughout the year, it was necessary to constantly adjust the length of the hours. Thus during the summer with its longer daylight period, the hours were longer than the hours during the period of darkness. During the winter months the reverse was true. In the thirteenth century, an Arabian mathematician suggested that the interval from one transit of the sun across the meridian to the next be divided into 24 equal hours. This suggestion was adopted approximately two centuries later.

Revolution

Although Copernicus and Kepler described the motion of the Earth around the sun, they were not able to prove that it was in fact a true motion. For this reason their theories were suspect, and it was not until some time later that indisputable proof of revolution became available.

One proof was provided by the discovery of the aberration of starlight. We have all experienced the effect of rain falling vertically when we stand still and at an angle when we move. When standing still we need to hold our umbrella directly upright to shield us from the rain, provided there is

---

*Celestial meridian: An imaginary line (great circle) on the celestial sphere, passing directly overhead (zenith) in a north-south direction and crossing the viewer's horizon at 90°.

Figure 7.11 A sidereal
day represents one 360°
rotation of the Earth with
respect to some distant
star. A solar day mea-
sured with respect to the
sun requires that the
Earth rotate about 1°
more than necessary for
a sidereal day.

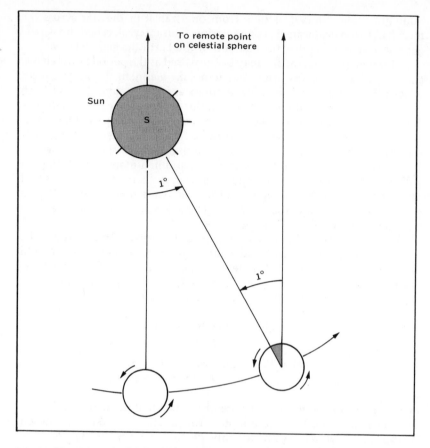

Figure 7.11 A sidereal day represents one 360° rotation of the Earth with respect to some distant star. A solar day measured with respect to the sun requires that the Earth rotate about 1° more than necessary for a sidereal day.

no wind to drive the rain. If we walk it is necessary to tilt the umbrella in the direction in which we are moving, and the degree with which we tilt the umbrella increases as we increase our speed. Light behaves in the same manner as the raindrops. However, we do not normally notice this phenomenon, because the speed with which we move relative to the speed of light is very small. This type of aberration was detected for light in the eighteenth century by James Bradley, the Astronomer Royal of England. He noted that the velocity of the Earth produced a measurable angular displacement in the position of the stars, requiring a slight adjustment in the positioning of the telescope to accommodate for the *aberration of starlight* (Figure 7.12). Since he recognized that motion was necessary for the aberration, he concluded that the phenomenon could only be ex- plained on the basis of the orbital motion of the Earth around the sun.

Another proof of the Earth's revolution is the annual parallax, or apparent yearly shift in the position of the nearby stars against the background of more distant stars. When Aristarchus in the third century B.C. first suggested that the Earth orbited the sun, his colleagues, who understood the nature of parallax, searched for this apparent shift, but were unable to find it and so abandoned Aristarchus' theory. Not until the early nineteenth century was the parallax of a nearby star measured (see chapter 13), thus providing another piece of evidence that the Earth orbited the sun.

It is also possible to demonstrate the Earth's motion in orbit by means of the Doppler effect, previously discussed in chapter 5. For a star on the

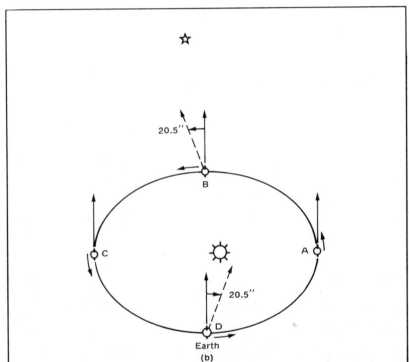

Figure 7.12 The principle of aberration may be illustrated in (a) with falling rain. The aberration of starlight is the apparent displacement of a star in the direction of Earth's motion in orbit when the Earth is moving at right angles to the direction of the star (b). Aberration would be evident at position B and D but not at A and C for the star shown.

ecliptic a Doppler shift is evident and will vary during the annual orbit, depending upon whether the Earth is traveling in its orbit toward or away from the star. The stars will exhibit their own motion through space, but this motion can be separated from that imposed by the Earth's orbital motion. Stars in a direction perpendicular to the ecliptic do not display any motion other than their own space motion, as is expected.

The interval of time required for the Earth to make one revolution around the sun is approximately 365¼ days. For this reason, three calendar years out of four have 365 days and the fourth year, *leap year*, has 366 to compensate for the accumulated quarter days. This time interval for the

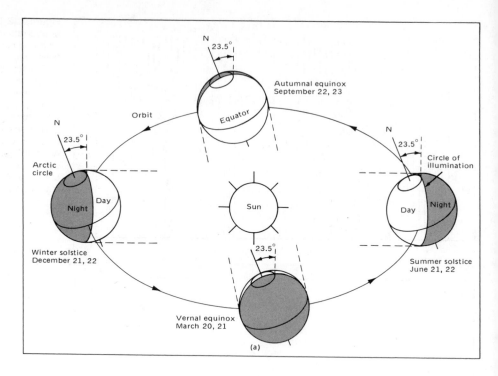

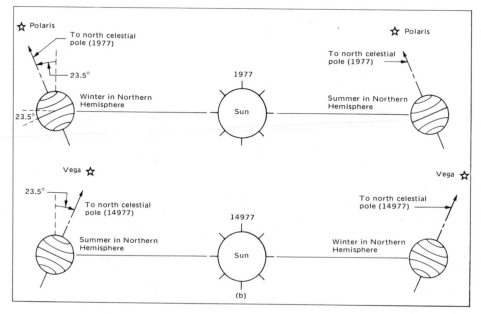

Figure 7.13 The Earth's equator is inclined 23.5° from the ecliptic plane (a). The sun is directly overhead in the southern hemisphere (23.5° S. latitude), and as the Earth revolves around the sun, the sun appears to move north being overhead on the equator on approximately March 20. Three months later the sun is overhead in the northern hemisphere. In about 13,000 years the Earth's axis will be pointing in the opposite direction due to precession (b).

year is approximate, and when we define the year more precisely we find that there are several ways in which the year may be measured. One complete 360° orbit around the sun with respect to the stars is the *sidereal year*. The length of the sidereal year is 365d 6h 9m 10s, and this is considered the true period of revolution.

The *tropical year*, used for calendar purposes, is measured between successive vernal equinoxes and is 365d 5h 48m 46s. The *vernal equinox* is the moment, on approximately March 20–21 of each year, when the sun is crossing the celestial equator and apparently moving from south to north. We briefly defined vernal equinox earlier and will now describe it more fully. We need first to recognize that the Earth's equator is inclined 23.5° from the ecliptic plane. As a result, when the Earth revolves around the sun, the sun appears to move in the sky so that on approximately March 20, the vernal equinox, the sun will be directly overhead over the equator apparently moving north (Figure 7.13). Because of the Earth's continued progress in orbit, the sun will continue its apparent northward movement to reach a position 23.5° north of the equator (Tropic of Cancer) about June 21, which is the *summer solstice*. At this time the apparent motion of the sun reverses itself, moving south, to recross the equator approximately September 22 for the *autumnal equinox*. The sun continues its apparent southward journey until December 21, the *winter solstice*, when the sun is directly overhead at 23.5°(Tropic of Capricorn) south of the equator. From here the sun reverses direction to repeat its apparent northward trek to the vernal equinox. It might be noted here that the intervals between the points specified are not uniform, due to the variation in the speed of the Earth in its orbit (Table 7.2). From Figure 7.14 it can be seen that the exact time and date for the four points will vary from year to year because of the accumulation of the quarter days between leap years. In addition, the variation in the orbital speed of the Earth will cause a slight variation in the day-to-day length of the rotational period of the Earth.

There is an additional means by which the year may be measured, and this is between successive perihelion passes. Planetary perturbation causes the Earth's perihelion to precess (in the same manner as Mercury's orbital precession) in the direction of the Earth's orbital motion. This year, called the *anomalistic year*, is 365d 6h 13m 53s in length. The length of all the years is given in solar days.

Variation in Quarterly Intervals of Earth in Orbit		TABLE 7.2
Quarterly Intervals	*Length of Interval*	
Vernal equinox to summer solstice	92d 19h 00m	
Summer solstice to autumnal equinox	93d 15h 17m	
Autumnal equinox to winter solstice	89d 19h 27m	
Winter solstice to vernal equinox	89d 00h 05m	

Throughout the discussion on revolution we have described the Earth's motion in orbit as if it were occurring around the center of the sun. This is not exactly true. Both the Earth and the sun have mass, and therefore both these bodies orbit around a point of balance between the two bodies which is the *center of mass* or *barycenter*. Since the mass of the sun is far greater than that of the Earth, the center of mass is located quite near the center of the sun and only very precise measurements enable the

Figure 7.14 The periods from equinox to solstice to equinox are not uniform. These periods vary depending upon the distance and velocity the Earth travels at different points in orbit. The difference in velocity also causes a slight variation in the length of the solar day.

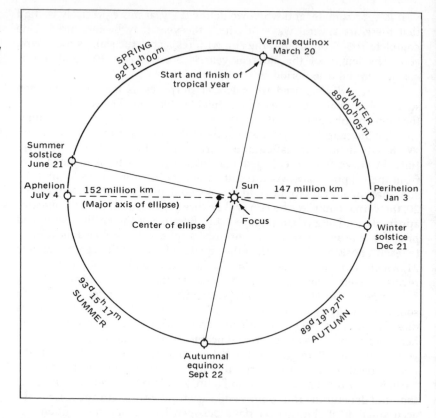

astronomer to separate the two points. The barycenter represents the point of focus of gravitational forces of the two bodies and therefore is also referred to as the *center of gravity.* Considering these relationships, we can say that the distance of the centers of the Earth and sun from the barycenter are inversely proportional to their respective masses (Figure 7.15). The same relationship applies to all objects orbiting the sun or for satellites orbiting the planets.

Precession

The precession of the Earth, a movement not readily discernible, resembles the gyrations of a spinning top or gyroscope (Figure 7.16). This movement is so slow that it is barely visible even to close observation, and then only over long periods of time. Recall that the axis of the Earth is inclined 23.5° from a normal to the ecliptic. Precession is the motion of the

Figure 7.15 Unequal masses (*M,m*) will revolve around a center of gravity (*CG*). The relationship of the distances of the objects to *CG* may be given as $M/m = R/r$.

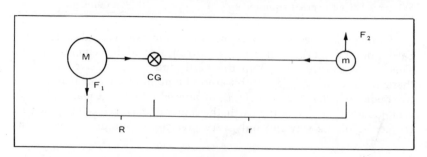

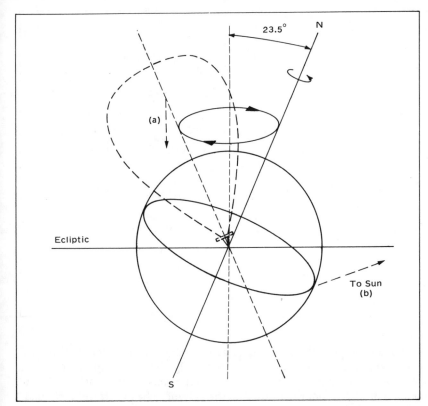

23.5°   N

(a)

Ecliptic

To Sun
(b)

S

Figure 7.16 The force of
gravity pulls downward
on the top causing it to
precess or change the
orientation of the axis by
a slow conical motion (a).
The force of gravity of
the sun, moon, and to a
lesser degree the planets
works on the Earth's
bulge to pull the axis
upright to a 90° angle
with the ecliptic, causing
the Earth to precess (b).

Earth's axis at this angle around the imaginary axis to the ecliptic. This movement causes a shifting of the Earth's axis relative to the stars, so that while Polaris is the present polar star, Alpha Draconis was the polar star in 3000 B.C. Hipparchus (second century B.C.) found Alpha Draconis to be 11° from the celestial pole, and this displacement led to his discovery of precession. Currently Alpha Draconis is about 25° from the celestial pole. Vega will be the polar star in 14,000 A.D. To return to the present position, after a complete cycle, will require about 25,900 years. Because of precession, the vernal equinox shifts slightly westward each year, causing the tropical year to be 20m 23.42s shorter than the sidereal year. This shifting is known as the *precession of the equinoxes.*

Why does the Earth precess? We have already described the Earth as an oblate spheroid, whose equatorial diameter is about 33.5 kilometres greater than the polar diameter. Also recall that the Earth's equator is inclined 23.5° from the ecliptic. The axis of rotation is therefore inclined 23.5° from a normal to the ecliptic, and the axis describes a slow, not readily discernible, conical motion around the normal, much as a spinning toy top or a gyro. This motion is caused by the gravitational force of the sun, moon, and to a lesser degree, the other planets on the equatorial bulge of the Earth. Although the moon's orbit is inclined 5° to the ecliptic, its average gravitational effect is centered on the ecliptic and therefore reinforces the gravitational influence of the sun. The moon's gravitational influence in this case is about twice that of the sun's influence, and the other planets contribute about 1/40 of the total.

In Figure 7.17 the bulge at (a) nearest to the sun and moon is influenced

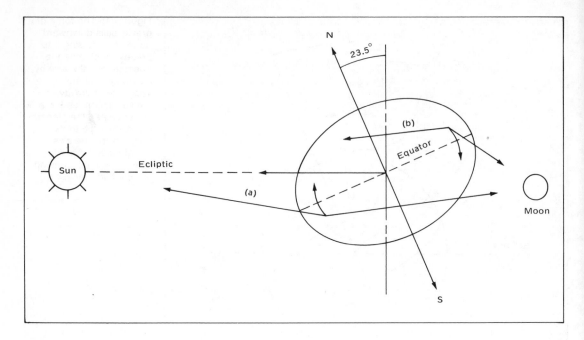

Figure 7.17 The diagram shows the component forces of the sun (applies also to the moon) as these forces act on the equatorial bulge of the Earth. The force at (a) has the greatest influence, being closer to the sun than (b), and acts to pull the Earth's bulge into line with the ecliptic, thus tending to pull the axis upright.

to a greater degree than the bulge at (b), resulting in differential forces which tend to pull the Earth's equatorial bulge into line with the ecliptic. As a result of these differential forces, the axis of the Earth remains inclined at 23.5° from the normal to the ecliptic. The axis traces a circular path around the ecliptic pole (point on the celestial sphere above the normal) once every 25,900 years.

7.5 TIMEKEEPING

We know that the solar day (24h) and the tropical year (365d 5h 48m 46s) are used for the purpose of recording the passage of calendar time. We also recognize a month of variable length (28–31 days), although originally the month was related to the lunar cycle of 29½ days (see chapter 8). The seven-day week represents approximately one quarter of the lunar cycle, and the days of the week were named for the seven planets as recognized in ancient times. None of these units for measuring time are compatible, that is, they are not evenly divisible into one another. As a result, developing a calendar universally suited for all purposes has been difficult. Each civilization developed a calendar according to its needs—some had several calendars.

The calendar currently used in the western world originated in Rome in about the seventh century B.C. The Roman calendar, adapted from an earlier Greek and Babylonian version, began the year in March, but this was changed to the first of January in the second century B.C. The year was composed of twelve months based on the lunar cycle. Seven of the months were 29 days in length, four were 31 days, and one was 28 days, totaling 355 days. This was 10¼ days short of one complete revolution around the sun, and as a result constant adjustments had to be made to maintain the calendar in phase with the seasons. Julius Caesar (c. 45 B.C.) was instrumental in calendar reform that resulted, except for small modifications made in the sixteenth century, in the calendar we use today. First, he established the period from one vernal equinox (March 21) to the next vernal equinox as the length of the year. He then had six months with 31 days, five months with 30 days, and February with 29 days (Table 7.3). This totaled 365 days, and he introduced leap year by adding an extra day to February each four years to account for the annual accumulation of quarter days. At the same time, the month Quintilis was changed to July in memory of Julius Caesar's reform.

MONTHLY DISTRIBUTION OF DAYS IN SEVERAL ROMAN CALENDARS   TABLE 7.3

Month	Number of Days		
	Pre-Julius Caesar	Julius Caesar	Caesar Augustus
Ianuarius	29	31	31
Februarius	28	29 - 30 l.y.*	28 - 29 l.y.*
Martius	31	31	31
Aprilis	29	30	30
Maius	31	31	31
Iunius	29	30	30
Quintilis (July)	31	31	31
Sextilis (August)	29	30	31
September	29	31	30
October	31	30	31
November	29	31	30
December	29	30	31

*l.y. = leap year

Caesar Augustus, about 7 B.C., is credited by some historians with making a few minor changes in the arrangement of days (Table 7.3). It is thought that he changed the name of the month Sextilis to August in honor of himself and took a day from February to add to August, thereby making it as long as July. He then moved a day from September to October and from November to December so that there would never be more than two consecutive months with 31 days. Some historians think these changes were accomplished at a later date, and the matter is in dispute.

The Julian calendar remained essentially unchanged for hundreds of years, but over time a deficiency became apparent. The calendar overestimated the length of the tropical year by 11m 15s, and this cumulative defect caused the vernal equinox to occur earlier each year. By the sixteenth century the vernal equinox occurred ten days early on March 11. In 1582, in consultation with a number of other governments, Pope Gregory XIII suggested a reform that reestablished the vernal equinox on March 21. He also had three leap year days eliminated from every 400 years to reduce the error. This was accomplished by having no leap year day on three century years out of four that normally are leap years. Thus the year 1600 had a leap year as it normally would but no leap year day was added to the years 1700, 1800, and 1900. The year 2000 will have a leap year day. Even these corrections have not been exact, and an adjustment of one day will be required every 3000 years.

Some of the Protestant countries did not accept the Gregorian calendar and did not make the correction until the eighteenth century. Great Britain changed to the new calendar in 1751, by which time the calendar was eleven days out of phase with the vernal equinox. To make the correction, eleven days were dropped from the calendar, an act misunderstood by many of the citizens. As a result riots ensued because the people felt they were being cheated, and they demanded the return of the eleven days.

In the Gregorian calendar the days of the week do not fall on the same date each year, which means that a new calendar must be issued annually. Some attempts have been made to establish a perpetual calendar in which the dates and days are the same for each year. One such scheme suggests four equal quarters of thirteen weeks each, with the first of the quarter beginning on a Sunday and ending on a Saturday. Each quarter would be three months, two with 30 days and one with 31 days. The four quarters total 364 days, and a year-end day, an extra Saturday, would be included to make the full year. On leap year an extra Saturday would be included at the end of the second quarter. This calendar was introduced in the United Nations Assembly but has not generated much interest toward adoption.

7.6 THE SEASONS

The seasons of the year—spring, summer, fall and winter—follow one another in endless cycle. Seasonal variations are due to the Earth's axial inclination of 23.5° from a normal to the ecliptic and are not the result of varying distances from the sun during the year. The Earth's elliptical orbit brings it closer to the sun at perihelion in January (147 million kilometres), and further away at aphelion in July (152 million kilometres). This results in a few percent more solar radiation being received in January than in July by the Earth as a whole, but it is not an important factor in climate.

The Earth's axial inclination causes the Northern Hemisphere to be subjected to direct solar radiation at one time of the year and the Southern Hemisphere six months later (Figure 7.18). The sun will be directly overhead 23.5° north of the equator on June 21 (summer solstice) in the Northern Hemisphere, resulting in this hemisphere receiving maximum solar radiation or *insolation*. In the southern latitudes, the surface receives the sun's rays at a slant. The amount of insolation received for a given area in the Northern Hemisphere would therefore be distributed over a greater area in the Southern Hemisphere and would result in less heat per unit of

Figure 7.18 Because of the Earth's axial inclination, the sun's rays are directly overhead alternately north of the equator and six months later south of the equator. The alternating concentration of solar radiation results in changing seasons.

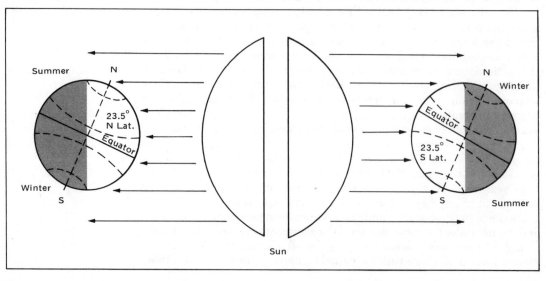

area (Figure 7.19). In addition, the direct rays in the Northern Hemisphere travel through less thickness of atmosphere than do the slanting rays south of the equator. Thus there is less radiant energy lost by atmospheric absorption and reflection in the north than in the south (Figure 7.19). The length of the day during which any area of the Earth's surface is subject to insolation is also a factor in the rhythm of the seasons. During the summer months in the Northern Hemisphere, days are longer than in the Southern Hemisphere, permitting a longer exposure to solar radiation (Figure 7.18). This allows a greater amount of heat to be delivered in a 24-hour period. Figure 7.18 reveals that when we have summer in the Northern Hemisphere, the north polar area has 24 hours of daylight and the south polar area has 24 hours of darkness. This combination of factors results in generally warmer temperatures north of the equator and colder temperatures in the southern latitudes.

As the Earth continues to progress in orbit, the sun will appear to travel south until it is directly overhead 23.5° south of the equator. At this time warm weather will be experienced in the Southern Hemisphere and cold weather in the Northern Hemisphere, or the reverse of what was occurring six months previously.

The occurrence of maximum insolation and day length, coincident with minimum atmospheric absorption and reflection of solar energy during June in the Northern Hemisphere (December in the Southern Hemisphere), would lead to the expectation that the highest average temperatures would be experienced at that time. This does not occur, because time is required for the atmosphere and land surface to accumulate heat, resulting in a seasonal lag of about one month. For this reason, maximum temperatures are generally encountered in July and August (January and February in the Southern Hemisphere). By the same token, minimum temperatures also generally lag the period of minimum insolation by approximately a month. Thus, January, on the average, is the coldest month north of the equator, and July the coldest month south of the equator. It must be recognized that, in addition to the normal seasonal cycle, there are other factors that modify climate, even along the same latitude and within short distances. If the Earth were a uniformly smooth sphere of similar surface material, then temperatures might be expected to

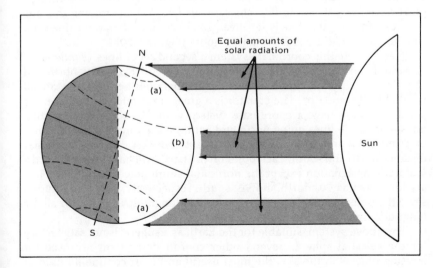

Figure 7.19 The sun's radiation is most concentrated where it strikes the Earth vertically and less concentrated where radiation strikes at an angle (a). The same amount of radiation is distributed over a greater area at (a) and is more concentrated at (b).

grade uniformly from equator to pole. The fact that the Earth is not a uniform sphere causes climate to deviate from a regular pattern of distribution.

7.7 COORDINATE SYSTEMS

We are familiar with the fact that it is possible to locate a point on a flat surface by using a grid system of two sets of parallel lines at right angles to each other. This represents a coordinate system. Most towns and cities are arranged in this fashion, with one set of streets oriented approximately north to south and the second set east and west. This enables us to locate a particular intersection or point by identifying two streets that cross at that site. Such a system works well on flat surfaces, and the area covered by even our large cities may be considered relatively flat by comparison with the Earth.

On a sphere such as the Earth, a grid system as just described is not suitable for locating specific points. To establish a coordinate system on the surface of the Earth, we take advantage of the rotation of the Earth to establish two points—the north and south poles. A line may be drawn from pole to pole, and we call this a *longitude line* or *meridian*. If the line is drawn completely around the Earth, thereby cutting the Earth into two equal halves through the poles, we have a *great circle*. By following a great circle on the Earth's surface, ships or planes may travel between two points by the shortest possible route. Twelve great circles drawn through the poles an equal distance apart would divide the Earth into 24 segments, each equal to 15° of longitude, and would represent a one-hour time zone. For this system to function in an orderly manner one meridian is designated as *prime meridian* and represents, in a manner of speaking, the starting point. The prime meridian is, by common consent, the line passing from the north pole to the south pole through a point on the old naval observatory in Greenwich, England, and is identified as 0° longitude. To the east of the prime meridian all longitude measured as angular distance is designated as east longitude up to 180°. To the west of the prime meridian the angular distance is given as west longitude. The 180° longitude line—on the opposite side of the Earth from the prime meridian and thereby forming a great circle with it—is the *International Date Line*. When crossing this line from east to west, the time is set ahead 24 hours, and when returning from west to east, the time is reversed 24 hours (Figure 7.20).

To complete this coordinate system a second set of lines— *latitude lines* or *parallels*—are established and measured as angular distance north and south of the equator. These lines are called parallels because they are all parallel to the equator. The equator is a great circle midway between the poles. We now have a coordinate system with two sets of intersecting lines—longitude lines going north and south, measuring angular distance east and west, and latitude lines going east and west, measuring angular distance north and south. Each point on the Earth will have a latitude and a longitude designation except the north and south pole. These two points are identified as 90° north and 90° south, respectively, and are latitudes, since the poles are 90° from the equator. No longitude is given, for all the longitude lines converge on the poles.

The above system, suitable for the Earth as a sphere, is not satisfactory for the celestial sphere. Several other coordinate systems are used for locating objects in the sky, the most useful and most commonly used of

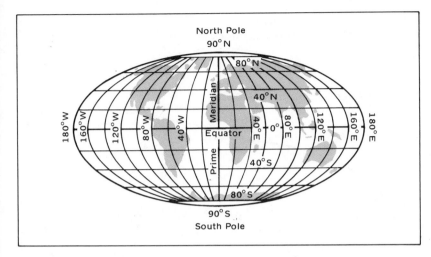

Figure 7.20 Parallels and meridians on the surface of the Earth.

which is the *equatorial system*. In this system the Earth is considered to be a small point and the basic reference points are the celestial poles, which are points in the sky directly above the earthly poles. The north celestial pole is identified approximately at the position of Polaris, the *North Star*. The equator serves as the basic great circle and is midway between the celestial poles. Lines corresponding to latitude and longitude on Earth provide us with a coordinate system in the sky. Angular distance from the celestial equator is the *declination*, which is designated as positive for a position north and negative for a position south of the celestial equator. The unit of measure that corresponds to longitude on Earth is right ascension. *Right ascension* is the displacement eastward from the vernal equinox and is measured in units of time, with each hour equal to 15° of angular displacement. Notice that right ascension is measured eastward only from the vernal equinox and not in both directions, as is the case with longitude on Earth (Figure 7.21).

The *horizon system*, although quite simple to use, has the disadvantage of having the basic frames of reference change with location and time. The basic reference point in this system is the *zenith*, which is a point directly overhead from the observer. Thus if the observer changes his position, the zenith changes as well. In addition, as the Earth rotates, the zenith moves eastward in the sky. The great circle is the true horizon, which is 90° from the zenith and not the uneven horizon seen on land. To determine the location of a celestial object, we draw a guide line from the zenith through the object to the horizon. Then the *altitude* of the object, or the angular displacement of the object from the horizon, is measured along this line. This represents one coordinate. From the north point on the horizon we can determine the *azimuth*, which is the angle formed between the north point and the point where the guide line touches the horizon going clockwise. Now we have the second coordinate and can fix the position of any object in the sky. Because of the disadvantage mentioned above, this coordinate system has limited application, being used primarily by navigators and surveyors (Figure 7.22).

An older system used in the past to study the movements of objects in the solar system is the *ecliptic system*. The ecliptic system is essentially the same as the equatorial system, measuring position by declination and right ascension. The system utilized the ecliptic as the basic great circle,

Figure 7.21 The equatorial system. The Earth is located at the center of the celestial sphere with the polar great circle and the equatorial great circle projected on the celestial sphere. Right ascension is measured in units of time eastward from the vernal equinox, and declination measured in degrees north and south of the celestial equator.

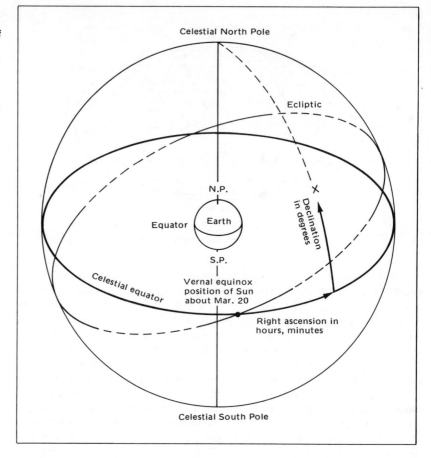

Figure 7.22 The horizon system. The frame of reference changes with a change in the observer's position. Angular displacement for altitude is measured from the horizon to zenith. From the north point on the horizon, azimuth is the angular displacement clockwise along the horizon.

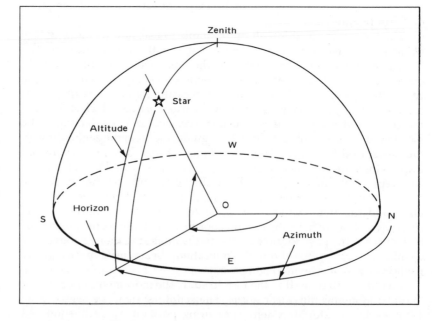

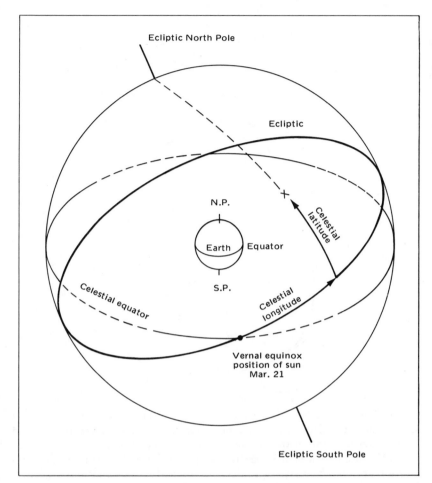

Figure 7.23 The ecliptic system. This system is similar to the equatorial system, utilizing the ecliptic great circle instead of the equatorial great circle. Celestial longitude and latitude are measured with respect to the ecliptic. The starting point for longitude is the vernal equinox.

and the ecliptic north and south poles measured 90° from the ecliptic (Figure 7.23).

To study the Milky Way astronomers make use of a coordinate system called the *galactic system*. The galactic system has as its fundamental great circle the central line of the Milky Way which forms a galactic circle. The *galactic latitude* is measured as the angular distance above (north) or below (south) of this line. *Galactic longitude* is measured in degrees along the galactic circle in an easterly direction. The starting point is the constellation of Sagittarius, which lies on the radius vector between the Earth and the galactic center (Figure 7.24). This constellation may be readily seen in the summer night sky.

7.8 LIFE ON EARTH

Before we talk of life in the universe we need first to know about life on Earth. When did it first occur? What were its first stages? Then we can ask, could these same events occur some other place? The history of abundant life on Earth can readily be traced back 600 million years—a period representing one-eighth of the Earth's age. Earlier than 600 million years

Figure 7.24 The galactic system. The fundamental great circle is the central plane of the Milky Way. Galactic longitude is measured from the galactic center eastward along the galactic plane. Galactic latitude is measured as angular displacement north and south of the galactic plane.

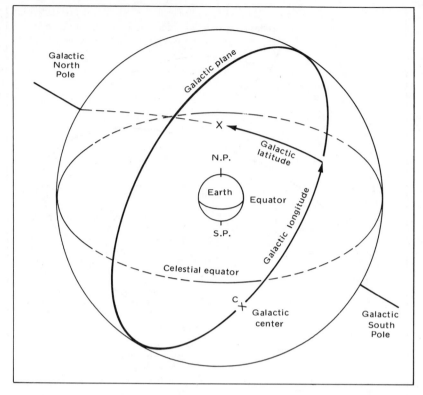

there is no immediately apparent trace of life in the Earth's rock, but it is recognized that there must have been some evolutionary precursors, and this represents one of the classic puzzles in paleontology.

Laminated rock structures that might be formed by microorganisms were first noted by Charles Walcott in the early 1900s and more recently by E. S. Barghoorn and William Schopf. These rocks have been shown to be fossils of blue-green algae and bacteria, which are themselves rather complex organisms. The oldest of the fossils were found in rock in South Africa and are now known to be about 3.5 billion years old. If we assume the 4.5 billion-year-old Earth to have been in a relatively molten or unstable state during the first half billion years of its existence, then these single-celled organisms developed during a period that was at most half a billion years in length. The question, then, is how did this come about?

The single-celled organisms are of the same material and have the same life processes common to all plants and animals, including man. They are composed of amino acids—the building blocks of protein—and the DNA molecule, which contains the genetic code. The DNA molecule—a nucleic acid—is composed of millions of pairs of nucleotides arranged as rungs in a spiral ladder. The order of the nucleotides represents the code necessary in arranging and assembling the amino acids. Variations in the code results in different protein.

This complex system had already evolved 3.5 billion years ago, but the question is how. Unlike the present atmosphere, the primitive atmosphere was composed of methane, ammonia, and other gases (section 7.3). These materials are thought to be the precursors of life. To test this idea Stanley Miller in 1952 set up apparatus composed of a model ocean and atmosphere containing water and the gases of the primitive environment. Energy was supplied to the model in the form of an electric discharge to simulate lightning, and in about one week the mixture turned a reddish brown. The "primordial soup," as it became known, was analyzed and found to contain amino acids, the precursors of protein. In order for the amino acids to be properly arranged to form protein, the nucleic acids carrying the genetic code were required. Leslie Orval at the Salk Institute in California was able to produce a chain of nucleotides about six units in length. This was not anywhere near the length of the million or more unit nucleic-acid chain, but it was a start. The question then raised was how the amino acids became arranged in an orderly sequence, and experimentation revealed that this could be accomplished by appropriate heating of the amino acids. As yet life has not been developed by this primitive process in the laboratory, nor is there an intention to do so. But the work provides a clue as to how life may have come about on the primitive Earth. Given the oceans that presumably existed, the nonoxidizing atmosphere, a source of energy in the form of lightning, and sufficient time—up to half billion years—it appears very likely that life could have developed in this fashion.

Could these same events have occurred elsewhere? Samples from the moon show this body to be devoid of life or the precursors to living materials. However, the Murchison meteorite (see chapter 11), an object that fell to Earth in Australia in 1969 and was picked up almost immediately after, provided scientists an opportunity to analyze a specimen without concern as to whether or not it had been contaminated by biological processes on Earth. A number of amino acids unlike any previously found were discovered, and it was concluded that these were produced during the process of meteorite formation. If this is true it has far-reaching implications, because it is an indication that life-forming processes are at work in other parts of the solar system.

7.9 SUMMARY

There are some difficulties with studying the Earth as an astronomical body when the observer is viewing the Earth from its surface, but despite this the Earth was recognized as a sphere at least 2500 years ago. Its size, mass, and density have been determined, and the Earth is found to be composed of a series of concentric spheres of progressively denser materials surrounding a very dense core. The outer surface of the Earth or the crust is composed of relatively lightweight rock, which comprises the continental masses, and slightly denser rock beneath the continental masses and covering the ocean floor. Separating the crust from the mantle immediately beneath is the Moho discontinuity, a zone where the velocity of the earthquake waves change. The mantle continues inward to the core, which is thought to be partially molten and composed of iron and nickel.

The atmosphere surrounding the Earth, now composed principally of nitrogen and oxygen, evolved from an earlier atmosphere of methane, ammonia, carbon dioxide, and other gases. The atmosphere occurs in three recognizable layers, each with distinguishing characteristics of temperature, turbulence, density, and composition. The Earth is also surrounded by a magnetic field within which are confined the Van Allen radiation belts. Knowledge of the Earth's atmosphere and magnetic field is important to an understanding of these features around other planets.

The Earth's motions—rotation, revolution, and precession—have been shown to be valid by various means and the periods determined with a fair degree of accuracy. There are several ways in which the periods of rotation and revolution may be determined, depending upon the frame of reference used to make the measurement. Precession, not so noticeable, may only be detected over long periods of time, and influences the position of the north celestial pole.

The motions of the Earth, particularly rotation and revolution, are useful in timekeeping. Rotation, based on the turning of the Earth with respect to the sun, is a measure for a day; the period required for the Earth to make one revolution around the sun is one year. The lunar period approximates one month, and the week about one quarter of the lunar period. From these various temporal periods a variety of calendars have been developed to keep a record of the passage of time. None are completely satisfactory and require alterations over the centuries. The Gregorian calendar, presently used in the western world, is sufficiently accurate to require only a one-day adjustment in 3000 years.

The seasonal changes on the Earth result from the Earth's axial inclination to the ecliptic. The amount of direct solar radiation (insolation) received by any portion of the globe at different times of the year varies as the Earth orbits the sun. This orientation also results in a change in day length as the Earth revolves around the sun, varying the length of time a portion of the Earth is subject to solar radiation. The angle at which the sun's rays enter the atmosphere will influence the amount of radiation that is absorbed by the atmosphere. In effect, the atmosphere acts as an insulator of greater or lesser efficiency, depending upon the angle at which the sun's rays enter the atmosphere.

A coordinate system is a necessary means for locating a point on a map or in space. Several systems are available but the one most generally used is the equatorial system, in which the north and south celestial poles are used as reference points and the celestial equator serves as the basic great circle. Declination is the angular displacement of an object north (positive) or south (negative) of the celestial equator. Right ascension is the displacement eastward from the vernal equinox and is measured in hours, minutes, and seconds, with each hour equal to 15° of angular displacement. Other coordinate systems exist but are not as adaptable for locating celestial objects as is the equatorial system.

Life on Earth may have formed over 3.5 billion years ago. The process is not fully understood, but life may have been formed from the primitive atmosphere containing methane, ammonia, and other gases in a primitive ocean. The application of energy by lightning is thought to have produced amino acids which by appropriate heating formed protein. The laboratory formation of the precursors of life by this process and the discovery of amino acids in meteorite material is a positive indication that life exists elsewhere in the universe.

QUESTIONS

1. What is the density of the Earth as calculated from Earth's mass and volume? Use the average value for the Earth's radius to determine volume.

2. Does the density of Earth's materials increase or decrease upon approaching the core?

3. Make a graph showing the profile of Earth's atmospheric temperatures.

4. Discuss an experiment by which it is possible to prove that the Earth rotates.

5. Discuss an experiment by which it is possible to prove that the Earth revolves around the sun.

6. Describe briefly why there is a difference in the length of the sidereal day and the solar day. Which one is used for the purpose of telling time?

7. Describe the basis for measuring the length of the sidereal year, the tropical year, and the anomalistic year.

8. Why is there an approximate 20-minute difference between the tropical year and the sidereal year?

9. What is the function of the leap year?

10. What was the calendar correction made in the Gregorian calendar? What influence did this have on the year 1900? What, if any, will be the effect on the year 2000?

11. What are the factors that bring about seasonal changes in temperature on various parts of the Earth?

12. Describe the equatorial system of coordinates.

13. Describe in outline form the possible way in which life may have evolved on Earth.

EXERCISES AND PROJECTS

1. How would you go about measuring the length of the year? Devise a procedure whereby the length of the year may be determined, and write it up in the form of a short report.

FOR FURTHER READING

GORDON, R. B., *Physics of the Earth*. New York: Holt, Rinehart and Winston, 1972.

HARTMANN, W. K., *Moons and Planets*. Belmont, Calif.: Wadsworth Publishing Co., 1972.

HESSE, W. H., and R. L. McDONALD, *Earth and Its Environment*. Encino, Calif.: Dickenson Publishing Co., 1974.

HOLTON, G., and D. H. D. ROLLER, *Foundations of Modern Physical Science*. Reading, Mass.: Addison-Wesley, 1958.

PONNAMPERUMA, C., and A. G. W. CAMERON, *Interstellar Communication: Scientific Prospectives*. Boston: Houghton Mifflin Co., 1974.

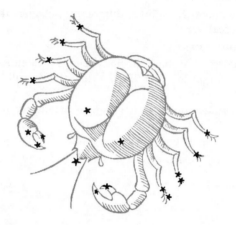

Cancer

CHAPTER 8   THE MOON

Which is more useful, the sun or the moon? The moon is the more useful
since it gives us light during the night, when it is dark, whereas the sun shines
only in the daytime, when it is light anyway.

—George Gamow

The Earth is accompanied in its travels around the sun by the moon. The
moon is known by several names, including Selene (from which *selenogra-
phy*, the study of the moon's surface, is derived), Phoeba, Cynthia, and
Luna. From Luna comes the word lunatic, for it was once believed that if
one slept under the light of the moon insanity would result. The moon has
been an object of great interest through the centuries, the subject of myth
and song and story—and little scientific information. With the invention of
the telescope, much was learned about its surface. Finally scientists
accomplished a lunar landing, with the intention of resolving all the
questions about the moon. While some questions were answered by the
landings many more have been raised, making the moon a subject for
study for many years to come (Figure 8.1).

8.1 LUNAR SIZE AND DISTANCE

The moon has a diameter a little more than one-quarter that of the Earth.
Although the moon ranks fifth in size compared to other satellites in the
solar system, it is more massive compared to the Earth than the other
satellites are in relation to their respective planets. As previously stated
(chapter 6), the Earth-moon system may be more characteristic of a double
planet than a planet-satellite system.

Aristarchus made the first recorded attempt to calculate the Earth-
lunar distance. However, the first approximately accurate determination
was not accomplished until 1751. The distance has been calculated many
times since, using the method of triangulation (see section 13.1). Modern
techniques now make use of radar and laser for such measurements, so the
calculated distance to the moon is now absolutely accurate.

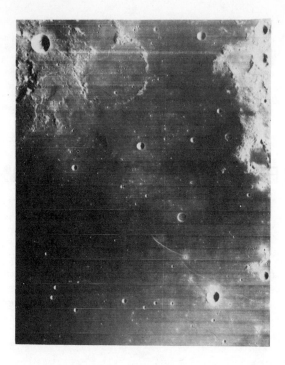

The mean distance from the center of the moon to the center of the Earth is 384,404 kilometres. The moon describes an elliptical orbit around the Earth; at its closest approach or *perigee* it is 356,555 kilometres from the Earth, and at its greatest distance or *apogee* it is 406,863 kilometres from the Earth.

Before scientists were able to determine the distance to the moon, they could measure the moon's angular diameter. Once the distance became known, this, together with the angular diameter, provided the data for calculating the actual diameter. The lunar diameter is now known to be 3476 kilometres.

8.2 LUNAR MOTION

The moon in orbit has the same relationship to the Earth as the Earth has to the sun, in that the Earth and moon revolve around a barycenter as do the sun and Earth. From the previous chapter we may have gained the impression that the Earth's center travels along the path of its orbit. This is not entirely true, since what is called the Earth's orbit is, in reality, the orbit of the center of mass or *barycenter* of the Earth-moon system. The true motion may be detected by closely observing the behavior of the nearby planets. Mars, for example, exhibits an apparent displacement that causes it to appear to move alternately ahead and behind its expected orbital motion. At its closest approach to the Earth, Mars shows a biweekly apparent displacement of 17 seconds of arc—a displacement that can be attributed to the Earth's movement around the barycenter. From such observations the barycenter has been calculated to be 4671 kilometres from the Earth's center on a line between the Earth and the moon. The centers of the Earth and the moon orbit this point about once a month (Figure 8.2), and the relationship of the distances, d_e/d_m, provides us with a ratio of the mass of the Earth to the mass of the moon, which is 81.5:1.

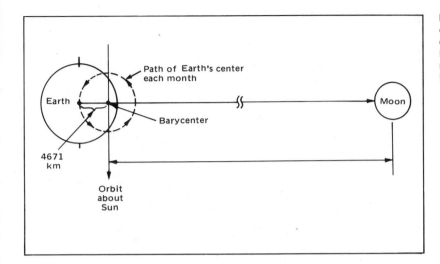

This data permits us to calculate the lunar mass if the mass of the Earth is known (it is). Then from the lunar mass and its volume, which can be determined once the diameter is known, it is possible to calculate the average lunar density. This has been found to be 3.3 grams per cubic centimetre. Information of this type serves as a basis for speculating on the lunar structure and origin as compared with the Earth.

We have described the lunar orbit with respect to the barycenter. However, it is convenient to describe the orbit of the moon with respect to the Earth's center when describing lunar motion as seen from the Earth. Such an orbit is known as a *relative orbit.*

In terms of the relative orbit, the moon's revolution around the Earth requires about one month, but there are two ways in which the period of the orbit may be measured. The *sidereal month* is the true period of revolution, being the interval between two successive conjunctions of the moon's center with the same star as seen from the Earth. The time required for the moon to travel its 360° orbit around the Earth is a sidereal month of 27d 7h 43m 11.5s, or about 27⅓ days. The *synodic month* is the interval between successive inferior conjunctions of the moon, or from new moon to new moon. This period is longer than the sidereal month by more than two days, being 29d 12h 44m 2.8s, or about 29½ days. The difference is due to the fact that during the moon's revolution about the Earth, the Earth in its turn has moved along its orbit around the sun, as illustrated in Figure 8.3.

The lunar orbit precesses once in slightly over 18 years. If the month is measured with respect to its perigee, we have an *anomalistic month* of slightly over 27½ days.

The moon's eastward motion, resulting from its movement in orbit around the Earth, causes the moon to fall behind the rotation of the Earth, so that the moon returns over the same meridian an average of approximately 50 minutes later each day. This is referred to as the *daily retardation* of the moon.

The moon rotates on its axis in the same length of time in which it revolves around the Earth, namely, the sidereal period of 27⅓ days. This is referred to as *synchronous rotation* (Figure 8.4). Because of this the moon presents the same hemisphere toward the Earth at all times, permitting viewers on Earth to see only a portion of the lunar surface. The far side is

Figure 8.3 The lunar sidereal month is measured as the time required for the moon to travel 360° around the Earth. The synodic month is the time it takes the moon to make a successive conjunction with the sun, and is slightly more than two days longer than the sidereal month.

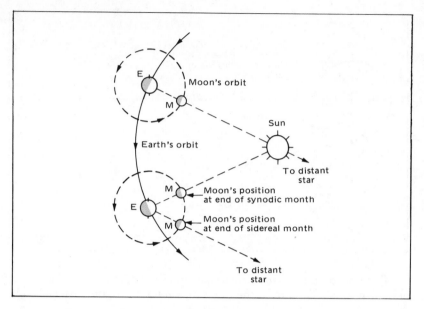

Figure 8.4 The moon accomplishes one rotation in the same length of time required to complete one sidereal period. This is called synchronous rotation.

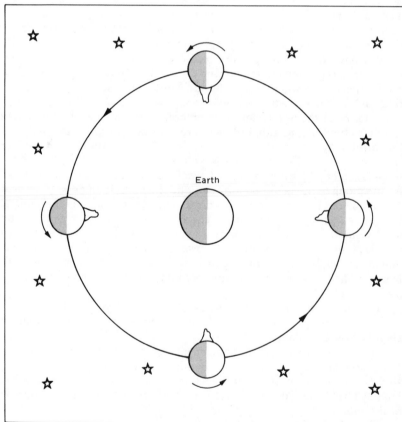

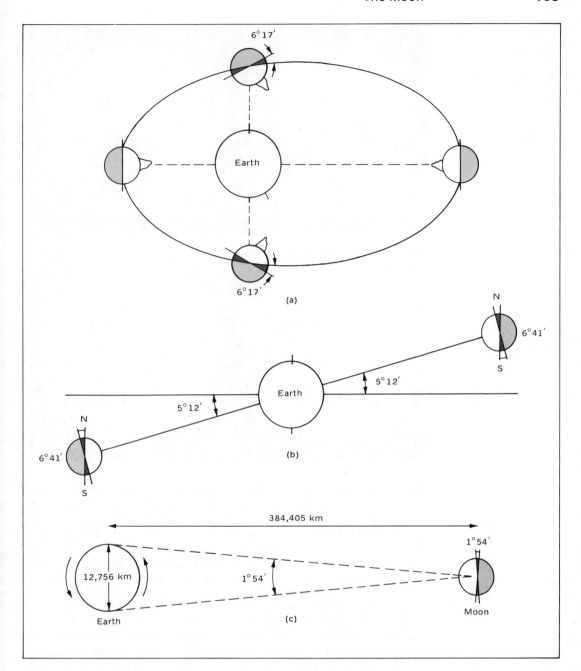

Figure 8.5 The moon's constant rate of rotation and nonuniform velocity in orbit results in 6° 17′ longitudinal libration (a). The 5° 12′ inclination of the moon's orbit to the ecliptic allows a 6° 41′ view over the poles. This is latitudinal libration (b). Due to the Earth's rotation a parallactic effect permits the viewer on Earth to benefit from a 57′ or 1° 54′ total diurnal libration (c).

Figure 8.6 Arrows point to similar features on the lunar surface to indicate the results of libration. Top photo: longitudinal libration. Bottom photo: latitudinal libration. (Yerkes Observatory photograph)

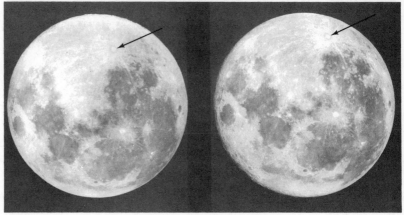

perpetually hidden from observers on Earth. However, although the moon rotates at a constant rate it moves at different speeds in its elliptical orbit. This leads to an apparent "rocking" of the moon, which permits the terrestrial viewer to see around the lunar limbs a few degrees each revolution (Figure 8.5). This is known as *longitudinal libration*. At the same time, the moon's orbital inclination to the ecliptic of about 5° causes an apparent "nodding," or *latitudinal libration*, which permits a view over the poles of a few degrees each revolution. Also we can observe the moon from a slightly different angle when it rises as from that seen when it sets, resulting in an additional degree of viewing due to *diurnal libration*. As a result of the various lunar librations it is possible to see approximately 59 percent of the lunar surface from the Earth. This was important to astronomers studying the moon from the Earth because of the restriction of synchronous rotation. However, with the capability of orbiting and landing on the moon, study of all parts of the lunar surface has been accomplished, and libration as a tool for viewing the moon is now of little relative importance (Figure 8.6).

8.3 LUNAR PHASES

One of the first celestial phenomena to be viewed and understood by astronomers was the phases of the moon (Figure 8.7). The moon itself is dark and only reflects light from the sun as the moon moves around in orbit. The alternately increasing and decreasing areas of the moon reflecting light toward the Earth are the phases of the moon. The moon always has one-half of its surface illuminated by the sun except during a lunar eclipse. The current phase of the moon will be determined by the position of the moon in orbit with respect to the Earth and the sun. When the moon passes between the Earth and the sun, the hemisphere of the moon facing the Earth is dark. This is the *new moon* phase. In a few days, as the moon progresses in its orbit and reaches a position where it is beyond the sphere of bright sunlight, the moon will appear as a thin crescent that may be seen in the afternoon and early evening. With each passing day the crescent increases in size until, at approximately one week, about half of the moon's face seen from Earth is illuminated, which represents the *first-*

Figure 8.7 The phases of the moon.

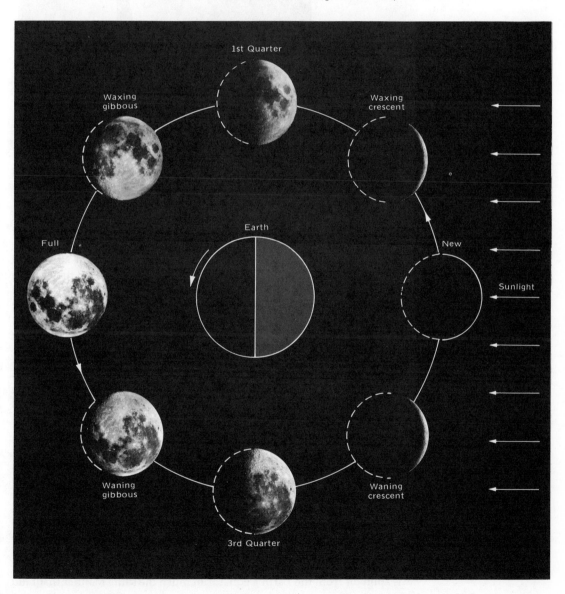

quarter phase. Then follows the *gibbous phase*, during which the lighted portion of the moon increases nightly until the moon is in opposition. At this point we have a *full moon.* The phases are then repeated in reverse order through gibbous, *third quarter,* and crescent to new moon again. The third-quarter moon has the same appearance as the first quarter, except the sunlight is coming from the east whereas in the first quarter the sunlight is coming from the west. The boundary between the bright and dark portion of the moon is called the *terminator.* It appears to the eye as a smooth regular line, but on close examination with a telescope the terminator is seen to be irregular, due to the mountainous nature of the lunar surface (Figure 8.8).

8.4 LUNAR ORIGIN

Once we understand the dynamics of lunar motion and the physical relationship of the moon to the Earth, we can ask the question, What is the origin of the moon? A number of proposals have been made to explain its origin, but none are as yet completely acceptable to all scientists.

Fission Theory

This theory, now out of favor, was originally proposed by George Darwin, son of Charles Darwin. In one version Darwin suggested that the moon was formed by the separation of a huge fragment from the Pacific Ocean at some time during the early history of the Earth. However, the frictional forces generated by such an event would have been too great to permit the moon to escape in this fashion. In addition, the moon fragment while within the Earth's Roche's limit would have been broken up into small fragments. Another version of the fissure theory suggests that the material was spun off from the Earth's equator in much the same manner that mud would fly off from a car wheel. To accomplish this, at some time in its history the Earth would have had to complete its rotation in two hours to enable material to spin off from the equator. While this is not an impossibility since the Earth may have had as short a rotation period as five hours at some point in its formative stage, no evidence exists to indicate a two-hour period. In addition, a satellite formed in this manner would have an equatorial orbit, and the moon does not have such an orbit.

Double-Planet Hypothesis

According to the double-planet theory, turbulence in dust and gas cloud in the vicinity of the evolving Earth (see chapter 7) caused the formation of a second body which became the moon. In other words, the Earth and the moon were formed as a double planet from essentially the same material. However, as we have noted, the Earth and moon vary in density, indicating some differences in the materials from which they were originally formed. If formed in the same vicinity it would be reasonable to expect that the Earth and moon would be almost identical in composition. Even when allowances are made for the greater compression of the Earth's interior due to its larger size, the average density of the Earth is still 50 percent greater than that of the moon.

Capture Hypothesis

To overcome the objection of the double-planet hypothesis, a proposal has been offered which suggests that the moon was formed in another part of the solar system and later captured by the Earth during a close encounter. Although not an impossibility, the requirements of such an encounter resulting in the moon becoming a satellite are so restrictive that the opportunity for capture would be extremely limited. A modification of the capture hypothesis has the Earth and moon forming in approximately the same orbit but in different parts of the orbit. According to this theory, the Earth and moon formed in a manner permitting the proto-earth to accumulate most of the materials, including a major portion of the heavier elements. Then, approximately 3.5 billion years ago, the Earth and the moon experienced a cataclysm of some kind in which the Earth's surface and the lunar surface became molten. Possibly this event signaled the capture of the moon by the Earth.

As previously stated, no theory thus far proposed on the origin of the moon is satisfactory in all respects. It was thought that a lunar landing and close examination of lunar surface material could solve the riddle, but this has not proved to be the case. If all the study to which the Earth has been subject has not yielded an acceptable theory on the origin of the Earth, it seems unlikely that a few samples of surface material would accomplish this for the moon.

8.5 LUNAR SURFACE FEATURES

Since the invention of the telescope the lunar surface has been most carefully studied and the gross lunar features have been fairly well defined as a result. The dark lunar areas were first mistakenly identified by Galileo as *maria*, or seas, but were soon recognized as broad, relatively smooth areas liberally sprinkled with small craters. The brighter portions of the moon were identified as the highlands and the heavily cratered regions. Many of the lunar craters are as much as 240 kilometres in diameter and are circled by ramparts reaching elevations of up to three kilometres. Despite the great amount of knowledge gained by the use of the telescope, close-up views of the moon from orbiter probes and Apollo landings excited great interest, even in the general public.

The maria are now seen as large, almost circular flows of basaltic lava, ranging from approximately 300 to over 1100 kilometres in diameter. The largest, the *Sea of Showers* or *Mare Imbrium*, measures slightly over 1100 kilometres in its greatest dimension. It is thought that the maria regions were originally huge impact craters that filled with lava during some catastrophic event early in the moon's history—possibly, according to one theory, when the moon was captured by the Earth. Hills at Fra Mauro, site of the Apollo 14 landing in 1971, are thought to have resulted from material ejected by the impact that formed Mare Imbrium. Samples returned by Apollo 14 astronauts led to the theory that the impact occurred about 700 million years after the formation of the moon. At that time, the theory holds, a *planetesimal,* possibly another Earth satellite about 150 kilometres in diameter, was swept up by the moon at an impact velocity of no more than 6 kilometres per second. The surface of the maria is generally pock-marked with small craters up to 25 metres in diameter, and occasionally a boulder field may be seen, deposited as a result of impact of a particle striking the moon. The predominant color of the maria landscape is a grey-brown, and the surface material is usually fine grained in texture (Figure 8.9).

Lunar orbiters, photographing the surface of the moon in 1968, were found to be deviating from their predicted flight paths, thereby revealing the presence of greater-than-normal gravitational forces. These positive

Figure 8.9 The lunar surface showing Mare Imbrium. The large crater in the lower left of the photograph is Copernicus. Craters Archimedes and Plato are also shown. (Courtesy of Hale Observatories)

gravitational anomalies appear to result from mass concentrations, or *mascons* as they became known, located in some of the ringed maria regions. Their exact nature and origin are not known, but one hypothesis is that they may have been formed by flows of lava from more dense subsurface regions when the moon was still volcanically active 3.5 billion years ago. Another theory suggests that the mascons are the remnants of huge meteorites, perhaps 100 kilometres or more in diameter, that impacted on the lunar surface to create the marias. Coincident with or shortly following the impact, lava flows filled the impact areas to form the existing marias. Some lunar geologists feel that the presence of the mascons is an indication that the moon is geologically inactive, that is, a cold dead planet. If the moon were geologically active, the reasoning goes, the mascons would slowly migrate toward the lunar center and disappear. The assumption that the moon is geologically dead seems to be supported by lack of extensive surface seismic activity and the age of the surface lunar rock.

Seismometers placed on the lunar surface during the recent Apollo landings have transmitted seismic data back to Earth. Information thus gained has indicated that energy released by moonquakes each year is only about one billionth or less that of the energy released by earthquakes. The strongest quake so far recorded on the moon had a magnitude of approximately 2 on the Richter scale. Such a quake would not normally be felt even at the epicenter. In addition, it was found that moonquakes originate at depths of 800 to 1000 kilometres below the lunar surface; most earthquakes originate within 100 kilometres of the Earth's surface. This is evidence that no molten rock or even plastic rock exists within 800 kilometres of the moon's surface. The conclusion reached is that the moon has a thick solid crust and is geologically quite inactive. It is thought that the moon was geologically active the first billion or two years of its existence and has been slowly dying, in contrast to the Earth which is still quite active.

Craters on the moon are easily recognizable features resembling volcanic cones on earth. The lunar craters occur in all sizes up to 240 kilometres in diameter, with the majority of the craters, especially the large ones, apparently having been formed in the early stages of the moon's existence.

Ever since the invention of the telescope, when astronomers had their first real view of the lunar surface, they speculated on the way the craters were formed. Two methods have been suggested—one is volcanic action, an internal process, the other is the impact of solid objects from space, an external process. In the latter process the impact fractures the surface material and causes it to be piled up to form a rim around the evacuated hole. An *ejecta blanket* of smaller debris is flung out beyond the rim (Figure 8.10).

One of the more prominent examples of a lunar crater is Copernicus. Eight times as wide as the Grand Canyon and twice as deep, Copernicus is typical of the large lunar craters (Figure 8.11). On the floor of the crater is a group of central mountains, and a series of terraces make up its inner walls. The outer flank of the crater has a hummocky appearance, and many small craters dot the area around the main crater. A system of light-colored streaks or *rays* extend in all directions for several hundred kilometres, suggesting material thrown out by impact. Was Copernicus formed by impact or is it of volcanic origin? The controversy has not yet been resolved, but the evidence strongly favors impact.

Figure 8.10 A vertical view of the north wall of Copernicus showing slumping of the wall and the ejecta blanket beyond the rim. (NASA)

Material brought back from the moon included specimens of lava, which indicated extensive volcanic activity. But also large numbers of tiny spheres of glass and broken fragments of glass were found in the soil. When these were examined in detail tiny craters were revealed, as if formed by high-speed particles from space. Traces of meteoric material were found in the craters, providing direct evidence that craters, ranging in size from a thousandth of a centimetre in diameter to many metres in diameter, were formed by impact. The distinct possibility exists that both volcanism and impact were responsible for crater formation on the moon. Volcanism may have been predominant during the first few billion years and then ceased. Impact may have occurred at a high rate of incidence in the early stages of lunar formation and is continuing today, but at a much reduced rate.

There are an estimated 200,000 craters on the moon over one kilometre in diameter, but only approximately 80 recognizable impact craters on Earth. Why is there this discrepancy? The Earth being larger, more massive, and therefore with greater gravitational force appears more vulnerable to impact than the moon. The answer lies in the fact that the Earth is subject to erosion by water, wind, and ice while the moon is not. It is estimated that features such as impact craters would be obliterated in 5 to 10 million years on Earth by erosion. Such features on the moon would not be so affected and would remain from the time the moon was first formed. This does not mean that erosion is totally unknown on the moon. Close viewing of crater rims reveals that the edges are not as sharp as might be expected. Gravitational erosion resulting in sliding and the gradual downslope movement of particles appears to occur. In addition, the constant bombardment of the lunar surface by micrometeoroids from space contributes to erosion in much the same way that wind erosion functions on Earth, but at a much reduced rate. The Earth is protected from this type of activity by the presence of the atmosphere.

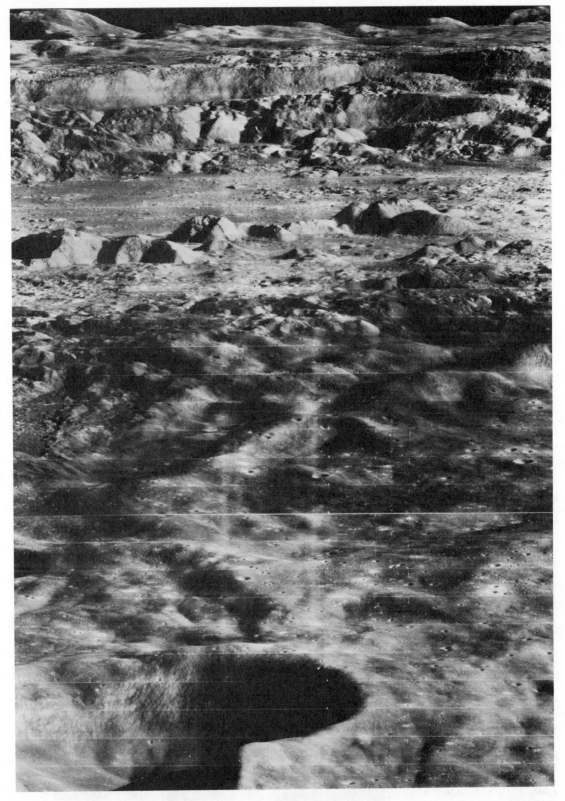

Figure 8.11 Crater Copernicus showing the inner wall and a group of mountains on the floor of the crater. (NASA)

In addition to craters, the moon also has several prominent mountain systems and valleys, clearly visible through the small telescope. Radioactive dating of lunar samples from mountain areas has revealed that the mountains are older than the maria and approximately as old as the solar system. The mountain ranges are extensive, some being hundreds of kilometres in length and reaching heights of 6000 metres. It is interesting to note that mountains and maria are features found mainly on the near side of the moon, while craters are the dominant feature of the far side (Figure 8.12). *Rilles*, another feature being studied, are well-defined, narrow trenches that appear to meander across the lunar surface. Hadley rille, examined during the Apollo 15 landing, roughly parallels the Apennine Mountains along the eastern border of Mare Imbrium (Figure 8.13). This rille is an elongated depression averaging about 1.5 kilometres in width and up to 365 metres in depth. The appearance of some rilles suggests that they were formed by the erosional action of water, but this is most unlikely in view of the total absence of water on the moon. Another theory proposes that rilles are the remains of long lava tubes from which molten lava has drained and the surface collapsed. This type of activity may be related to the formation of *domes*, which are blisterlike structures on the floor of maria and may be formed by the upwelling of molten lava from beneath the lunar surface (Figure 8.14). Although common on the moon, a satisfactory explanation of the origin of such features has not yet been found.

A profile view of the Hadley rille reveals definite layering, such as seen in sedimentary deposits on Earth. On Earth such deposits are commonly formed by water action, but again this is not possible on the moon due to lack of water. There is a likelihood that the layers represent successive lava flows, or they may have formed by the impact of large particles on the

Figure 8.12 Craters on the lunar farside. Photograph was taken from Apollo 10 Lunar Module. (NASA)

Figure 8.13 Hadley rille parallels the Apennine Mountains on the eastern border of Mare Imbrium. (NASA)

lunar surface. Each impact throwing out an ejecta blanket would build up a stratified profile layer by layer over eons of time (Figure 8.15).

The lunar surface material has been the object of intensive study since the lunar landings made possible the return of lunar samples. The surface is composed primarily of loose scattered rock and boulder material, produced mainly by impact, and a layer of generally fine-textured sand up to 10 metres thick on an igneous bedrock. The soil material contained microscopic beads of glass or obsidian (Figure 8.16), ranging through a variety of colors. In addition, plagioclase was plentiful and olivine,

Figure 8.14 Volcanic type domes in the Marius Hills region in Oceanus Pro-cellarum. (NASA)

Figure 8.15 Stratified profile built up layer by layer over eons of time is revealed in Hadley rille. (NASA)

pyroxene, troilite, ilmenite, sanidine, and iron were positively identified as minerals in the rock. The rock material was mainly crystalline igneous and some breccia but varied somewhat in age and composition from landing site to landing site. For example, samples from Mare Imbrium contained more iron, magnesium, and nickel than samples from Mare Tranquillitatis, or Sea of Tranquility, but less titanium, potassium, and rubidium. The age of rock at Mare Imbrium (1.7 to 2.7 billion years) was about one billion years less than the ages determined for rock at Mare Tranquillitatis. However, one rock from Mare Imbrium contained sig-

Figure 8.16 The lunar soil contains microscopic beads of glass or obsidian. (NASA)

nificantly higher quantities of uranium, thorium, and potassium than other samples collected at that site, and radioactive-dating technique revealed the sample to be 4.6 billion years old.

The moon has an extremely tenuous atmosphere because it has too low a gravitational field to retain any volcanic vapors that may be outgassed. In addition, the solar winds are quite effective in sweeping away traces of gas that may appear at the surface, so that they generally dissipate within a few months. Measurements show that the mass of the entire atmosphere may be no more than 10 tons. It has been pointed out that if the moon's atmosphere were 10,000 times more dense than it is, solar winds would not be able to sweep it away in a few months. Rather, the heavier gases such as neon would remain for thousands of years. The concern has been expressed that such a buildup of gases could result from repeated landings on the moon. Each Apollo mission, it is estimated, added about 10 tons of rocket exhaust gases to the lunar environment. Extensive exploration and colonization, with the attendant discharge of gaseous wastes, could build up an atmosphere of pollutants that would forever contaminate the lunar surface.

The placing of magnetometers on the moon has revealed the presence of a magnetic field so weak as to be essentially nonexistent.

The absence of a lunar atmosphere is responsible for the wide range in temperatures on the moon's surface. Temperatures vary from 100°C at lunar noon to –150°C during the lunar night. High temperatures and the lack of an atmosphere and water have precluded the development of life on the moon. No organisms were found in the lunar samples, nor was there any evidence of fossil material. Thus it would appear that the first extraterrestrial body in the solar system to be explored by man is devoid of life. This may be an advantage, for if the moon is colonized at some time in the future, the lack of alien organisms will be one less hazard man will have to contend with.

8.6 THE ECLIPSE

Watching the occurrence of a total eclipse, with the gradual covering of the sun by the moon until only the corona of the sun can be seen, is an awe-inspiring experience. Small wonder that ancient people, who little understood what was taking place, were frightened by the event. At the time of totality the area from which the eclipse is being viewed appears strange in the dim light and the temperature drops rapidly and perceptibly. Birds, insects, and animals react as though night were falling. In a few minutes the moon moves on and the sun is exposed, with the light first appearing through valleys on the lunar surface providing what is known as the *diamond ring effect*. (Figure 8.17).

Until the recent invention of the coronagraph it was not possible to view the outer edges of the sun except during a total eclipse. The occurrence of a total eclipse made possible the examination of prominences and corona of the sun. In the pattern of movements of the sun, Earth, and moon, there are always two to three solar eclipses possible during the year. Some are *partial eclipses,* when only a portion of the sun is covered by the moon. Some are *annular eclipses,* when a thin ring of the solar disk is visible around the covering moon. This takes place when the moon is at apogee and therefore too far from the Earth for the umbra of the moon's shadow to touch the Earth. Some eclipses are *total eclipses,* when

Figure 8.17 The diamond
ring effect during a solar
eclipse. (Courtesy of A.
W. Johnson, C. Halver-
son, D. Duncan)

the entire sun is covered. On the average, only two total eclipses may be seen on Earth every three years.

A total eclipse of the sun is possible because the angular diameter of the moon is slightly larger than the angular diameter of the sun when the moon is near perigee. It is then possible to see a total eclipse when the true shadow of the moon, the *umbra*, passes overhead (Figure 8.18). The umbra produces a round shadow on the Earth's surface never more than 275 kilometres in diameter at or near the point on the Earth directly beneath (i.e., nearest) the sun. As the umbra approaches the position of contact with the earth, the shadow is more elongated. The *penumbra*, which surrounds the umbra like an inverted cone, does not completely hide the sun, and observers located at any point in the lightly shaded area of Figure 8.14 will see only a partial eclipse. The penumbra forms a circle about 6500 kilometres in diameter where it touches the Earth.

A total eclipse of the sun occurs only when the moon is directly between the Earth and the sun or at a new-moon position. It would seem,

Figure 8.18 A total
eclipse will be seen by
the observer within the
umbra portion of the
moon's shadow. A partial
eclipse is seen in the
penumbra.

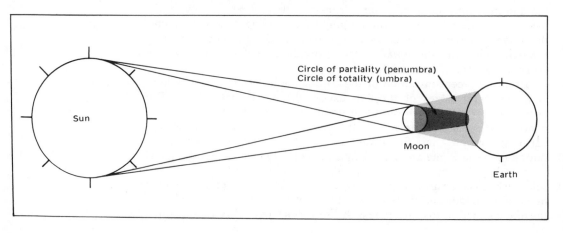

Circle of partiality (penumbra)
Circle of totality (umbra)

Sun

Moon

Earth

then, that we should have an eclipse each month, and the fact that we do not indicates that one or more additional conditions must be met in order for such an event to take place. The moon's dark shadow, the umbra, is 373,520 kilometres long, which is less than the mean distance from the Earth to the moon. This means that in order for an eclipse to occur the moon must be near perigee at the same time the moon reaches new-moon phase. If the moon is close to apogee at this time an annular eclipse would result. One other factor that must be considered is the proximity of the moon to the orbital plane (ecliptic) of the Earth. The moon's orbit is inclined 5° 12′ to the ecliptic, which means that during some new-moon episodes the lunar shadow will fall above or below the Earth (Figure 8.19). Thus a total eclipse can occur only when the moon passes through the ecliptic at what are called the *nodes* of the orbit. The nodes are those points where the lunar orbit intersects the ecliptic, and a line connecting the nodes is known as the *line of nodes*. In order for a total eclipse to occur the line of nodes must be pointed to the sun, a condition that exists approximately twice each year during what is called the semiannual *eclipse season.*

A *lunar eclipse*, when the Earth's shadow falls on the moon, has essentially the same requirements as a solar eclipse. In a lunar eclipse the moon must be in a full-moon position at one of the orbital nodes. The distance from the moon to the Earth is not a consideration, because the Earth's shadow extends far beyond lunar apogee. The moon is not completely dark during such an eclipse, because the moon receives some light as a result of the refraction of sunlight by the Earth's atmosphere. The

Figure 8.19 The inclination of the plane of the lunar orbit with respect to the ecliptic permits a possible eclipse only twice annually when the line of nodes *AB* is pointed toward the sun (a). When the moon is not at the line of nodes position the lunar shadow may pass over or under the Earth and no eclipse can occur (b).

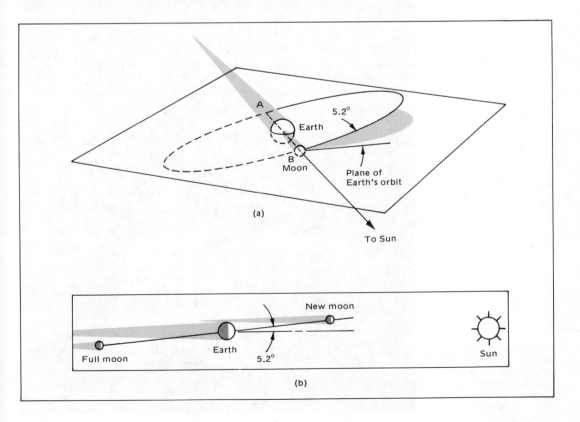

blue component of the light passing through the atmosphere is scattered and is therefore reduced in intensity. However, the red light passes relatively uninterrupted through the atmosphere, resulting in the moon reflecting a dull red color during the eclipse period (Figure 8.20).

Eclipses can be predicted quite accurately for some time into the future. This can be accomplished through detailed knowledge of the motions of the moon and Earth relative to the sun. Predictions for solar eclipses in the near future are:

Oct. 12, 1977: Mid-North Pacific, southeastward into northern South America.

Feb. 26, 1979: North Pacific Ocean, northwest tip of United States, across Canada into central Greenland.

Feb. 16, 1980: Atlantic Ocean, across central Africa, Indian Ocean, India, southern China.

July 31, 1981: Southeast Europe across Siberia to mid North Pacific.

June 11, 1983: South Indian Ocean to East Indies to western Pacific.

May 30, 1984: Pacific Ocean across Mexico, southern United States across Atlantic to North Africa.

Nov. 12, 1985: Antarctic Ocean.

Figure 8.20 A lunar eclipse on November 18, 1975, shown in several stages over the Capitol Building, Washington, D. C. (World Wide Photos)

8.7 TIDES

Tides are the alternate rise and fall of the sea level—a phenomenon that can be readily observed along any shoreline. Although tides were recognized in antiquity, their cause was unknown until Isaac Newton developed the universal law of gravitation. It then became apparent that the gravitational force of the moon—and to a lesser extent the sun—attracted the waters of the Earth's surface and, together with the Earth's rotation, created the tides.

The moon's influence is more than twice that of the sun because of the moon's proximity to the Earth. There is a variation in tidal magnitude according to the movement of the moon and its position relative to the sun. The greatest or *spring tides* occur during the full-moon phase, when the moon and the sun are on opposite sides of the Earth, and during the new-moon phase, when the moon is directly between the Earth and the sun. In these positions the gravitational forces of the moon and the sun reinforce each other. Variations in spring tides will result depending upon whether the moon is at perigee or apogee. Lesser or *neap tides* occur when the moon is generally in the first- or third-quarter phase (Figure 8.7).

The dominance of the moon in generating tides permits us to discuss tidal action in terms of lunar movement and the rotation of the Earth. The tides are measured on the basis of a lunar day of 24 hours and 50 minutes. This is the time required for the moon to arrive at the zenith the following day, due to its eastward movement in orbit and the Earth's rotation (see section 8.2). Ideally, a high tide traveling across the Earth would be attracted by, and aligned directly under, the moon and is known as the *direct tide*. Friction, if considered infinite between the Earth's surface and the ocean waters, would cause the tidal bulges to shift eastward at the same rate as the Earth rotates. Actually, the direct tidal bulge occurs at an equilibrium point between the maximum force of the moon's gravity and the Earth's friction (Figure 8.21). Another tidal bulge, known as the *opposite tide*, occurs simultaneously on the opposite side of the Earth. The result is two daily high tides approximately 12 hours and 25 minutes apart. Low tides are 90° behind high tides and occur approximately 6 hours and 12 minutes behind each high tide.

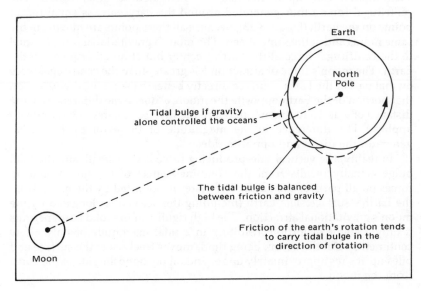

Figure 8.21 Friction of the rotating Earth tends to drag the ocean's tidal bulge in the direction of rotation.

Figure 8.22 The Earth
and moon revolve around
the barycenter (*B*). The
Earth's center (*C*) traces
out a path around (*B*).
Each point on the Earth
will trace out a similar
circular path of the same
radius. This movement is
not related to Earth's
rotation.

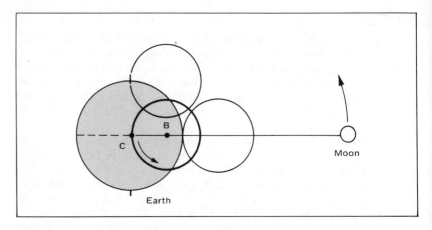

The following explanation for the simultaneous occurrence of two high tides on opposite sides of the Earth is somewhat oversimplified. It is first necessary to refer to the Earth-moon motion around the barycenter (section 8.2) and to emphasize that the Earth revolves, and does not rotate, around this point. Therefore all points on the Earth describe equal circles with respect to the barycenter (Figure 8.22), and each circle thus formed has a radius equal to the distance from the center of the Earth *C* to the barycenter *B*. The Earth revolves around *B*, as does the moon, because of their mutual force of attraction due to gravity. Now let us consider the waters of the Earth as a shell around the Earth. Due to the revolution of the Earth around the barycenter, a motion is imparted to the ocean, causing it to move with respect to the Earth in a linear path in a direction opposite to that of the moon. At the same time the Earth's gravity causes the ocean to be drawn toward the Earth's center, but equally from all directions, so for purposes of explaining tidal action we can ignore Earth's gravity.

For convenience, let us ascribe the motion imparted to the ocean to centrifugal force. A centrifugal force is one which seemingly causes an object to move outward from some central point. It is not a real force, as such motion can, upon analysis, be attributed to other forces. The centrifugal force *f* thus generated around the barycenter is equal for all points on the Earth (Figure 8.23), because all these points are moving in the same radius and at the same speed. The moon's gravitational force is equal to the centrifugal force at the Earth's center but at no other point on the Earth. The moon's force of attraction *F* is greater than the centrifugal force on that part of the Earth's surface directly beneath the moon, *d*, and less on that portion of the Earth opposite the moon, *o*, due to the differences in the distance of *d* and *o* to the moon. (Here the inverse squares law for gravity applies.) The difference in the magnitude of these forces, *t*, is what generates the direct and opposite tides.

In reality, the vertical tide-producing force is very small, and the tidal bulge actually results from the combined effect of the tide-producing forces on all parts of the Earth. The water is not lifted by the moon from the Earth's surface. The Earth (including the oceans) is elongated by the moon's gravitational attraction. The high rigidity of the solid Earth permits it only a slight distortion, resulting in a tidal magnitude of only a few centimetres. But the ocean, being fluid, moves freely over the surface and piles up at sites approximately under and on the opposite side of the Earth from the moon.

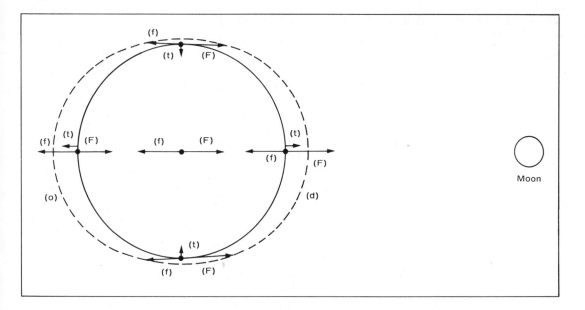

The magnitude of the ocean tides may be less than a metre in most instances, but the contour of the land masses and the funneling of the tides into widemouth bays (for example, the Bay of Fundy) may cause sea-level changes of up to 20 metres. Tides also occur in lakes, but the amplitude of the water-level change is not great. Even in the Mediterranean Sea, tides vary the sea level only about 30 centimetres.

Tides are of some astronomical significance, in that they act to slow down the Earth's time mechanism. As the tide moves over the shallow ocean bottom, the friction acts as a brake and slows the Earth's rate of rotation. As a result the day has increased in length about 1/1000 of a second each century. This means that the day is about one second longer than it was 100,000 years ago. The rate of slowdown appears to be an extremely short . length of time, but over millions of years the amount becomes significant. Evidence from the study of growth rings of certain fossil coral formed during the Silurian period, 400 million years ago, indicates that the year may have been approximately 400 days long. It is assumed that the Earth's annual period of revolution has remained unchanged. The greater number of days during the same annual period indicates that the days were once shorter.

The moon's distance from the Earth is also affected by this type of tidal activity and is gradually receding. Half a billion years ago the moon was half its present distance from the Earth. Now it is estimated to be receding approximately one centimetre per year. Ultimately, the moon will reach a point where it will become an independent object in orbit around the sun.

What is the magnitude of the force responsible for slowing the Earth's rotation and causing the moon to recede? It is difficult to determine the total, but K. F. Bowden, a British oceanographer, estimates that only 13 percent of the tidal energy coming into the English Channel from the Atlantic reaches the shore. The other 87 percent, or an estimated 210 million horsepower, is lost through friction on the shallow floor of the channel. In reality, such energy is not lost but accomplishes work—which in this instance results in the slow constant change in the Earth-moon relationship.

Figure 8.23 The difference in magnitude of centrifugal force (*f*) and gravitational force (*F*) generates direct and opposite tides (*t*).

8.8 ARTIFICIAL SATELLITES

Although launching and orbiting artificial satellites are engineering problems, there is some benefit to be gained from examining the motion of satellites. From this we can learn something more about the motion of natural satellites around their respective planets and of planets around the sun.

Orbital motion of an artificial satellite is actually the result of two distinct forces. To define these forces we may make use of Newton's illustration, wherein he describes the launching of a projectile from the top of a mountain in a direction parallel to the Earth's surface (Figure 8.24). The force applied at launching would cause the projectile to move with a constant speed, and if no other force were applied the projectile would move straight off into space in uniform linear fashion, in accord with Newton's first law of motion (see page 48). However, in reality the projectile does not move in a straight path but rather travels in a curved path around the Earth, because gravity acts to overcome the inertia of the projectile. What this means is that, in addition to the linear motion, a projectile, starting at zero acceleration, will by the end of one second be accelerating 9.8 metres per second^2 toward the Earth as a result of the force of gravity. By the end of one second the projectile would have fallen a distance of 4.9 metres. If a horizontal speed of approximately 8 kilometres per second were reached, the projectile would remain at a uniform distance above the Earth's surface, because the 4.9-metre fall in an 8-kilometre distance during one second coincides with the curvature of the Earth. The magnitude of the force of gravity necessary to overcome the projectile's inertia is given by Newton's second law (see page 49).

The resulting action of uniform linear motion and the force of gravity can be illustrated in the following manner (Figure 8.25). Let us consider the projectile located at C and the force of gravity as being centered at G. The

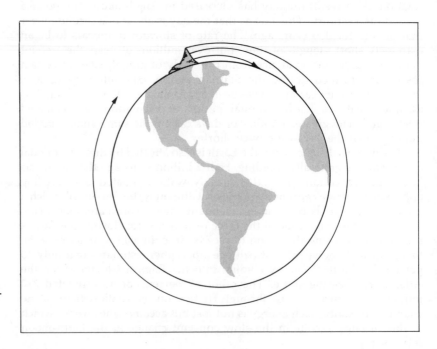

Figure 8.24 Newton described the paths of projectiles launched horizontally from various altitudes above the Earth's surface.

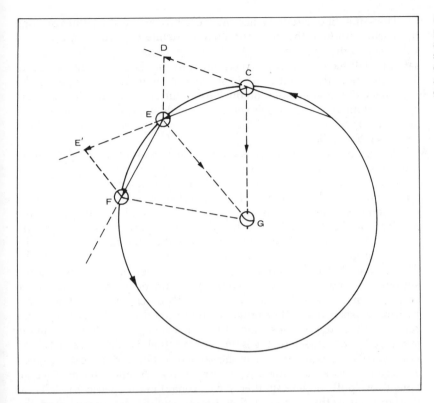

Figure 8.25 The orbital path of a satellite in accordance with Kepler's second law is the result of the satellite's inertia and the influence of gravity.

projectile is set in uniform linear motion along *CD*, and while so traveling is influenced by gravity, which attracts the projectile toward the center of gravity at *G*. The net effect of this is to cause the projectile to move in a new direction *DE*. At *E* the projectile is moving toward *E'*, but again gravity exerts its influence, causing the projectile to move toward *F*. If the motions illustrated were reduced to very small increments, the net result of the force of gravity, called the *centripetal force*, would cause the projectile to follow a uniform curve. The centripetal force is the unbalancing force mentioned in Newton's second law, and the acceleration produced by this force is directed toward the center of gravity.

The motion of natural satellites following the behavior just described, but orbiting the Earth or other planets at different velocities, produces the rather unusual effect of satellites appearing to move in opposite directions. For example, the two natural satellites of Mars orbit that planet in the same direction as Mars rotates on its axis. Phobos, closest to the planet, is about 6000 kilometres from the Martian surface and requires seven hours and thirty-nine minutes to complete its orbit. This is approximately 0.3 of the Martian day of 24 hours and 37 minutes. From the surface of Mars, Phobos is seen to rise in the west and set in the east, requiring about 4.5 hours to cross the Martian sky. Deimos, the more distant of the two satellites at approximately 20,000 kilometres from the surface, completes an orbit in 30 hours. This period being longer than the Martian period of rotation causes Deimos to appear to rise in the east and set in the west, as does the moon seen from the Earth. Because of the difference in velocities the satellites appear to be traveling in opposite directions when seen from the Martian surface, but are in reality traveling in the same direction.

The same set of circumstances may exist for artificial satellites orbiting the Earth, although the need for such a satellite to revolve in orbit at a velocity less than the speed of the Earth's rotation is rare. On occasion a satellite will have an orbital speed equal to the Earth's rotational speed. The satellite is then said to be in *synchronous orbit* and will appear to be stationary above some point on the Earth's surface. Astronauts in separate spacecraft moving at different velocities would appear to each other to be going in opposite directions if their respective motions were noted without reference to the Earth.

At this point in the discussion of orbital motion it may be well to inquire into the somewhat misunderstood phenomenon of *weightlessness*. First, let us regard *weight* as a force resulting from the mutual gravitational attraction of a body and the Earth or any other celestial object. The weight of a body is determined by its mass and the pull resulting from the acceleration due to gravity. If a body is not supported, these factors will result in free fall. When a body is supported by the Earth's surface, the body is supported by a force equal and opposite to the weight of the body. This means that the forces acting at the interface of the body and the supporting surface are equal and opposite. If the body and the surface are in motion in the same direction, as in a rapidly descending elevator, so that no force exists between them, then the body is considered to be weightless relative to the supporting surface. Once in a stable orbit all objects in the spacecraft are subject to the same forces, that is, linear velocity and gravity. All objects are moving in the same direction and at the same speed and are therefore motionless with respect to each other and produce no "push" or "pull" on one another. An astronaut in the spacecraft "falls" at the same rate as the spacecraft; therefore he experiences no push from the "floor" and with the slightest effort can propel himself around the craft. For him a condition of weightlessness is said to exist. We may take this one step further by saying that the astronaut is weightless relative to the spacecraft, but the spacecraft-astronaut system is not weightless relative to the Earth. They are in a constant state of free fall. Only when the spacecraft and the astronaut return to a supporting surface on Earth can the respective weights be determined.

Having reviewed the motion of a projectile in orbit, we may now examine the types of orbits a projectile may travel. The same forces that act on the orbital motion of an artificial object also determine the movement of the moon around the Earth and the planets around the sun. Most orbits of bodies in space are elliptical. A perfect circular orbit is a rarity because of the very precise conditions required to achieve such an orbit. Types of orbits and the conditions necessary to produce them may be illustrated by describing the launching of a spacecraft from Earth into orbit around Earth and examining the type of orbit attained. It should be recognized that natural bodies in space are not "launched" in the same sense as a spacecraft, but the types of orbits are comparable (Figure 8.26).

One of the factors which determines the orbital configuration of a spacecraft is the spacecraft's velocity when the rocket ceases firing. A tangential (tangent to Earth's surface) velocity is imparted to the spacecraft when it reaches the desired altitude, and centripetal force (gravity) changes the direction of the spacecraft so that it orbits the Earth. The spacecraft's velocity and orbital radius must be such that the force of gravity at that radius will produce an orbital path which does not intersect the Earth. When such a condition is satisfied the spacecraft is said to be in a stable orbit.

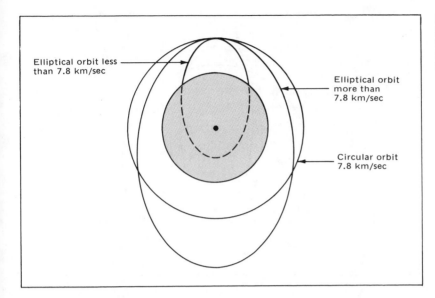

Elliptical orbit less than 7.8 km/sec

Elliptical orbit more than 7.8 km/sec

Circular orbit 7.8 km/sec

Figure 8.26 The orbital path followed by a space vehicle will be determined by its launch velocity.

The velocity suitable for a particular orbital radius of an artificial satellite may be easily determined. It will be recalled from Kepler's third law that if the radius a of a planet's orbit is known, the period P can be computed and vice versa. The same applies to satellites orbiting planets. The radius of a spacecraft's orbit is predetermined, depending on the purpose of the satellite. The orbital radius (measured from the center of the Earth) and the period are then used to calculate the velocity required for the orbit to be achieved. For example, if a is equal to 6.4×10^6 metres and P is equal to 5.1×10^3 seconds (85 minutes), it is possible to calculate the orbital velocity v from

$$v = \frac{2\pi a}{P} \qquad \text{(see 6.3)}$$

$$v = \frac{2 \times 3.1416 \times 6.4 \times 10^6 \text{m}}{5.1 \times 10^3 \text{sec}}$$

$$v = 7.92 \times 10^3 \text{m/sec}$$

The above example applies to a circular orbit, but the same principle is appropriate for an elliptical orbit, although the computations are more complex.

To achieve a circular orbit at an altitude of approximately 160 kilometres, the spacecraft must be given a tangential velocity at burnout of 7.8 kilometres per second (Figure 8.27). A velocity in excess of 7.8 kilometres per second would produce an elliptical orbit with the ellipse outside the circle. Entry by a spacecraft into such an orbit would be at perigee, the point of closest approach to the Earth by a spacecraft in an elliptical orbit. On the other hand, when the velocity is less than 7.8 kilometres per second an elliptical orbit inside the circle would be achieved and the spacecraft would strike the Earth. The point of entry of

Figure 8.27 The orbit
achieved by a space ve-
hicle launched at a given
velocity may be varied
depending upon condi-
tions of injection into
orbit.

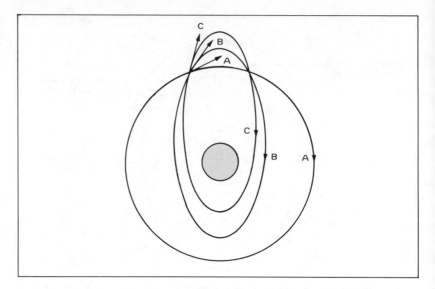

Figure 8.27 The orbit achieved by a space vehicle launched at a given velocity may be varied depending upon conditions of injection into orbit.

the spacecraft into such an orbit would be at apogee, the most distant point on the ellipse from the Earth. Such would be the fate of projectiles fired from the mountain at a velocity less than that required to achieve a circular orbit (Figure 8.24).

The orbital configuration is also dependent upon the spacecraft's mode of injection into orbit. A tangential injection at 7.8 kilometres per second would produce a circular orbit. However, an injection at this velocity that deviates from the tangential mode would produce an elliptical orbit.

How can we describe the trajectory of probes to the planets? Elliptical orbits to Venus and Mars are the most efficient orbits to achieve in that they require the least amount of energy. The elliptical path of a probe to Venus, for example, would be tangent to the Earth's orbit at aphelion and tangent to Venus' orbit at perihelion (Figure 8.28). The probe would leave the Earth with a velocity of approximately 27 kilometres per second with respect to the sun (slightly less than the Earth's orbital velocity of 30 kilometres per second) and would reach Venus in about 5 months. The same conditions would exist for a probe to Mars, except here the orbital pathway would be tangent to the Earth's orbit at perihelion (probe speed 33 km/sec) and to the Martian orbit at aphelion (Figure 8.28). Then with a velocity of 32.8 kilometres per second, slightly greater than the Earth's solar velocity, the probe would arrive at Mars in about 8.5 months.

Because of the distances to be traveled and the lengths of time involved, careful planning is necessary to make certain that the probe is launched so as to arrive at its destination simultaneously with the planet.

Orbits thus far discussed have been closed pathways, and are of the type followed by natural bodies in space and certain artificial satellites around the Earth. Spacecraft may also follow an open course, achieved by imparting to it an *escape velocity*, a velocity sufficient to permit the spacecraft to move continuously away from the Earth and not return. Under these circumstances the spacecraft will travel along a *parabolic* pathway with respect to the Earth, instead of a circle or an ellipse as in Figure 8.27. This does not imply that the spacecraft has escaped the Earth's gravitational influence, because this influence extends out to infinity. In this case the

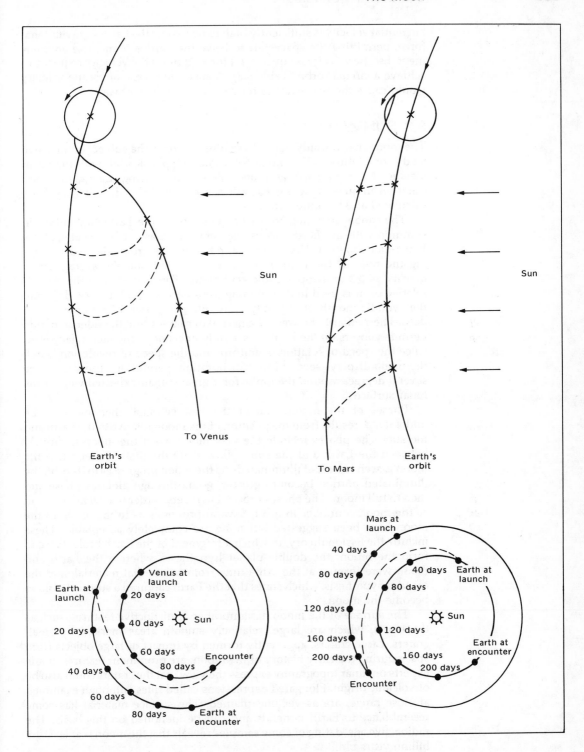

Figure 8.28 Elliptical orbits are the most efficient paths for a space vehicle to follow in going to Venus or Mars.

tangential velocity is sufficiently high to overcome the Earth's gravitational force, permitting the spacecraft to leave the Earth's immediate environment. Escape velocity is equal to 1.4 (or $\sqrt{2}$) times the velocity required to achieve a circular orbit (Table 6.2). A speed in excess of escape velocity would cause the spacecraft to follow a hyperbolic pathway.

8.9 SUMMARY

The moon, Earth's only natural satellite, has been the subject of intensive study for centuries. The moon follows an elliptical orbit, with the mean distance from the Earth accurately determined, using triangulation and other techniques, to be 384,404 kilometres. The lunar diameter has been calculated as 3476 kilometres.

The moon does not, in a true sense, orbit the Earth but rather, in common with the Earth, orbits the center of mass or barycenter of the Earth-moon system. This relationship has been instrumental in determining the ratio of the Earth-moon mass as 81.5:1 and the average lunar density as 3.3 grams per cubic centimetre. For convenience the moon's relative orbit is used in determining the sidereal month of 27⅓ days and the synodic month of 29½ days. The moon's period of rotation is 27⅓ days. The similarity between the period of rotation and the sidereal period permits viewers on the Earth to see only one side of the moon. However, since the speed of rotation is uniform and the speed of revolution is not, the moon displays several forms of libration, permitting observers to see several degrees around the limbs for a greater-than-expected view of the lunar surface.

Phases of the moon, one of the first celestial phenomena to be understood, results from the position of the moon relative to the Earth and the sun. The phases include the new moon when the moon is directly between the Earth and the sun, followed by the first quarter when the moon is seen to be half illuminated. As the moon progresses in its orbit the illuminated portion becomes greater, going through gibbous phase and then to full moon. The phases proceed in reverse order to new moon again as the moon continues in orbit. Several proposals as to the origin of the moon have been suggested but none are completely acceptable. These include the fission theory, in which a fragment of the Earth broke loose to form the moon; the double-planet hypothesis, wherein the Earth and moon were formed at the same time from primordial material; and the capture hypothesis, which states that the Earth captured a stray asteroid to become the moon.

The surface of the moon has a number of distinctive features, such as the maria, which are large, relatively smooth areas marked by small craters. Large craters, apparently formed by impact of large objects from space during the early history of the moon, are prominent features. Some experts on lunar topography express the view that the craters could also be of volcanic origin. Elongated depressions called rilles, although examined at close range, are as yet unexplained. The surface material has some resemblance to Earth minerals and can be identified on this basis. The radioactive age-dating of some samples reveals that the moon is at least 4.5 billion years old.

Surface temperatures on the moon fluctuate widely, ranging from 100°C down to −150°C. This is due principally to the total absence of any kind of atmosphere. The moon also lacks water and has no magnetic field. No traces of life have been found on the moon.

A solar eclipse is an event that occurs when the moon comes directly between the Earth and the sun. This can only occur if the moon is in a new-moon position as it crosses one of the orbital nodes and is near perigee in orbit. Because of these restrictions a solar eclipse occurs on the average only twice in three years. A lunar eclipse occurs when the Earth is between the moon and the sun.

The alternate rise and fall of the tides are the result of gravitational forces of the moon and, to a lesser extent, the sun. Two daily tides, approximately 12 hours and 25 minutes apart, alternate with low tides each day, although there is some deviation from this pattern. Tidal friction resulting from this activity is causing the Earth's rotation to slow down by one second per day each 100,000 years.

The orbital motion of one body around another is the result of uniform linear motion and the force of gravity. The combination of these two forces will result in a body following a uniform curved path. The body is, in effect, continually falling toward the Earth at the same time it is moving tangent to the surface. This so-called falling phenomenon will result in weightlessness when two bodies, as for example an astronaut in a spacecraft, are falling simultaneously through the same path. Two space-craft moving in the same direction in orbit but at different speeds would appear to be moving in opposite directions.

The types of orbits described by artificial satellites will depend upon the velocity and the manner in which the satellite was injected into orbit. It may follow an elliptical path, a circular path, a parabolic path, or a hyperbolic path. Elliptical orbits are attained by injecting the satellite into orbit at less than escape velocity, a parabolic orbit at escape velocity, and a hyperbolic orbit at a speed in excess of escape velocity.

QUESTIONS

1. Describe the motion of the moon and Earth with respect to the barycenter. Where is the barycenter of the Earth-moon system located?

2. Why is there a difference of over two days between the synodic and sidereal month?

3. Why does the moon always present the same "face" toward the Earth?

4. How is it possible to see more than 50 percent of the lunar surface from Earth?

5. At what time would you expect to see the moon cross the meridian during full-moon phase; during third-quarter phase?

6. What is the relative position of the moon with respect to the Earth and sun during the new-moon phase?

7. In one sentence each describe the following lunar features: craters, maria, rilles, mascons, ejecta blanket, rays.

8. Briefly define what is meant by: a total eclipse, a partial eclipse, an annular eclipse.

9. In what phase is the moon during a total solar eclipse? Why does a total eclipse not occur each time the moon is in this phase?

10. What conditions must be met in order for a total solar eclipse to occur?

11. What is the difference between a spring tide and a neap tide? Diagram the relative positions of the sun and moon with respect to Earth during these tides.

12. Define the following terms: apogee, perigee, barycenter, terminator, libration, daily retardation, synchronous orbit.

EXERCISES AND PROJECTS

1. A communications satellite in synchronous equatorial orbit has an orbital radius of 42,250 kilometres. What is its orbital velocity?

2. Make a star chart of the sky as it appears during the evening hours. See the star chart of the month in Appendix 6 for guidance. As soon after new moon as practical, plot the position of the moon against the background of stars. Make your plot at the same time each evening and make the diagram large enough to show the change in lunar phase.

FOR FURTHER READING

GORDON, R. B., *Physics of the Earth.* New York: Holt, Rinehart and Winston, 1972.

HARTMANN, W. K., *Moons and Planets.* Belmont, Calif.: Wadsworth Publishing Co., 1972.

McDONALD, R. L., and W. H. HESSE, *Space Science.* Columbus, Ohio: Charles E. Merrill Publishing Co., 1970.

WHIPPLE, F., *Earth, Moon and Planets.* Cambridge, Mass.: Harvard University Press, 1968.

Gemini

CHAPTER 9   **THE TERRESTRIAL PLANETS**

Love seldom haunts the breast where learning lies,
And Venus sets ere Mercury can rise.

—Alexander Pope

The terrestrial or earthlike planets include Mercury, Venus, and Mars, which together with Earth are the planets found closest to the sun and, compared to the gas giants, have relatively low masses and high densities. Interest in planetary properties has been heightened because of the recent space probes. As a result of the space program, more has been learned about the planets in the past decade than in the preceding 150 years.

9.1 MERCURY

Physical Properties

Mercury, named after the messenger of the gods in Roman mythology due to its speed in orbit, is the closest planet to the sun, with a mean distance of 0.4 AU or 58 million kilometres. (See Figure 9.1.) its distance at perihelion is 46 million kilometres, and the distance at aphelion is 69 million kilometres, indicating a highly eccentric orbit. Mercury is difficult to see because of its closeness to the sun—the greatest elongation at perihelion is 18° and at aphelion, 28°. This is equivalent to the angle seen when the hands of a clock stand at five minutes to twelve. The proximity of the planet to the sun permits us to view it just prior to sunrise and just after sunset. At these times the reflected light from the planet passes through the thickest part of the atmosphere, causing a maximum distortion of the image. Therefore astronomers attempt to study Mercury in broad daylight, despite the sun's brilliant glare, by making use of special techniques.

Mercury's period of revolution around the sun is 88 days. Its sidereal period of rotation, thought in 1889 by G. V. Schiaparelli to be a synchronous rotation period of 88 days, is now measured by radar techniques to be 58d 15h 30m. This means that Mercury rotates on its axis

Figure 9.1 View from Ma-
riner 10 over the limb of
Mercury. (JPL/NASA)

three times for every two orbits around the sun, a phenomenon resulting from the influence of the sun's gravitational force on Mercury. The Earth is not influenced in the same manner because of its greater distance from the sun. But the moon, being so near the Earth, is affected to a much greater degree by the Earth's gravitational field. This has resulted in a one-to-one relationship between the moon's period of rotation and revolution.

Mercury's solar period (or the period from one noon to the next) is 176 days (Figure 9.2). This provides us with an interesting relationship, in that we have one solar period and three sidereal periods during two periods of revolution. The motions of Mercury also result in a unique situation with respect to sunrise on that planet. Mercury's elliptical orbit brings it quite close to the sun, while traveling at a fast rate. If we could stand on the appropriate place on Mercury so that sunrise occurred as the planet went through perihelion, we would see the sun rise and then a short time later set again over the same horizon. This is due to the fact that the orbital speed at perihelion is faster than the speed with which the planet is rotating. As the planet moves further along in orbit, the sun will rise again.

Mercury, the smallest planet in the solar system, has a diameter of 4864 kilometres. Its mass is 0.055 that of Earth's mass and Mercury's density is $5.45\,g/cm^3$ which is slightly less than the density of the Earth.

Surface Features and Interior

Little was known about Mercury's surface features until the 1974 Mariner 10 probe. Prior to this, radar studies indicated a rough surface but did not yield too much detail. Several circular features about 50 kilometres in diameter were reported by radar astronomers as being possible craters. When the first pictures of Mercury were returned to Earth by Mariner 10, the initial reaction was that Mercury looked exactly like the moon, although careful study soon revealed some differences (Figure 9.3). The

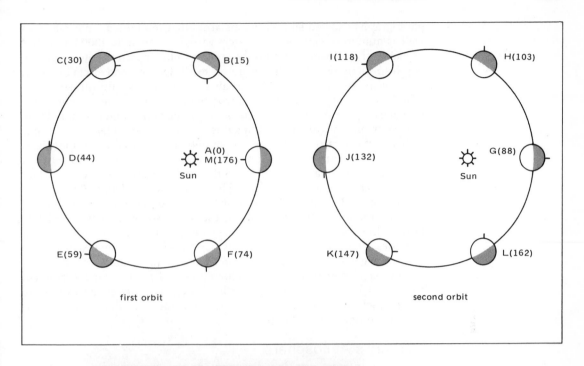

first orbit

second orbit

Figure 9.2 Diagram showing the relationship of Mercury's sidereal day (58.6 days), solar day (176 days), and period of revolution (88 days).

Figure 9.3 Photomosaic of Mercury from eighteen photos taken at 42-second intervals by Mariner 10 when 210,000 kilometres from Mercury. North pole is at top, and its equator extends from left to right about two-thirds distance from top. A large circular basin 1300 kilometres in diameter is emerging from the terminator slightly above left center. Rayed craters are prominent. (JPL/NASA)

photographs showed smooth plains areas that resembled lunar maria, as well as numerous craters. The craters varied in size up to 1300 kilometres in diameter, and were found to be similar to lunar craters with respect to depth-to-diameter ratios. Some craters exhibited rays and ejecta blankets. Rilles were visible on the Mercurian surface, but not the straight rilles found on the moon. Irregular scarps were visible, some as much as a kilometre or more high, which cut across some of the larger craters. There is nothing on the moon resembling this feature. Some evidence of volcanic activity was detected, such as the smooth material and wrinkle ridges found on the crater floors and plains areas (Figure 9.4). It is felt that volcanic activity may have occurred during the early phases of planetary development following a period of heavy impacting (Figure 9.5).

Although the surface features of Mercury bear some resemblance to those on the moon, the internal structure of these two bodies is probably quite different. Some inferences may be drawn from the difference in densities. The density of Mercury, which closely approximates that of the Earth, suggests the possibility that Mercury has a metallic core. This metallic core may be about 2900 kilometres in diameter and composed of an iron-nickel alloy, as is the Earth's. There is a presumed overlay of lighter rock, similar in composition to the rock of which the moon is thought to be entirely composed.

Mercury's Atmosphere

Of major interest was the discovery of an atmosphere on what was previously thought to be an airless planet. The principal constituent is helium, thought to originate from the solar winds or possibly as a byproduct of the radioactive decay of minerals on Mercury. The helium is detectable as far as 500 kilometres from the surface. Other gases detected include neon from solar winds; argon, a decay product; and a faint trace of hydrogen near the surface.

Figure 9.4 A relatively uncratered region on Mercury from 86,800 kilometres above the planet. Older cratering followed by volcanic filling appears in the region above center and slightly to the left. The prominent crater with central peak in center of photo is 30 kilometres in diameter. The bright halo crater to its right is 10 kilometres across. Sun is from the right. (JPL/NASA)

Figure 9.5 New crater in center of older crater is seen at a distance of 20,700 kilometres above Mercury. The new crater is about 12 kilometres in diameter. (JPL/NASA)

One of the interesting features of Mercury's atmosphere is the helium "tail" streaming out from the planet in a direction away from the sun. It appears to be shaped by a weak and unexpected magnetic field associated with Mercury and discovered during the first encounter between Mariner 10 and Mercury on March 29, 1974. The magnetic field, about one-thousandth as strong as Earth's, is a puzzle because the slow rotation of the planet seems to rule out an internal mechanism as the fields' source. The exact source of the magnetic field is not yet known, but it was established as intrinsic to Mercury by the third and final encounter of Mariner 10 with Mercury in March, 1975.

Prior to Mariner 10, sensors on Earth provided reliable data showing temperature readings up to 425°C at high noon on Mercury. This bright-side temperature was confirmed by Mariner 10, which also detected temperatures of 185°C at the terminator and –175°C on the dark side. The range of 600°C between the bright side and the dark side of Mercury is by far the widest in the solar system. This is due in part to Mercury's proximity to the sun and in part to the tenuous atmosphere which, if much denser, would help to retain the heat on the dark side of the planet.

9.2 VENUS

Physical Properties

Venus, named for the Roman goddess of beauty, is sometimes called the Earth's twin because of the similarity in size and mass. From radar measurements Venus' diameter has been found to be 12,110 kilometres and, unlike the Earth, Venus does not exhibit flattening at the poles. Venus' mass is 0.815 (Earth = 1.00) and its density 5.25 grams per cubic centimetre. Venus has the most nearly circular orbit of any of the planets, with a mean distance of 0.72 astronomical units from the sun. The orbital period is 225 days, and the planet rotates slowly on its axis in a period of 243 days in a retrograde direction (opposite that of the Earth). Although

Venus comes closer to the Earth than any other planet, little is known about the surface due to the dense cloud layer surrounding it. Reflections of sunlight from the cloud layer causes Venus to appear as one of the most brilliant objects in the night sky.

Because of the similarity between Earth and Venus, it had long been hoped that beneath the Cytherean* cloud layer, life in some form would be found flourishing. Due to the cloud layer the surface of the planet is completely hidden, but despite this, several models for Venus had been proposed in the past. An early model assumed the clouds to be water vapor, much like the clouds on Earth. Because Venus is close to the sun, the surface was thought to be a hot, humid, swampy jungle, similar in aspect to that pictured for the Earth during the time of the dinosaurs 100 to 200 million years ago. When it was discovered that the clouds were not water vapor and that the atmosphere was composed mainly of carbon dioxide, a new model was constructed that viewed the surface of Venus as a hot, dry desert with hurricanelike winds blowing dust high into the atmosphere. The fact that high concentrations of carbon dioxide were found in the atmosphere troubled some astronomers. Carbon dioxide in great quantities exists on the Earth but is tied up in carbonate rock. Why was this not so on Venus? Possibly, it was suggested, because the entire surface is covered by water. This model later had to be discarded when radio astronomers in 1960 discovered that surface temperatures on Venus far exceeded the boiling point of water.

Venus' Atmosphere

Several probes sent to Venus by the United States and Russia since 1967 have confirmed its high surface temperatures, as well as making several other significant discoveries. The temperature of the outer cloud surface was found to be about -25°C. Data returned to Earth by the Soviet Venera 8, which made a soft landing on the Cytherean surface on July 22, 1972 and remained functional for about 50 minutes, indicated a surface temperature of 400°C (900°F) and a surface pressure of 90 atmospheres (90 times the sea level pressure of Earth's atmosphere) (Figure 9.6).

The surface temperatures were found to be as hot on the dark side as on the bright side, despite the slow rotation of the planet. In addition, there appears to be little difference between cloud temperatures at the poles and the equator. Photographs taken by Mariner 10 (1974) in ultraviolet light revealed details of cloud structure not evident in the visible range of the spectrum (Figure 9.7). The turbulent cloud layers extended from 30 to 65 kilometres above the surface of the planet, and seemed to be stratified and moving at rates exceeding 300 kilometres per hour. At these rates the clouds circle Venus in about four days, compared to the planet's rotation of 243 days. It is thought that the turbulence seen at the cloud surface is not duplicated in the lower atmosphere. The atmospheric movement is probably responsible for distributing heat energy uniformly over the planet by convection cells, or air masses, which descend to lower levels at the poles and ascend to the upper reaches of the atmosphere at the equatorial regions. Only a very moderate flow of air, 6 to 8 kilometres per hour, is assumed at the ground surface because of the high density of the atmosphere there.

---

*Currently accepted adjective form for Venus.

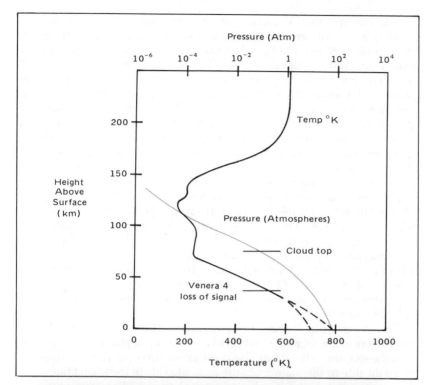

Figure 9.6 Temperature and pressure profile of Venus' atmosphere. Data from Mariner and Venera probes. The Venera probe ceased functioning 30 kilometres above the planet's surface; therefore temperature to surface is extrapolated.

Figure 9.7 View of Venus in ultraviolet light taken from 720,000 kilometres by Mariner 10. The individual frames were computer-enhanced at JPL's Image Processing Laboratory, then mosaicked and retouched at the Division of Astrogeology, U. S. Geologic Survey, Flagstaff, Arizona. (JPL/NASA)

The atmosphere of Venus is composed primarily of carbon dioxide (97 percent), a fact known before 1960 and confirmed more recently by the Venus probes. A little nitrogen, less than 2 percent, has been detected, and a small amount of atomic oxygen has accumulated in the upper atmosphere. Hydrogen has been discovered in the upper atmosphere but almost no deuterium. Deuterium is a heavy isotope of hydrogen found in a small number of water molecules. Since deuterium cannot escape into space as readily as hydrogen because of the planet's gravity, one would expect to find deuterium in the upper atmosphere if the source of hydrogen were water. The absence of deuterium gives a clue to the planet's history. If we assume the same evolution for Venus as for the Earth, we could expect water as one of the products of this process. The lack of deuterium implies a lack of water on Venus; it is possible that the solar winds removed the hydrogen from the inner portion of the solar nebula from which Venus was formed. A lack of hydrogen in the early stages of planetary formation would result in the absence of water at a later time. Hydrogen now present is thought to come from the solar winds, rich in protons but lacking deuterium. The solar winds strike the planet unimpeded because, unlike the Earth, Venus has no magnetic field, and in the Cytherean atmosphere the protons become neutralized to form hydrogen atoms.

The solar winds, aside from adding hydrogen, influence Venus' atmosphere in another way. Acting like a supersonic plasma streaming out from the sun, the solar winds also produce an atmospheric shock wave on the sunlit side of the planet. On the opposite side of the planet there is an elongated wake which extends about 250 Cytherean diameters or approximately 3 million kilometres out from the planet. Venus' wake seems to be much smaller in diameter than the Earth's, which is influenced by its magnetic field. Mariner 10, approaching the dark side of Venus, detected the wake about five days before arriving in the vicinity of the planet (Figure 9.8).

Surface Features

Very little is known about the surface of Venus, except that there is sufficient light to take photographs. Venera 8 provided data indicating that the light on the surface is about 10 percent of that on an average day on Earth. Venera 8 data also revealed that the surface material on Venus is similar to granite. Plans are in progress for sending additional probes to Venus in the late 1970s that will be capable of making soft landings (Figure 9.9).

Why High Temperatures?

Several questions may be asked about Venus, an important one being why the planet is so hot. Because Venus is closer to the sun than is the Earth, it receives about twice the solar radiation, but, all other things being equal, this would not account for the extreme temperatures. The answer seems to lie in the fact that most of Venus' atmosphere is composed of carbon dioxide. Carbon dioxide can absorb heat radiated by the planet. The heat may initially come from solar energy or escape to the surface from the planet's interior. The Earth's surface receives heat from the sun and from its interior, but most of the heat escapes into space because of the small percentage of carbon dioxide in the atmosphere. In the case of Venus, with huge amounts of atmospheric carbon dioxide (70,000 times more than

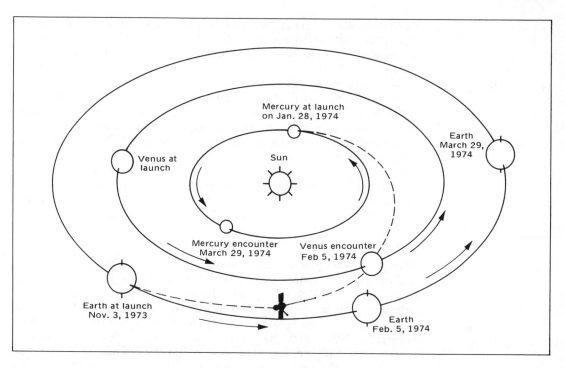

Figure 9.8 The Mariner 10 flight path to Venus and Mercury.

Figure 9.9 Venus in crescent phase photographed in blue light. (Courtesy of Hale Observatory)

Earth) to absorb planetary infrared radiation, the heat is trapped, thereby raising the surface temperature considerably above the level attainable if Venus were an airless body. This heating process is called the "greenhouse effect."

Next, one might ask about the source of so much atmospheric carbon dioxide on Venus. Earth also has a considerable amount of carbon dioxide, but it is tied up as carbonate rock. The carbon dioxide, outgassed by volcanic activity into the Earth's atmosphere, reacted with calcium and magnesium released from silicate rock by the action of water. Since, as Mariner 10 indicated, there appears to be very little water on Venus, this reaction could not take place, and the carbon dioxide remained as a gas in the atmosphere. At the high temperatures that exist, Venus must be devoid of life and an unlikely prospect for future manned exploration.

9.3 MARS

Mars, the fourth planet from the sun (Figure 9.10) named after the Greek god of war, is a favorite with science fiction writers. The Red Planet, as Mars is sometimes called, has been the object of more speculation about extraterrestrial life than any other planet and, consequently, the focus of many men-from-outerspace science fiction stories. An incident that occurred in 1938 reveals the extent to which the public will accept the idea of extraterrestrial life. A radio drama included a realistic account, in the form of a newscast, of a landing by Martians in the New Jersey countryside. This caused the switchboards of radio stations, newspapers, and police departments to be inundated with phone calls from the public who believed the invasion to be real.

Mars Seen From Earth

Mars, during a favorable opposition, comes within 55 million kilometres of Earth, approaching closer than any planet except Venus. Even at this close approach Mars's telescopic image is generally blurred because of variable turbulence in Earth's atmosphere which interrupts "good seeing." However, Mars has relatively little atmosphere, and on rare occasions when both atmospheres display minimum turbulence it is possible to observe surface features with a telescope. Under ideal conditions astrono-

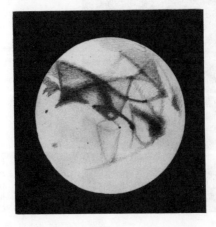 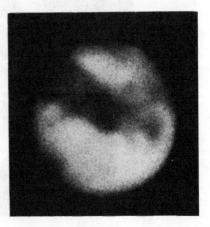

Figure 9.10 Comparison of a drawing (left photo) of the Martian surface with a photograph of Mars (right photo) made at the same time, 1926. (Lick Observatory photograph)

mers have observed such features as seasonal darkening areas, spreading from equator to the poles as the Martian summer approaches. In the past this action has been interpreted as the growth of vegetation when water was released from the polar caps during the Martian spring and summer, or as a change in the color of certain minerals as moisture from the polar caps was released into the atmosphere.

Physical Properties

Mars is 6788 kilometres in diameter, which places it approximately midway between Earth and the moon in size. Mars has a density of 3.97 g/cm^3, the lowest for the terrestrial planets and slightly greater than the lunar density. This suggests the possibility that Mars does not have a metallic core as does Earth. Evidence from Mariner 6 and 7 probes indicates the presence of a partially molten core on Mars, but with the mantle and core not yet completely separated.

Mars is 1.52 astronomical units from the sun, revolves around the sun in 687 days, and rotates in 24h 37m 22.6s on an axis inclined 25°12' to the orbital plane. The period of rotation and axial inclination would permit astronauts to feel at home on Mars, but the year is twice that of Earth's. Because the axial inclination of Mars is quite similar to that of Earth's, Mars also has seasonal variations which can be observed from Earth. These seasonal variations are displayed by the alternate advance and retreat of the polar caps and darkened areas.

From earthbound studies as well as the several Mariner probes, temperatures on the Martian surface have been found to be quite variable. The temperatures range from –125°C over the polar caps to 15°C at noon at the equator. Nighttime temperatures at the equator, unobservable from the Earth, are as low as –75°C. Mariner probes have found the atmospheric pressure to be less dense than that deduced from terrestrial observations. Pressure on Mars is 4 to 7 millibars compared to atmospheric pressure on Earth at sea level of 1013.2 millibars.

The composition of the Martian atmosphere is primarily carbon dioxide with small traces of water vapor, nitrogen, and ozone. Under average Martian surface conditions water can exist only as ice or vapor. The very cold temperatures experienced and low atmospheric densities permit very little water vapor in the atmosphere, but even the small amounts of water recorded can produce saturation and cloud formations at proper temperature and pressure. However, if all the atmospheric water were condensed into rain it would barely wet the surface. Cloud formations are also thought to result from condensation of carbon dioxide and from dust (Figure 9.11).

Other constituents detected in the Martian atmosphere are products of carbon dioxide and water. Photodissociation by ultraviolet light from the sun causes the separation of water into hydrogen and oxygen and the dissociation of carbon dioxide into carbon monoxide and oxygen. Atomic hydrogen was detected in excess of 20,000 kilometres above the Martian surface, well beyond the high point of the Mariner 9 orbit. The hydrogen concentration is stable, which is an indication that the average amount of water vapor in the atmosphere is constant. The concentration observed means that about one million gallons of water are dissociated each day. This is enough water to form a layer 6 metres deep if this process has been continuous for the entire life of Mars.

Figure 9.11 Mars during favorable opposition in 1956. Four views are in different colored light as indicated. (Lick Observatory photograph)

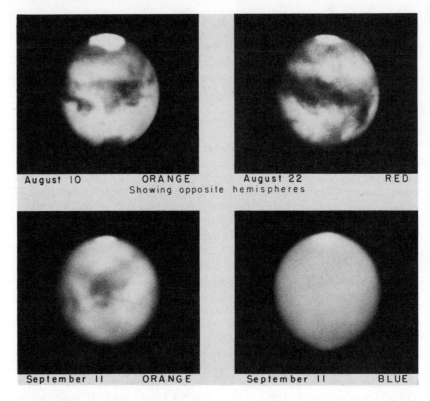

August 10   ORANGE
Showing opposite hemispheres
August 22   RED

September 11   ORANGE
September 11   BLUE

The Mariner Probes to Mars

Several probes to Mars have added to the store of knowledge of Mars. The first probe, Mariner 4, encountered Mars on July 15, 1965, after a flight of 307 days. The probe came within 10^4 kilometres and transmitted its findings 2.17×10^8 kilometres back to Earth. The 21 Mariner 4 pictures clearly showed about 100 craters along the narrow strip of the planet it surveyed (Figure 9.12). Some of the craters even had the peculiar mounds in the center that are characteristic of many lunar impact craters. The Martian surface appeared very old and seemed to confirm what some planetary astronomers had believed all along—that Mars was a geologically dead world. Canals were not evident, although some straight-line features in these Mariner 4 photographs could be associated with canal-like structures seen from Earth. The linear features seemed to be well-weathered, pock-marked geologic features, possibly natural cracks in the planet's crust caused by meteoroid impact. Nothing was observed suggesting any form of life on Mars, but it must be pointed out that neither could life be detected on Earth from a distance of 10,000 kilometres.

Other experiments carried on by Mariner 4 disclosed additional important features on the planet. Mars has no significant magnetic field of its own and no radiation belts.

In 1969, two simultaneous probes, Mariner 6 and 7, completed a successful voyage to Mars and returned more detailed data to Earth than did the previous probe. Mariner 6 took photographs across the equatorial zones of the planet, which included many known light and dark features of its surface. Mariner 7 photographed the surface in an approximately north-south course, intersecting that of Mariner 6 and continuing over the

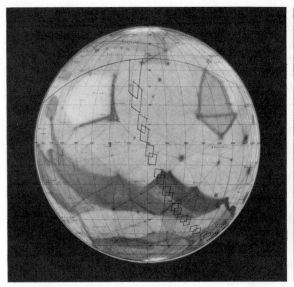 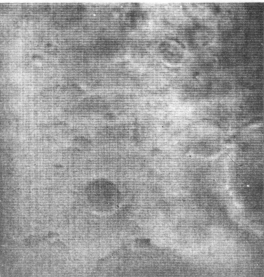

south polar cap. The two probes were designed to cross at different times of the day to obtain a range of lighting conditions in the same area.

Mariner 6 and 7 revealed several different types of Martian terrain. One type included many craters from 50 to 80 kilometres in diameter and a few as large as 500 kilometres. Another form of surface was a chaotic region described as irregular and jumbled, with a topography similar to that of a large landslide on Earth. A featureless area akin to a dry lake bed on Earth was also photographed. No mountain systems like those found on Earth were detected.

As in the case of the Mariner 4 probe, some of the surface marking seen through Earth-based telescopes was not observed in the Mariner 6 and 7 photographs. For example, the seasonal darkening, often visible from Earth, was not evident, nor were there any topographic features visible in the photographs that could be related to seasonal darkening. Only in a few instances were any linear features visible that might conceivably be identified as canal markings from the long distance to Earth. However, nothing that might indicate the presence of a vast network of canals was seen in the Mariner photographs.

The first three probes to Mars indicated nothing other than that Mars appeared to be a dead planet like the moon. The terrain was monotonous, with little evidence of volcanic activity and no mountain ranges or fault systems.

Mariner 9, an orbiter instead of a flyby, was inserted into Martian orbit on November 14, 1971, during a planetwide dust storm that had previously been observed from Earth. The orbiter was better equipped and came closer to Mars than previous probes, thereby revealing details of the planet that surprised even the most optimistic astronomer. Over 7300 photographs with a resolution of one kilometre were taken covering the entire planet, and some areas were photographed to a resolution of 100 metres for greater detail. None of the photographs revealed the presence of life on the planet, but they did disclose that Mars certainly was not a dead planet.

Photographs taken as Mariner 9 approached the planet showed only

Figure 9.12 Left photo: Area of Mars covered by 21 photographs taken by Mariner 4. (Prepared by Army Map Service, Corps of Engineers, U. S. Army) Right photo: Photograph number 10 of the Mariner 4 sequence. The region covered represents an area of 275 kilometres by 255 kilometres. Photo was taken July 14, 1965, from an altitude of 12,900 kilometres. (JPL/NASA)

five distinct features through the dust. These were the south polar cap and four dark spots. When the storm cleared in late December the four spots were revealed to be volcanic mountains (Figure 9.13). One of these, *Olympus Mons* (Figure 9.14), is the largest volcanic formation ever seen by man. Its area could cover the entire state of Colorado. The diameter of the volcanic shield is more than 500 kilometres, and its height above the surrounding plains to the highest point on the lip of the caldera is 24 kilometres. The summit caldera is about 70 kilometres across, and the volcano is three times the height and twice the diameter of the island of Hawaii, the largest volcanic feature on Earth. The other three volcanic structures labeled *North Spot, Middle Spot,* and *South Spot* are not as large as Olympus Mons but are huge compared to most volcanos on Earth.

Does the presence of these volcanos indicate extensive volcanic activity on Mars? Lava flow patterns on the flank of Olympus Mons are characteristic of fluid basaltic eruptions presently forming on the Hawaiian Islands (Figure 9.15). Fringing lava flows surrounding the South Spot caldera are indicative of lava having a high silica and aluminum content and would have been therefore less fluid than Olympus Mons flows. Fifteen additional large volcanic cones have been discovered that were formed by the repeated flows of low-viscosity lavas. This is confirmation of past volcanic activity, but there is at present no direct evidence that these volcanos are still active. Mariner's infrared radiometer did not detect unusual hot spots in any of the calderas. Some indirect evidence of current volcanic activity exists. In the past earthbound observers had found that certain locations on Mars seemed to brighten in the Martian afternoons. These sites have now been associated with clouds developing around volcanic cones, and the clouds may be water vapor condensate vented by the volcanos. On the other hand, the clouds may be formed by moisture directly from the atmosphere condensing as it rises toward the volcanic peaks. Much of the southern hemisphere of Mars is covered by craters, but many of these may be impact craters. Scattered throughout the cratered region are some plains areas similar to the lunar maria.

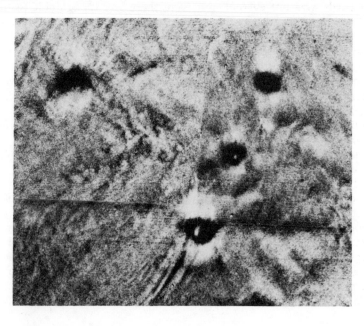

Figure 9.13 Four dark spots appear through the dust storm during the early phases of the Mariner 9 mission. The large spot to the left is Olympus Mons. The other three are North Spot (Ascraeus Lacus), Middle Spot (Pavonis Lacus), and South Spot (Arsia Silva). (JPL/NASA)

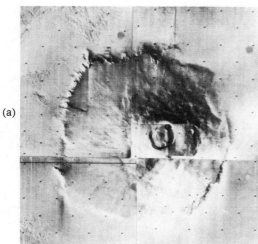

(a)

Figure 9.14 (a) Olympus Mons as revealed after the dust storm had subsided. The clearing atmosphere revealed a mountain 500 kilometres across at the base. The main crater (summit caldera) is 70 kilometres in diameter. (JPL/NASA) (b) Olympus Mons superimposed on California shows the volcano's enormous size.

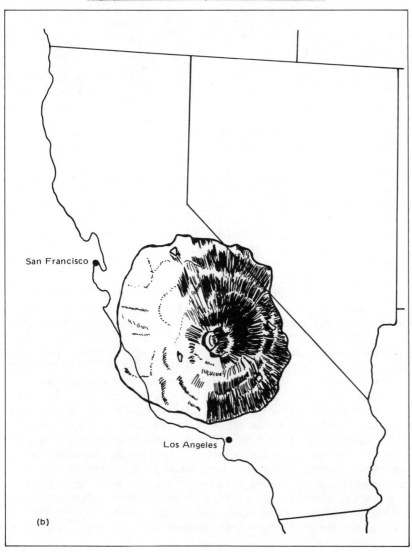

San Francisco

Los Angeles

(b)

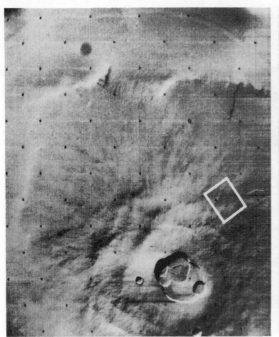
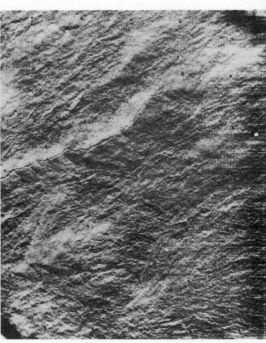

Figure 9.15 Photo shows detail from flank of Olympus Mons. The pattern suggests flowage of material downslope similar in appearance to that seen on terrestrial lava flows. (JPL/NASA)

Evidence of internal stress is recorded in the region near the four large volcanos. In one area fracturing occurs in three major directions, with uplifted and tilted blocks of the crust disclosing the extent of the stress. Farther to the east lies a lattice of canyons with more extensive fracturing, in appearance more rugged than the Earth's Grand Canyon (Figure 9.16). Extending eastward from that complex fault system, the faults coalesce and form a giant rift valley that is as wide as 125 kilometres, extending over 4000 kilometres across the Martian surface, and may be as much as 6 kilometres deep (Figure 9.17). The only structure on Earth that is comparable in size is the East African rift zone that extends from the tip of South Africa up through the Dead Sea.

A number of streamlike channels have been observed on Mars, but as yet no satisfactory explanation for their formation is available. Channels such as those in Figure 9.18 are sinuous and have tributaries that coalesce into well-defined streams. The topography is such that these stream channels would discharge downstream and appear much like major stream channels on Earth. The detail shown in Figure 9.19 of a Martian channel shows a well-defined, braided stream pattern such as those generally formed by intermittently flowing streams on Earth. Every argument and analysis of these flow patterns would lead to the conclusion that they were formed by the action of flowing water. Water, in considerable quantities, has been found to exist in the polar caps and possibly frozen in the subsoil. Astronomers are also seeking alternate ways by which the channels could have been formed. Such suggestions include collapsed lava tubes, erosion by liquid carbon dioxide, deflation by wind, and faulting. However, all of these have been found inadequate for one reason or another. Alternatives that would allow water to form in quantities sufficient to form these channels have also been sought. One

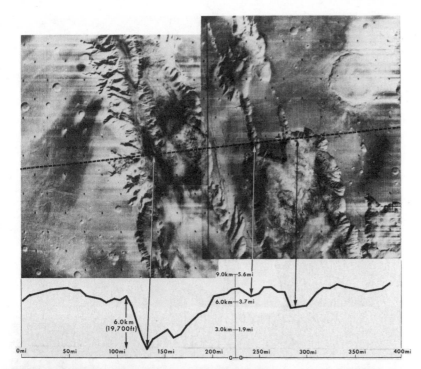

Figure 9.16 Martian canyon reaches a depth of 6 kilometres compared to Grand Canyon depth of 1.6 kilometres. The branching canyons appear to be a landform unique to Mars. Subsidence along lines of weakness in the crust and sculpturing by winds are believed to have formed the features. The arrows relate to low points along the dotted line. North is to the right. (JPL/NASA)

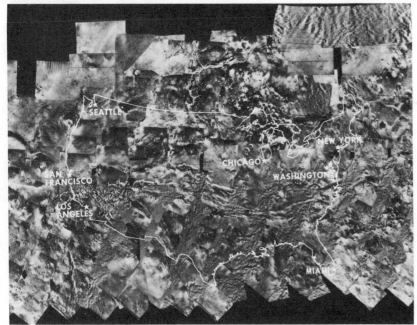

Figure 9.17 The giant rift valley Coprates extends 4000 kilometres across the Martian landscape. To show the relative size an outline of the United States has been superimposed over the rift valley. (JPL/NASA)

explanation that would allow for water erosion assumes that the polar caps are formed of water ice covered by a thin layer of carbon dioxide ice. The polar caps alternate in size every 2.5×10^4 years due to the 5×10^4 precession cycle of the planet. As the larger polar cap becomes oriented toward the sun at perihelion it melts, releasing water into the atmosphere

Figure 9.18 A braided
channel sweeping past a
Martian crater suggests
former presence of fluid
erosion of the Martian
surface. A detailed study
will be required to determine whether water or
some other fluid is the
active agent. The crater
is 19 kilometres in diameter. (JPL/NASA)

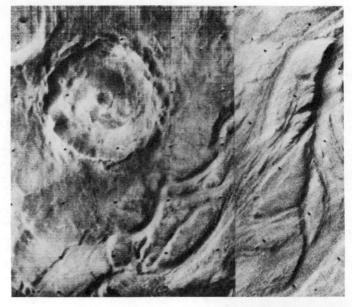

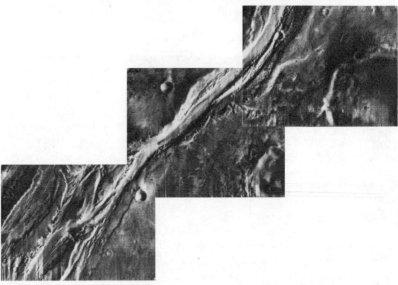

Figure 9.19 A mosaic
showing detail of a Mar-
tian channel thought to
be formed by running
water. Direction of flow is
from lower left to upper
right and the segment of
the channel is about 75
kilometres long. (JPL/
NASA)

which falls as rain. Running water from rivers or directly from the melted
ice would bring about the erosion of the surface and the accumulation of
water as ice on the alternate polar cap (Figure 9.20). It is felt by some
astronomers that in areas where atmospheric pressure is in excess of 6.1
millibars, in the warmer portions of Mars, water may exist temporarily in
liquid form. In the surface soil water may remain liquid for several hours
and below the surface up to several years.

A sequence of pictures taken by Mariner 9 of the polar caps reveals
that the polar caps retreat very rapidly during the onset of the Martian
summer, and then the rate of retreat abruptly slows down almost to a halt.

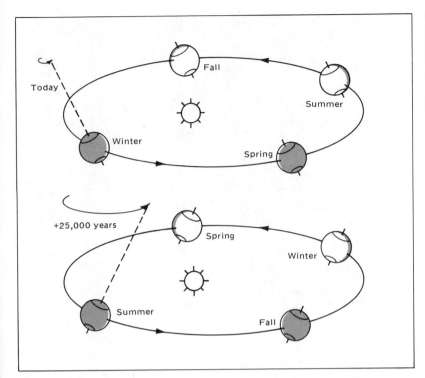

Figure 9.20 As the larger polar cap, oriented away from the sun at perihelion, gradually becomes exposed to the sun at perihelion due to Martain precession, the ice melts and water in liquid form may temporarily exist on the Martian surface.

One hypothesis to explain this phenomenon suggests that the upper layer of carbon dioxide is rapidly evaporated or sublimated, exposing a body that is mostly water ice. At the temperatures existing at the Martian poles, melting of water ice would be extremely slow. It has been found that massive amounts of water ice are concealed beneath the surface near the polar regions. Figure 9.21 shows terrain near the south pole that appears layered and may be composed of substantial amounts of water ice. The ice appears overlayed with soil and dust, possibly carried there by dust storms similar to the giant Mars dust storm of 1971 and built up over many millions of years.

Variable surface changes detected during the Mariner 9 mission led astronomers to believe that light-colored surface areas are covered by fine particulate matter. This material is subject to movement by the high-velocity winds occurring on the Martian surface. The dark areas are covered by coarse granular material or rocks which are not so readily moved by wind action. In all cases where surface changes were observed, the boundaries of light and dark regions changed, so that light areas always retreated and the dark areas were enlarged. It is thought that a somewhat uniform dust layer may be deposited over the surface of the planet during planetwide dust storms such as seen in 1971. Later this dust is gradually removed by local wind storms, exposing the darker surface material underneath. In some areas the dark portions on the floors of craters showed a similar pattern to sand dune fields on Earth. This indicated the presence of particles larger than that of which the light-colored dust is composed.

Figure 9.21 The oval tableland near the south pole of Mars conceals massive amounts of water ice trapped beneath the surface. The light and dark contoured surface is thought to represent layered deposits of dust and volcanic ash and possibly carbon dioxide and water ices. Sun is shining from the left about 10° above horizon. (JPL/NASA)

The Viking Probe to Mars

On July 20, 1976, Viking 1 landed on Mars in man's first successfully functioning, unmanned landing on one of the nine major planets in the solar system. The Russians had made two previous attempts but both were failures. The Viking space craft was equipped to perform a number of scientific experiments designed to acquire biological, chemical, and environmental data about Mars.

Figure 9.22 View from Viking space craft showing Martian surface strewn with rocks and fine sand. At left is sand piled up in a manner reminiscent of sand dunes. The horizon features are approximately 3 kilometres away. (JPL/NASA)

Biological investigations—the search for the presence of Martian organisms—has generated by far the greatest interest of any of the experiments. Three distinct investigations to incubate surface soil samples were conducted, each investigation based on different fundamental assumptions about the requirements of Martian life.

Data on the Martian atmosphere received from the Viking lander

indicated much higher concentrations of nitrogen and argon than anticipated. Nitrogen amounted to about 3 percent of the total atmosphere and argon about 1½ percent. Nitrogen is considered essential in the life process but its presence does not guarantee the existence of life on the planet. All the elements necessary for life are present on Mars; nothing has been detected that is detrimental to the development of life. The question at this point in time is—did it happen? Nothing from the Viking experiments to date precludes the existence of life. The results are exciting and encouraging. As yet the data is inconclusive.

Temperaures in the *Chryse Planitia* basin, where the Viking lander is located, range from approximately –31°C in the daytime to –85°C at night. Wind velocity has been measured at 30 to 65 kilometres per hour.

Perhaps the most incredible information received thus far comes from the photographs of the Martian surface. These remarkably clear photographs show a rolling surface strewn with rocks of all sizes. The scene is quite similar to the deserts in the southwestern United States. The rocks show angular facets and in some cases are deeply pitted. The rocks are lying on the surface or are partially buried in a finely granulated material of sand and dust. There is evidence of wind-caused drifting of the fine-grained material similar to that observed in desert regions on Earth. Some photographs show features reminiscent of sand dunes. (See Figures 9.22, 9.23, and 9.24.)

The color of the rocks is generally grey to black and they are partially coated with a reddish dust. The sand is tinted red, which is thought to be the reason for the planet's predominantly red color. The color forms as a result of water combining chemically with the iron-rich Martian soil. The chemical reaction, occurring under the influence of unfiltered ultraviolet light from the sun, causes the soil to oxidize and acquire its rusty color. This reaction could not occur on Earth because our atmospheric ozone layer excludes most of the ultraviolet light from the sun. The photographs also showed the Martian sky to be pink to orange in color.

Figure 9.23 Mars view with large dark boulders dominating the scene. The largest boulder is about 3 metres wide and one metre high. The rocks may have been derived from lava flows or stream action. The fine material between the rocks appears to have been deposited by wind.(JPL/NASA)

Figure 9.24 Near view of Martian surface. The object at the lower left is the housing on the Viking lander containing the surface sampler scoop.(JPL/NASA)

The Martian Satellites

In the mid-eighteenth century, both Voltaire and Jonathan Swift referred to two Martian moons in their writings, although the moons were not discovered until 1877. In that year Mars was at a favorable opposition when Asaph Hall, an American astronomer, discovered the two tiny satellites. Phobus (Fear), closest to the planet, is 6000 kilometres above the surface and has a period of revolution of 7 hours 39 minutes. Because of its rapid period of revolution it rises in the west and sets in the east twice each day, although it revolves around the planet in the same direction as the planet rotates. Deimos (Panic), 20,200 kilometres from the Martian surface, requires 30 hours 18 minutes to orbit the planet. It rises in the east and takes 2½ days to reach the western horizon. Both satellites are in synchronous rotation, each turning once on its own axis during one revolution about the planet. The two satellites, probably the smallest thus far discovered, are too small to be seen other than as a point source of light through a telescope. Mariner 9 photographs have disclosed that Phobus and Deimos are irregularly shaped (Figure 9.25). Phobus, the larger, is 23 by 16 kilometres, while Deimos is 13 by 9 kilometres. Both satellites are so heavily cratered that new impacts overlay older craters. Although subject to the same impact activity as Mars, crater density may be up to one hundred times greater on the satellites. This is probably due to the satellites being unaltered by volcanism, faulting, or erosion by wind, ice, and water. The satellites may represent material little changed since the solar system was first formed.

Figure 9.25 The larger of the two Martian satellites, Phobus, seen from a distance of 5540 kilometres. The profusion of craters suggests that the satellite is old and possesses considerable structural strength. (JPL/NASA)

9.4 SUMMARY

Mercury, the smallest planet and the one nearest to the sun, was until 1974 almost a complete unknown. The planet's motions were known, but its surface characteristics were not visible because of the planet's proximity to the sun. Radar studies revealed a rough surface but not too much detail. The Mariner 10 probe in 1974 provided some excellent photographs of Mercury's surface, disclosing a planet very similar to the moon in appearance. A helium atmosphere was detected on Mercury which previously was thought to be airless.

Venus resembles the Earth in mass and size, but there the similarity ends. The Cytherean surface is completely hidden by a cloud layer and therefore has never been seen. The atmospheric pressure is about 90 atmospheres and is composed mainly of carbon dioxide with small amounts of nitrogen, oxygen, and hydrogen. No water appears to be present, and it is possible that water has never been present on Venus. Extremely high temperatures, up to 470°C, have been detected on the surface, making Venus a totally uninhabitable planet. Venus has a retrograde rotation of 243 days and orbits the sun in 225 days.

Mars is the planet that has stimulated a great amount of interest, particularly since the return of the Mariner 9 photographs. What was once thought to be essentially a dead planet is now known to be a planet with volcanic activity and possibly a planet occasionally eroded by flowing water. The magnificent photographs returned by Mariner 9 reveal the largest volcanic structure ever seen by man and a huge rift zone on Mars. Water appears to be more plentiful than previously thought, and some scientists feel that the prospects for the detection of life have gone from poor to fair. Viking 1 and 2 have provided more data on the physical properties of Mars but have yet to resolve the problem of whether or not life exists on Mars.

Photographs of the two Martian satellites show them to be irregularly shaped objects, seemingly unaltered by any force other than impact. The satellites appear relatively unaltered with time.

QUESTIONS

1. What is the relationship between Mercury's sidereal period of rotation, its period of revolution, and its solar period?
2. Is it possible to see Mercury at superior conjunction? Why?
3. In what ways does Mercury's surface differ from the lunar surface?
4. What evidence is there that Venus lacks water? How may this have come about?
5. Describe how an abundance of carbon dioxide in the atmosphere may have resulted in the high temperature on Venus.
6. What is a reason given for the high carbon dioxide content of the atmosphere of Venus?
7. What has been the history of "canal" observations on Mars?
8. Which Martian configuration brings Mars closest to Earth: quadrature, opposition, or conjunction?
9. Describe the attempt to determine if life exists on Mars. Was it successful?
10. Photographs of Mars show erosional features on Mars that are difficult to explain due to the near absence of water. What suggestions have been made to account for the erosional features?
11. What would the motions of the Martian moons be like as seen from the surface of Mars?

EXERCISES AND PROJECTS

1. What would be your weight on Mercury, Venus, and Mars?
2. If Mars, Jupiter, or Saturn are approaching opposition during the term you are studying astronomy, plot the position of the planet nightly against the background of stars. Provide yourself with a suitable star chart and see if you are able to plot the planet's movements through retrograde motion.

FOR FURTHER READING

GINGERICH, O., ed., *New Frontiers in Astronomy*. San Francisco: W. H. Freeman and Company, 1975.

HARTMANN, W. K., *Moons and Planets*. Belmont, Calif.: Wadsworth Publishing Co., 1972.

KAULA, W. M., *An Introduction to Planetary Physics*. New York: John Wiley & Sons, 1968.

WHIPPLE, F., *Earth, Moon and Planets*. Cambridge, Mass.: Harvard University Press, 1968.

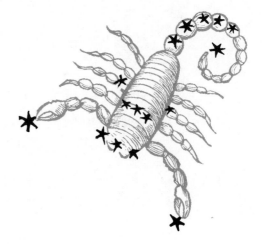

Scorpio

CHAPTER 10 THE GIANT PLANETS AND PLUTO

The inhabitants of Jupiter . . . it would seem, be cartilaginous and glutinous masses. If life be there, it does not seem in any way likely that the living things can be anything higher in the scale of being than such boneless, watery, pulpy creatures . . .

—William Whewell, 1854

The giant planets include Jupiter, Saturn, Uranus, and Neptune. These planets have great mass and low densities compared to the terrestrial planets. Pluto, on the other hand, appears to resemble the terrestrial planets, but there is a great deal of uncertainty about its physical characteristics. No doubt future probes will enlighten us as to the nature of this distant planet.

10.1 JUPITER

Physical Properties

Jupiter, the largest planet in the solar system (Figure 10.1), revolves around the sun in an orbit 5.2 astronomical units from the sun. The large planet is a giant compared to Earth, being eleven times its diameter and 318 times its mass. Jupiter comprises about 70 percent of the solar system's mass not incorporated in the sun. Jupiter is extremely bright when seen from Earth. On the other hand, the terrestrial planets would be difficult to see from Jupiter. Mercury and Venus would probably not be visible, since they are too close to the bright sun that, seen from Jupiter, is only about 20 percent of the size as seen from Earth.

Jupiter requires 11.8 years to complete one revolution of its orbit. As seen from Earth the giant planet's period of rotation varies with latitude, being 9 hours and 50 minutes at the equator and slightly slower toward the poles, indicating the extensive nature of the gaseous atmosphere and the fluidity of the planet. The disk of Jupiter is somewhat flattened or oblate, causing the equatorial diameter of 137,400 kilometres to be approximately

Figure 10.1 Jupiter as photographed from Pioneer 10 shows clouds spun into parallel bands by the planet's rapid rotation and internal heat. The Great Red Spot is the top of a gigantic storm system raging in the Jovian atmosphere. (NASA photograph)

9200 kilometres greater than the polar diameter. Along with the fact that Jupiter's density is 1.32 grams per cubic centimetre, this is further evidence of the partially gaseous structure of the planet.

Planetary Structure

The terrestrial planets have lost most of their primordial gas, due to their low gravitational attraction and the action of the solar winds emanating from the sun. Jupiter, because of its great size, has retained the original products from the formative period of the solar system, thus causing the planet to be of great interest to astronomers.

Data from Pioneer 10 reveals that Jupiter's atmosphere is composed of 84 percent hydrogen and 15 percent helium, with the remaining 1 percent comprising ammonia compounds, methane, and some traces of water, ethane, acetylene, and other gases. Ethane and acetylene result from the breakdown of methane and ammonia by ultraviolet radiation from the sun.

Jupiter's disk, viewed from Earth, consists of clouds spun into a series of parallel bands by the planet's rapid rotation and internal heat. The clouds whip around the planet at speeds up to 500 kilometres per hour, forming greyish-white zones similar to cirrus clouds on Earth. These clouds are composed of ammonia crystals at a temperature of about –120°C and a pressure of 0.7 atmospheres. Alternating with the white clouds and approximately 20 kilometres lower are darker orange-brown belts of ammonium hydrosulfide crystals at temperatures of approximately –110°C (Figure 10.2). The white clouds appear to be convection currents billowing up from the planet's interior, and the darker belts are the descending portion of the circulation pattern (Figure 10.3). Above the cloud layer, which for convenience we may consider the giant planet's surface, is a region of rarefied gas which includes some ethane and

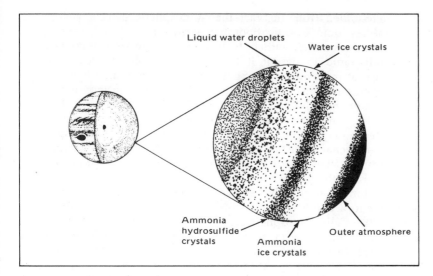

Figure 10.2 The upper Jovian atmosphere from cloud tops to depth of about 90 kilometres. The upper surface is ammonia crystals and the lower darker bands are ammonium hydrosulfide. Below this water exists as ice crystals and liquid droplets.

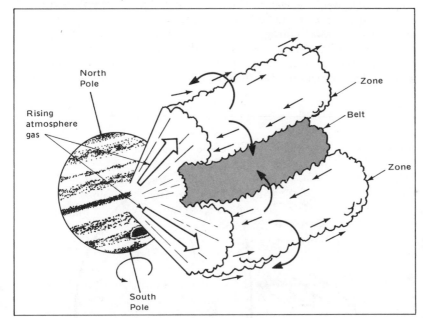

Figure 10.3 Jupiter's atmospheric gases well up to the surface due to convection circulation. Gases moving upward toward the equator from north and south hemisphere move against direction of rotation due to coriolis force. Gases moving toward the poles move with direction of rotation. (From NASA diagram)

acetylene. This region extends about 200 kilometres above the cloud layer where temperatures reach −140°C. Below the clouds, where temperatures increase, ammonia in liquid form mingles with water ice crystals, and further down water exists in liquid form with ammonia in solution. This region, or that immediately below, may be relatively clear, with sufficient sunlight to permit photographing by a probe planned to enter the Jovian atmosphere in the 1990s.

The Great Red Spot is a feature that has been visible in Jupiter's atmosphere for the past several centuries. It appears to be the spinning vortex of a violent storm that is about 40,000 kilometres in its greatest dimension and rises 8 to 10 kilometres above the cloud layer. This has been determined from the fact that atmospheric density is less and temperatures cooler over the Spot than at the tops of the adjacent clouds. The Spot occurs in the southern hemisphere, and rotation of the planet is sufficiently rapid to permit it to maintain its shape even amid the turbulence of the clouds. The Spot appears to be located between two wind currents moving in opposite directions, with shear winds of up to 600 kilometres per hour holding the Spot in place like a ball bearing between two moving surfaces. A similar but smaller Red Spot, over 10,000 kilometres in diameter, has been detected in the northern hemisphere. It has the same characteristics as its larger counterpart, lending support to the theory that these features are temporary meteorological phenomena that arise from the planet's interior.

Data from Pioneer 10 has shown that the Jovian gaseous atmosphere continues to a depth of about 1000 kilometres. Here, at temperatures around 2000°C, the atmosphere grades into near liquidity in what appears to be an atmosphere-ocean continuum. At 2900 kilometres from the top of the atmosphere, temperatures and pressures increase to 5500°C and 90,000 atmospheres, respectively. At this level hydrogen exists as a liquid about 25 percent as dense as water. When the 24,000-kilometre level is reached, temperatures of over 11,000°C occur and the pressure is about 3 million atmospheres. Here hydrogen no longer exists as a liquid but attains a liquid metallic quality. At the center of the planet, 68,700 kilometres from the equatorial surface, a small, rocky core exists under incredible pressures and at temperatures of 30,000° C (Figure 10.4). This is five times the

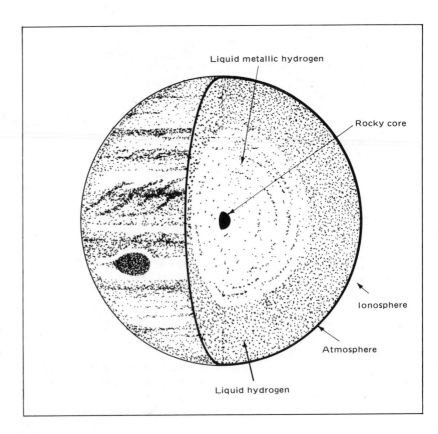

Figure 10.4 A cross-section of Jupiter showing possible rocky core at center. Temperatures and pressures increase with depth causing hydrogen to assume a liquid metallic characteristic.

temperature of the sun's surface, but not sufficiently high to initiate a thermonuclear reaction that would cause Jupiter to be a star. It is estimated that Jupiter would need to be approximately 80 times more massive to enable it to generate a star's self-sustaining nuclear reaction.

Heat and Radiation

Jupiter receives only about 4 percent of the energy received by Earth from the sun. This amount is not nearly sufficient to account for the internal heat detected on Jupiter, nor does it account for the energy radiated by the giant planet. Jupiter's internal heat is most probably residual primordial heat from the time the planet was first formed. The existence of high heat during the formative stages of Jupiter is borne out by the fact that the two large satellites closest to the planet, Io and Europa, have rocky surfaces, whereas the two outer large satellites, Ganymede and Callisto, are covered by frozen gases or water ice. It is believed that Jupiter radiated enough primordial heat to prevent ice from forming on Io and Europa.

Alternate suggestions have been made to account for Jupiter's high internal heat. One theory states that the source of energy could be the fractionation or separation of hydrogen from helium near the center of the planet. Another suggests that the source of energy could result from the slow contraction of the planet. An amount as small as one millimetre per year would be sufficient to maintain the present level of heat at the planet's interior.

The Jovian magnetic field is enormous, having a total energy of about 400 million times that of the Earth's magnetic field. Strength of the magnetic field at Jupiter's surface (cloud tops) is about ten times the strength of Earth's field at its surface. The inner portion of Jupiter's magnetic field is doughnut-shaped and extends over one million kilometres beyond Jupiter's surface. A disk-shaped outer magnetic field continues out over 3 million kilometres from the surface. Polarity of the Jovian magnetic field is the reverse of Earth's.

Jupiter's magnetic field is tilted approximately 10° from the planet's rotational axis. The center of the magnetic field is about 2100 kilometres north of Jupiter's center, and is offset from the axis parallel to the equator almost 8000 kilometres (Figure 10.5). The displacement of the magnetic field had been predicted prior to the Pioneer 10 flight. However, the displacement revealed by Pioneer 10 proved to be much less than predicted. The magnetic field is thought to result from the "dynamo effect," caused by eddy currents within the planet's liquid interior. The eddy currents generate electric currents and, as a consequence, a magnetic field. It is felt by some scientists that only a large planet with an extraordinarily active interior could produce a magnetic field so far offset from the planetary center as is Jupiter's.

Trapped within Jupiter's inner magnetic field is a radiation belt of high-energy particles. The radiation belts are of the highest intensity thus far observed in the solar system and were measured by Pioneer 10 to be about 100 times the lethal dose for humans. Radiation is most intense along the equatorial plane.

Jupiter's Satellites

Jupiter has fourteen satellites. Four were discovered by Galileo in 1610 and are quite large, from 2900 to 5000 kilometres in diameter, and ten range in

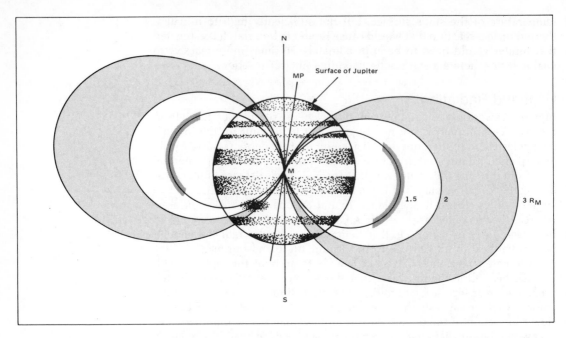

Figure 10.5 Diagram shows the center of Jupiter's strong magnetic field to be offset from the planet's center.

size from 25 to 250 kilometres in diameter. Only those discovered by Galileo were officially named. The others, with the exception of the thirteenth discovered in 1973 and the fourteenth discovered in 1975, have unofficial names (see Appendix 3). All the satellites are identified with roman numerals—the Galileon satellites numbered in order of their distance from the planet while the rest are numbered in order of their discovery.

The Jovian satellites appear to occur in three distinct groups. The first group, including Amalthea and the Galileon satellites, range from 4×10^5 up to 1.5 million kilometres from the planet. After a considerable gap of almost 10 million kilometres four additional satellites are found, including those numbered VI, VII, X, and XIII. The outermost group at approximately 22 million kilometres includes XII, XI, VIII, and IX. This group revolves around Jupiter in retrograde direction opposite that of the other satellites. In 1975, satellite XIV was discovered.

An astronaut standing on one of the five innermost of the satellites would have an unparalleled view of Jupiter. From Amalthea, closest to Jupiter, the planet would fill one-quarter of the sky and be 1200 times brighter than the full moon. Unfortunately no astronaut will see this view, for the inner satellites are within the deadly radiation belts surrounding Jupiter. Manned landings on the outer satellites will be more feasible for they would circumvent this hazard as well as avoid the high gravitational field of the planet and the larger satellites. The larger Jovian satellites will probably be explored by unmanned probes in the early part of the 1980s.

10.2 SATURN

Physical Properties

Saturn, the most distant planet capable of being seen by ancient astronomers, is particularly noted for its rings. Saturn's equatorial diameter is

115,000 kilometres, almost 9.5 times Earth's. The huge planet's mass is 95 times Earth's. Saturn is the second largest planet in the solar system but has the lowest density—0.71 g/cm³—of any planet. This density is less than that of water, and if an infinite ocean were available Saturn could float. Saturn is 9.54 AU from the sun, which appears in the sky as a bright object only about 10 percent the diameter as seen from Earth. At this distance the terrestrial planets are within 11° from the sun and are probably not visible.

Saturn revolves around the sun in 29.6 years in a moderately elliptical orbit. Like Jupiter, Saturn's period of rotation, on an axis with an angle of obliquity of 26.7°, varies with latitude. By observing distinctive markings on Saturn's surface, the rotational period at the equator has been found to be 10h 14m while at higher latitudes it is 10h 38m. This characteristic, along with the fact that the disk of Saturn is somewhat oblate, indicates that the planet has an internal structure similar to that of Jupiter (Figure 10.6).

Very little is known about the internal structure of Saturn, but it is presumed to be similar to that of Jupiter. Hydrogen and helium predominate and methane has been observed spectroscopically. Ammonia has not been detected, but this is consistent with the low atmospheric temperature which probably causes ammonia to condense into "snow." Below the outer visible atmosphere and cloud layer, there is presumably a region where gases become liquid. Radio observations have revealed the presence of a strong magnetic field and radiation belts, a further indication of Saturn's rotating fluid interior. The solid core, if one exists, must be quite small to account, in part, for the planet's low density.

Saturn's Rings

A feature unique to Saturn is its spectacular rings system, composed of billions of particles in an equatorial orbit (Figure 10.7). The rings were first seen by Galileo in 1610 but were more accurately described by Christian Huygens in 1655. The system is composed of three concentric rings with an overall diameter of 276,000 kilometres. The inner ring is viewed by some as two rings, the *crepe ring* and the *subcrepe ring*. These are faint, reflecting little light, and come within 13,500 kilometres of Saturn's surface. The center ring is the largest and the brightest and therefore

Figure 10.6 Photograph of Saturn (100-inch telescope) shows a banded atmosphere and the structure of the ring system. (Courtesy of Hale Observatories)

Figure 10.7 Details of the ring structure of Saturn.

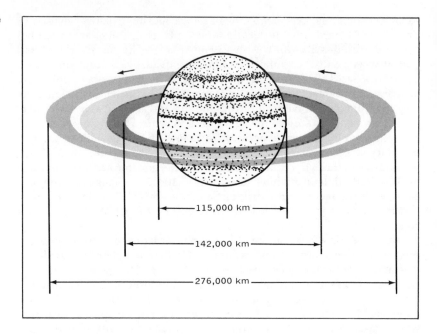

simply called the *bright ring*. The *outer ring* is separated from the bright ring by a 4800 kilometre gap, the most prominent of several such gaps. This space is known as *Cassini's gap*.

When seen from the Earth the aspect of the rings changes as Saturn orbits the sun. Because of Saturn's angle of obliquity the rings at one point in the orbit will be edge-on to Earth and practically invisible. About seven and one-half years later the rings may be seen inclined 26.7° to the line of sight and provide a spectacular view through the telescope. This will be followed in seven and one-half years by an edge-on view again (Figure 10.8). The rings were at their maximum visibility in 1974 and will appear edge-on about 1981.

The rings are not solid, but rather are formed by billions of small particles. These particles are thought to be from one millimetre up to several metres in diameter, a fact deduced from radar studies of Saturn's rings made early in 1973. Doppler shifts in the spectrum of the rings reveal that the particles follow orbits prescribed by Kepler's laws, with particles in the inner ring moving at higher velocities than those farther out. The exact composition of the particles is not known. They reflect sunlight in the manner of solids, and it is thought that they may be primarily water ice or ice-coated rock material and frozen gas, possibly ammonia.

We are not certain of the origin of the rings, but several theories have been suggested. One holds that the rings represent the remnant of the gas and particle cloud surrounding the newly forming planet and from which satellites were formed. However, according to this theory the satellites never took shape, because the material was inside Roche's limit for Saturn and could not consolidate into satellites by gravitational force. Another version proposes that an existing satellite or several bodies wandered within Roche's limit for Saturn and disintegrated. This produced the swarm of particles that, due to tidal effects and the gravitational influence of other satellites, formed into a flat disk. It is now felt that the gaps in the rings result from perturbations of the ring material by other satellites of Saturn. This is the same influence thought to be responsible for the flat structure of the rings as they now exist.

Saturn's Satellites

Saturn has ten satellites, of which eight are thought to be true satellites and two, the most distant from the planet, probably captured. The latter two, Iapetus (S VIII) and Phoebe (S IX), have orbits that are inclined 14.7° and 30°, respectively, to the equatorial plane, and Phoebe revolves around the planet in retrograde direction. Of the eight remaining satellites, six are within 525,000 kilometres and two are 1.2 and 1.4 million kilometres from the center of Saturn. All eight have orbits inclined 1.5° or less from the equatorial plane. (See Appendix 3.)

Of particular interest to astronomers is Titan (S VI), the largest of the Saturnian satellites. Not only is Titan the largest satellite associated with Saturn, but on the basis of data obtained in 1974 Titan is approximately 5800 kilometres in diameter, making it the largest satellite in the solar system. It is about the same size as Pluto and slightly larger than Mercury.

Titan has an atmosphere, a fact known since 1944 when methane was detected. Since then molecular hydrogen has also been discovered on Titan. In 1974, evidence of a strong temperature inversion was found, in which temperatures rise with distance above the satellite's surface instead of falling as is generally expected. This "greenhouse effect" is accomplished on Venus and Earth by the presence of water and carbon dioxide in their atmospheres, but for Titan the phenomenon may be due to the existence of molecular hydrogen in its atmosphere. Although small quantities of hydrogen have been detected, it is suggested that greater amounts may be masked by a dense cloud cover. The breakdown of volcanic products such as water, methane, and ammonia may result in molecular hydrogen and some organic compounds, according to Carl Sagan of Cornell University. The possible presence of such compounds

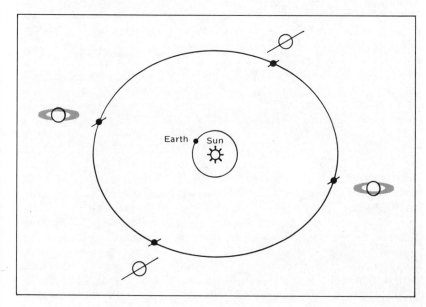

Figure 10.8 Saturn's rings are alternately seen inclined 26.7° to the line of sight and edge-on seven and one-half years later.

and the possibility that Titan may be sufficiently warm has led Sagan to suggest that some form of primitive life may exist on Titan.

10.3 URANUS

The planets discussed thus far were known to ancient astronomers, since they are readily visible to the naked eye. Uranus is faintly visible to the naked eye and had been plotted on star maps as early as 1690, having been mistaken for a star. In 1781, William Herschel, a musician and amateur astronomer of some note, saw a disk-shaped object through his telescope, which he first mistook for a comet. After observing this new object for several months, he concluded that this was indeed a new planet, since it appeared as a well-defined orb, not hazy like a comet. In addition, the orbit was much more circular than the normal long elliptical orbit of a comet. Thus Uranus gained the distinction of being the first planet discovered after the invention of the telescope. The name Uranus, after the Greek god of the heavens, was suggested by Bode (Figure 10.9).

There was not sufficient data from previous sightings to establish an orbit for the new planet, so new data had to be accumulated for the purpose. By 1821, an orbit had been determined, but in a few years Uranus began to deviate from its prescribed course. This indicated the presence of an as-yet-unseen object. Uranus had been diverted from its calculated course by two minutes of arc, an error considered too large to be observational.

Uranus is approximately 19.2 AU from the sun and requires slightly more than 84 years to complete one revolution. Of all the planets the plane of the orbit corresponds most nearly to the ecliptic, being inclined only

Figure 10.9 Uranus and its satellites.(Lick Observatory photograph)

0.77°. Uranus has about the same density as Jupiter, but is slightly more than ⅓ the diameter of the huge planet—50,100 kilometres in diameter.

The period of rotation of 10h 49m was determined by observing the Doppler shift in the spectrum of Uranus. An unusual feature is that the axis of rotation is almost horizontal to the orbital plane (Figure 10.10). The equatorial plane is inclined 98° to Uranus' orbital plane, thus causing the planet to rotate in retrograde direction. This orientation permits astronomers to view all parts of the planet during one complete orbit. If one pole is currently pointed toward Earth, then in 21 years the equatorial plane would be seen edge-on, and 42 years from now the opposite pole would be pointed toward Earth (Figure 10.10).

Uranus appears greenish in color, which has been attributed to the presence of methane in its atmosphere. Molecular hydrogen has also been detected, but little or no ammonia appears to be present because of the low atmospheric temperature. It is presently assumed that the internal structure of Uranus resembles that of Jupiter and Saturn. Because of the inclination of Uranus' rotational axis, the atmospheric circulation patterns probably vary periodically as the planet moves around in orbit. Little is known of the magnetic field or radiation belts, if any, surrounding Uranus.

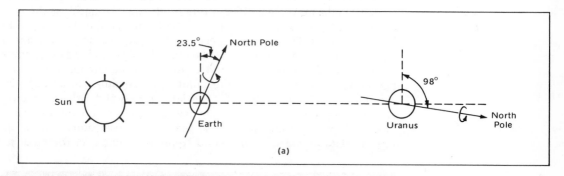

(a)

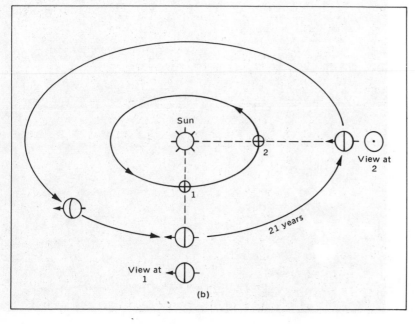

(b)

Figure 10.10 (a) Uranus' axis of rotation is almost horizontal to the orbital plane. To describe Uranus' rotation as counterclockwise it is necessary to identify the angle of obliquity as 98°. (b) Because Uranus' axis of rotation is nearly horizontal to the orbital plane, we see the equatorial plane edge-on at 1 and 21 years later we see the planet pole-on at 2.

Uranus has five known satellites in equatorial orbits. Their direction of revolution corresponds to the direction of rotation of the planet and is therefore retrograde. The satellites orbit quite close to the planet, with the most distant about 586,000 kilometres from the planet's center. The largest of the satellites, Titania, is about 1760 kilometres in diameter and the smallest, Miranda, has a diameter of 550 kilometres.

10.4 NEPTUNE

Little is known about Neptune because of its great distance from Earth, but it is sometimes referred to as "Uranus' twin" because of their similarity in size and atmospheric composition. Neptune is 49,500 kilometres in diameter and presumed to be composed principally of hydrogen and helium, although only methane and hydrogen have been detected. The planet is 30 AU from the sun and requires 164.8 years to complete an orbit. Neptune has traversed approximately 75 percent of its orbit since its discovery, and will complete one revolution in 2011. The period of rotation, determined by the Doppler effect, is 15h 40m, and the axis is inclined 29° to the planet's orbital plane (Figure 10.11).

The existence of Neptune was predicted on the basis of deviations detected in the movement of Uranus in orbit. The relationship is shown in Figure 10.12. Uranus (a), on an inner orbit from Neptune, is attracted by Neptune's mass. Thus Uranus speeds up as it moves along in its orbit. In this way Uranus arrived at its 1821 position (b) slightly ahead of schedule. At this point Neptune fell behind Uranus, causing Uranus to slow and permitting it to arrive at its starting point (a) on schedule. The perturbation exhibited by Uranus caused J. C. Adams (1843) and Urbain Leverrier (1846) independently to suspect the existence of the more distant planet. Leverrier contacted Johann Galle at the Berlin Observatory, and on the basis of data provided, Galle found Neptune within 1° of the predicted

Figure 10.11 Neptune and its satellites.(Lick Observatory photograph)

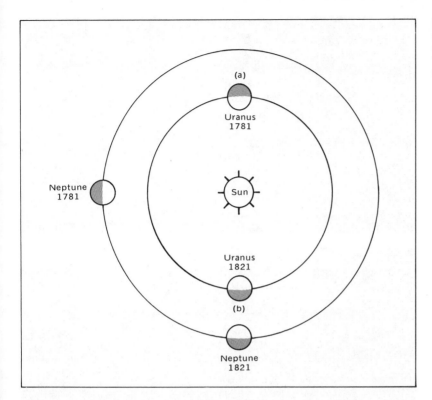

position. All three of the participants in the discovery were given appropriate credit for their work.

Neptune has two satellites, the larger of which, Triton, is closest to the planet and orbits the planet in retrograde direction. The reason for the retrograde motion of such a large satellite is a mystery. One explanation is that Pluto and Triton were once satellites of Neptune, revolving around the planet in direct orbits. As a result of some interaction with Triton, Pluto escaped the influence of Neptune to become a planet orbiting the sun. This interaction caused Triton to be inserted into a retrograde orbit. It has been predicted that Triton will eventually crash into Neptune or break up to form a ring system similar to those around Saturn. Triton is now about 350,000 kilometres from Neptune. The second satellite, Nereid, travels in a direct but very eccentric orbit from 1.5 to 9 million kilometres from the planet.

10.5 PLUTO

Certain small irregularities in the movement of Uranus could not be explained entirely by the presence of Neptune, and this led some astronomers to consider the possibility of a "trans-Neptunian" planet (Figure 10.13). The existence of such a planet was first mentioned in print in 1879 by a Frenchman, Camille Flammarion, although it was Percival Lowell in the early twentieth century who, from his Arizona observatory, made the first intensive search for the object. By studying the behavior of Uranus and Neptune, Lowell hoped to locate the object he called "planet X." His search lasted almost ten years (from 1905 until his death), during

Figure 10.13 Pluto, two photographs showing motion of the planet in twenty-four hours. (Courtesy of Hale Observatories)

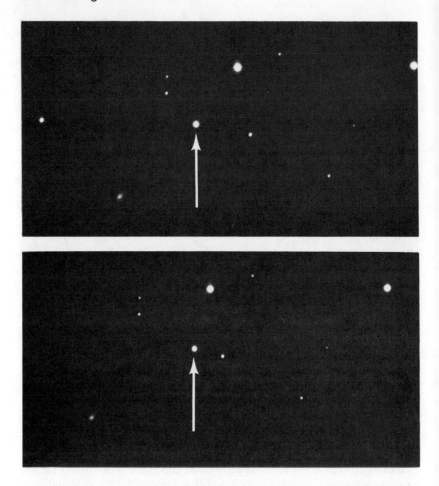

which time he suggested two possible locations for the perturbing planet.

Meanwhile a similar investigation was being conducted by W. H. Pickering. He mathematically computed a position for the unseen planet that approximately coincided with one of the positions suggested by Lowell. Photographs taken at the Mount Wilson Observatory in 1919 based on Pickering's determinations failed to reveal the new planet, but only because the image of the planet happened to fall directly on a small defect on the photographic plate.

Finally in 1930, twenty-five years after Lowell initiated the search for Planet X, Pluto was discovered by Clyde Tombaugh, a young American astronomer at Lowell Observatory. Tombaugh made use of a "blink" microscope, whereby two photographs, taken at different times of the same region of the sky, are rapidly shifted back and forth so the observer sees a constant but flickering picture. If one object in the sky has moved during the interval between the time the photographs were taken, the object will seem to appear and disappear or "blink" as the photographs are shifted. In this way, Tombaugh recognized the new planet quite close to one of Lowell's suggested positions. Ironically, Lowell was in the process of comparing two photographs with the blink microscope at the time of his death in 1915. Had he been able to finish, he might have discovered the planet he searched for so long.

In the final analysis, the discovery of Pluto may not have been the result of meticulous measurement and patient pursuit, but rather of a search covering the right area of the sky which was chosen because of incorrect data. Several years after its discovery a mass approximately 0.9 of Earth's mass was calculated for Pluto. This, along with its size, yielded a density far higher than was considered possible, since the amount exceeded the value that would have been obtained had the planet been made of lead. More recently, with improved techniques, a mass of 0.11 of Earth's mass has been calculated which, along with a more accurate figure for Pluto's size, yields a density of 4.85 g/cc. While this value seems more plausible for a planetary density, the mass upon which the density is based is too small to account for the perturbation of the inner planets upon which Pluto's discovery is based. As a result many astronomers consider the finding of Pluto at that point in space a lucky coincidence.

Pluto is the most distant planet, being 39.5 AU (on the average) from the sun. However, Pluto's orbit has the highest eccentricity of any of the planets, causing Pluto to travel inside Neptune's orbit at perihelion. At this point Pluto is 29.5 AU from the sun, while at aphelion its distance is 49.4 AU. There is no likelihood of Neptune and Pluto colliding because of their orbital characteristics. The inclination of Pluto's orbit is 17° to the ecliptic, and theoretical calculations indicate that the two planets will never approach each other closer than 18 AU. This is about the same as the distance from Earth to Uranus.

Pluto appears to have no visible atmosphere and is extremely cold, possibly –210°C or lower. Pluto is the second smallest planet, being about 6400 kilometres in diameter, only slightly smaller than Mars. It has physical characteristics similar to that of the terrestrial planets with respect to size, density, and mass. The period of rotation is 6.39 days, a finding based on a slight regular variation in brightness of the planet. Pluto's period of revolution is 248.4 years.

10.6 ADDITIONAL PLANETS?

The question that comes to mind at this point is, Are there additional planets as yet undiscovered in the solar system? Some attempt was made by Clyde Tombaugh to answer this question, but his efforts have been fruitless. He had explored a region of the sky far beyond the orbit of Pluto, out to 270 AU, without results. Because of the vastness of space, it is easy to overlook a tiny, dark object in the remote areas of the solar system.

In 1972, evidence of the existence of a tenth planet was obtained from what appeared to be the erratic behavior of Halley's comet. Halley's comet is one of the most studied of the comets that periodically revolve around the sun (see section 11.1). Deviations from its normal orbit indicated that the comet was affected by a mass equivalent to three times that of Saturn, located approximately 65 AU from the sun. The hypothetical planet would be in a retrograde orbit, requiring 512 years to complete one revolution, and the orbit would be inclined 60° to the ecliptic. The proposed direction of revolution and the inclination of the orbital plane to the ecliptic is at variance with that of the other planets, making this planet an unusual member of the solar system, if it does indeed exist. A subsequent search for the planet in the predicted region has thus far failed to reveal its existence.

Soviet astronomers have also investigated the possibility of planets beyond Pluto. Mathematical analysis of cometary movement during the past century has shown that a planet may be located at 54 AU, and another planet about the size of Earth may be approximately 100 AU, from the sun.

10.7 SUMMARY

New insights are being gained with respect to the large planets as a result of the recent probes to Jupiter. Although only Jupiter has so far been scanned by probes, it is felt that the data obtained may be, to some extent, applied to the other giant planets as well. The giant planets appear to be composed primarily of hydrogen and helium with minor amounts of methane, ammonia, and other gases. Strong magnetic fields and radiation belts have been detected around Jupiter. It will not be surprising if these exist in association with other large planets as well.

Thirty of the solar system's 34 satellites are arrayed around the large planets; a few of them are actually larger than the smaller planets. These satellites, formerly ignored for the most part, are now stimulating considerable interest. Atmospheres, possible snowcaps, and warmer temperature than their distance from the sun would indicate have caused astronomers to speculate on the possibility of life occurring on some satellites.

Pluto is still an enigma. Similar in physical attributes to the terrestrial planets, Pluto is thought to be a former satellite that has escaped from an orbit around Neptune. Little is known about the planet other than its motions and orbit. The planet's size and mass are still in doubt.

Additional planets beyond Pluto have been sought, thus far without success. Speculation on the existence of an object is based on perturbations of Halley's comet in orbit, but the object has never been visually detected.

QUESTIONS

1. How do the giant planets differ from the terrestrial planets?
2. What is the nature of the Great Red Spot on Jupiter?
3. What suggestions have been made to account for Jupiter's high internal heat?
4. What physical property of Saturn would enable Saturn to float on water? What is the numerical value of this property?
5. Describe the ring structure of Saturn. What is a possible origin of Saturn's ring structure?
6. What do we know about Saturn's satellite Titan that makes it of particular interest to astronomers?
7. What is distinctive about Uranus, Neptune, and Pluto as compared to the other six planets?
8. Describe the unusual rotational feature of Uranus. In view of this and considering Uranus' period of revolution, what would be the length of the "day" at Uranus' north pole?
9. What is unusual about the discovery of Neptune?

10. What may be the reason for Triton's retrograde revolution around Neptune?
11. Briefly describe the search for and discovery of Pluto.

EXERCISES AND PROJECTS

1. If planets were located at 54 AU and 100 AU as suggested by Soviet astronomers, what would be the period of revolution of each of these planets?
2. With a good pair of binoculars observe the Galileon moons of Jupiter. Plot their motion with relation to Jupiter over several hours and over several nights. Make a chart of their motion by plotting time versus position.

FOR FURTHER READING

ALEXANDER, A. F. O'D., *The Planet Saturn.* London: Faber and Faber, 1962.

ALEXANDER, A. F. O'D., *The Planet Uranus.* New York: American Elsevier Publishing Co., 1965.

HARTMANN, W. K., *Moons and Planets.* Belmont, Calif.: Wadsworth Publishing Co., 1973.

KUIPER, G. P., ed., *Planets and Satellites.* Chicago: University of Chicago Press, 1961.

WHIPPLE, F. L., *Earth, Moon and Planets,* 3rd ed. Cambridge, Mass.: Harvard University Press, 1968.

Equuleus

CHAPTER 11

MINOR COMPONENTS OF THE SOLAR SYSTEM

"The literary world," said he, "is made up of little confederacies, each looking upon its own members as the lights of the universe; and considering all others as mere transient meteors, doomed soon to fall and be forgotten, while its own luminaries are to shine steadily on to immortality."

—Washington Irving

Aside from the planets and their respective satellites, there are numerous objects circling the sun in more or less fixed orbits. These objects include the comets, meteoroids, and asteroids, briefly mentioned in chapter 6 and about which more details will be given here. Most comets, following regular orbits around the sun, are seen only with a telescope. Occasionally a large comet will appear that is visible to all, but often at an inconvenient time and therefore missed by most. Those with patience may see a meteoroid enter the Earth's atmosphere as a flash of light, commonly referred to as a "shooting star." Meteoroids, the small particles found in interplanetary space, collide with the Earth by the millions, but only those a few millimetres or more in diameter create the flash of light which traces their path as they enter the Earth's atmosphere. The asteroids are a group of small objects, the majority of which orbit around the sun between Mars and Jupiter. The largest of the asteroids, Ceres, is about 770 kilometres in diameter and, in 1801, was the first asteroid to be discovered.

The total mass of comets, meteoroids, and asteroids is insignificant compared to the rest of the solar system, but they are important in helping us understand how the solar system was formed.

11.1 COMETS

History

Throughout history comets have been viewed with foreboding. Ancient peoples considered them harbingers of disaster. Almost every visible comet since before the time of Christ and well into the Renaissance was associated with the death of some prominent figure. The bright comet of 43 B.C. was supposedly the soul of Julius Caesar carried to the heavens, and in 453 A.D. a comet's appearance announced the death of Attila the Hun. So prevalent was the belief that comets never seen were recorded, such as the one that allegedly heralded the death of Charlemagne in 814 A.D. Other disasters were also attributed to the appearance of comets. In 1066, a great comet (later identified as Halley's comet) supposedly foretold the Norman victory at the Battle of Hastings (Figure 11.1). Almost 900 years later Halley's comet was pointed to as the precursor of World War I. In the Western Hemisphere the Aztecs considered the comets as a sign that Quetzalcoatl, the serpent god, would return to Mexico. When Cortez arrived in 1519, he was greeted by the Aztecs as the white-bearded god, associated with Quetzalcoatl, whose coming was foretold by a comet's appearance twenty years earlier. Although we don't completely understand all aspects of comet structure and origin, we know that comets are integral parts of the solar system, just as are the planets, and not some manifestation of the supernatural.

Structure

We tend to think of comets as composed of a head and a tail, but this is only partly true. A large bright comet consists of four features: the *nucleus, coma, tail,* and a giant tenuous cloud of *hydrogen* gas. The latter feature was discovered in 1970 when the Orbiting Astronomical Observatory, OAO-2, returned information on comet Tago-Sato-Kosaka and comet Bennett to Earth. These observations, verified by the Orbiting Geophysical Observatory, showed the hydrogen cloud around comet Tago-Sato-Kosaka to be

Figure 11.1 Segment of the Bayeux Tapestry showing crowd awed by the appearance of a comet (Halley's comet) in the eleventh century. (Bettmann Archive)

as large as the sun and that around comet Bennett to be much larger—
actually over 12 million kilometres in diameter. It is believed that the
hydrogen is formed when ultraviolet light from the sun breaks up water
molecules, producing hydrogen and free hydroxyl radicals (Figure 11.2).
The hydrogen emits light only in the ultraviolet portion of the spectrum,
which is absorbed by the atmosphere. For this reason the hydrogen cloud
around comets can be detected only from above the atmosphere.

The nucleus and coma make up what is generally referred to as the
head of the comet. The exact nature of the nucleus is still a mystery, and
there is some question as to whether it exists at all as a separate entity. An

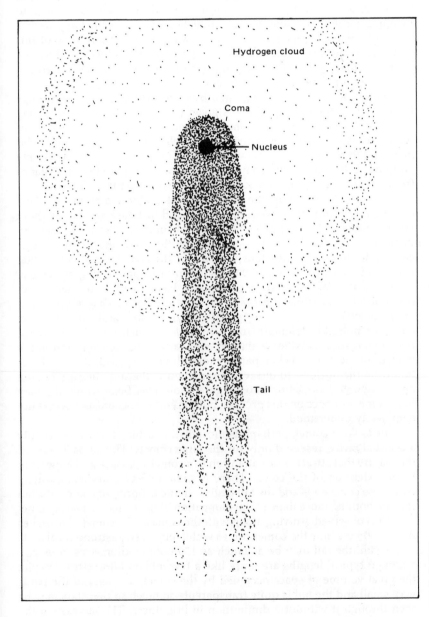

Figure 11.2 The basic
structure of a typical
comet includes the nu-
cleus, coma, tail, and
hydrogen cloud.

upper limit for the diameter of the nucleus has been placed at 3 to 4 kilometres for the short-period comets—those that orbit the sun with a period of less than 200 years—and 7 to 8 kilometres for long-period comets. There have also been some estimates that cometary nucleii may be as great as 200 kilometres in diameter.

Two possible models for the structure of the nucleus are recognized. One, proposed by F. L. Whipple, a Harvard astronomer, suggests that the nucleus is a solid mass of frozen water and gases, such as ammonia, methane, and carbon dioxide. In this frozen mass are imbedded solid particles of various sizes, giving rise to the metaphor that a comet is, in effect, a "dirty snowball." Whipple views the nucleus as being rather porous and therefore a poor conductor of heat. This, according to Whipple, permits the outer surface to be heated as the comet approaches perihelion while the interior remains relatively cool. A variation of this structure is the "layer cake model," proposed by E. J. Opik, in which the nucleus is made up of alternating layers of meteoroid material and frozen gases around a core of solid hydrogen.

The second model was suggested by R. A. Lyttleton from Cambridge University. In his model, which is given the picturesque title of "flying sandbank," the nucleus and coma are a continuous structure. The head of the comet is composed of individual particles of frozen gases and solid material orbiting the sun as a compact mass. The particles are more closely grouped at the core of the comet but more widely spaced, with distance from the core giving the illusion of a solid nucleus at the center.

The coma, surrounding the nucleus, is a cloud formed by solar energy, causing the sublimation (passing from a solid to a gaseous state without going through the liquid phase) of materials in the nucleus. This activity increases as a comet nears the sun, causing the coma, especially of large comets, to expand tremendously. Some of the comets have comas that expand in diameter up to 500,000 kilometres on approaching perihelion.

The coma of a number of comets have been examined spectroscopically, revealing the presence of a number of *daughter molecules*, often ionized, such as hydrogen, hydroxyl, carbon, and nitrogen. The daughter molecules originate from *parent molecules* such as water, ammonia, methane, carbon dioxide, cyanogen, and carbon monoxide found in the nucleus of the comet. When parent molecules are exposed, after sublimation from the nucleus, to ultraviolet radiation of the sun, they are broken apart by photodissociation into daughter molecules. Solar winds also may have some influence on this process, but the entire mechanism is as yet not completely understood.

The tail of a comet, perhaps its most distinguishing feature, is actually a celestial rarity, reserved only for the large comets. The tail is, however, the feature that attracts most attention to a comet's passage and appears to be an extension of the coma. There is some difficulty in distinguishing where the coma ends and the tail begins. In the majority of comets the tail may be nothing more than a faint elongation of the coma. In a large comet the tail is observed growing along with the coma as the comet approaches the sun. By the time the comet comes within one or two astronomical units of the sun, the tail may be as much as 150 million kilometres in length, although typical lengths are more like a few million kilometres. Despite the great volume of space occupied by the comet, the mass of the tail is very small and the tail is quite transparent—so much so that stars may be seen through it without a diminution in brightness. The increase in the length of the tail results from the increase in sublimation of the materials that make up the nucleus as the comet comes closer to the sun.

Without exception, the tails point away from the sun (Figure 11.3). This phenomenon was first noted by the Chinese in the ninth century, but it wasn't until the sixteenth century in Europe that Peter Apian commented on the fact that the tail of a comet then visible appeared to be repulsed by the sun. The fact that the tail is repulsed may seem peculiar in view of the strong gravitational forces exerted by the sun. However, there are two other forces also emanating from the sun that are sufficiently strong to cause the repulsion of cometary tails.

One such force is the force of light itself. Recall that a light beam consists of a stream of photons. A stream of photons reflecting off one side of a small particle of dust exerts a small force on it, causing the particle to be moved. As the comet approaches the sun, intense solar radiation will push tiny particles from the coma to produce a *dust tail* that points away from the sun. In Figure 11.4 a photograph of comet Bennett reveals the existence of two tails. One tail is smooth and slightly curved, tending to lag behind the radius vector. This tail results from the action of solar radiation. The second tail is straight and displays a slight degree of turbulence. This tail is influenced by a second force that is generated by the solar winds, consisting of protons and free electrons ejected from the sun. These subatomic solar particles travel at speeds up to 2000 kilometres per second and interact with the coma to produce a *gas tail*, consisting of ionized gases. A comet is not restricted to two tails. A comet may have one

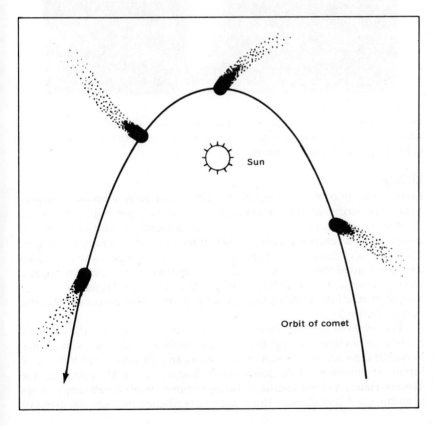

Sun

Orbit of comet

Figure 11.3 Sunlight and the solar winds cause a comet's tail to point away from the sun as the comet passes perihelion.

Figure 11.4 Comet Bennett, April 4, 1970, showing two tails. (Lick Observatory photograph)

kind of tail or the other or several of each. Long-period comets tend to have more complex tails whereas short-period comets have narrow tails consisting mainly of ionized gases.

Orbits

Comets follow orbits that are conic sections—usually ellipses or hyperbolas. The stringent requirements for a circular or parabolic orbit make these extremely rare. A comet following an elliptical orbit is a periodic comet, since it returns at fairly regular intervals to the vicinity of the sun. Those comets moving in hyperbolic paths are nonperiodic and once having rounded the sun will return to deep space never to be seen again. Occasionally such a comet has its orbit perturbed by the larger planets and altered to an elliptical orbit, thus causing the comet to become a periodic comet (Figure 11.5).

The orbits of periodic comets are far more elliptical than the orbits of planets. For example, comet Encke, with a period of 3.3 years, approaches to within 0.33 AU of the sun at perihelion and returns to the vicinity of Jupiter at aphelion. In so doing comet Encke crosses the orbits of five planets. Halley's comet, perhaps the best known of all comets, approaches to within 0.6 AU of the sun, thus reaching perihelion between the orbits of

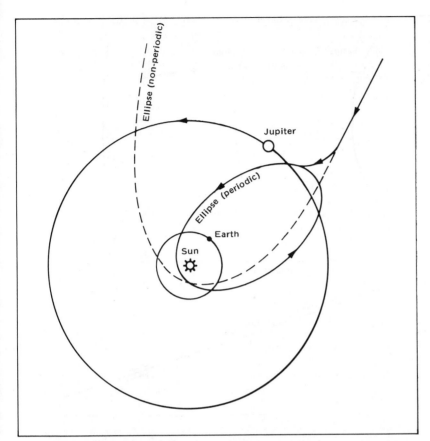

Ellipse (non-periodic)

Jupiter

Ellipse (periodic)

Earth

Sun

Figure 11.5 The orbit of a nonperiodic comet is altered by a large planet such as Jupiter, causing the comet to become a periodic comet.

Mercury and Venus. Halley's comet goes out beyond the orbit of Neptune at aphelion, requiring approximately 76 years to complete one revolution (Figure 11.6). The last perihelion occurred in 1910, and we may expect to see this comet again in early 1986. During the 1910 return the Earth passed through the comet's tail, creating some panic because of the dangerous gases thought to be present. However, no one was harmed and only a meteor shower resulted.

Comets occasionally come close to the Earth. One of the closest approaches in modern times (1927) was comet Pons/Winnecke, which came within 5.6 million kilometres of Earth. Do comets collide with Earth? There has never been a recorded incident of such an event, but this does not exclude the possibility that it has ever occurred in the past or that it will ever occur in the future. It has been hypothesized that the nucleus of a small comet may have collided with the Earth on June 30, 1908, in Tunguska, Siberia. A brilliant fireball was observed in broad daylight, and the impact produced seismic shock waves that were detected in Europe. Trees were knocked down pointing away from the impact site for a distance of 30 kilometres from impact. No crater was ever found, nor were any particles from the mass recovered. It was estimated that the mass weighed 10^5 tons and was about 60 metres in diameter. If this were a comet the body exploding in the air would have caused the gases to be dissipated in the atmosphere.

Unlike the planets, comets move in orbits that are inclined to the

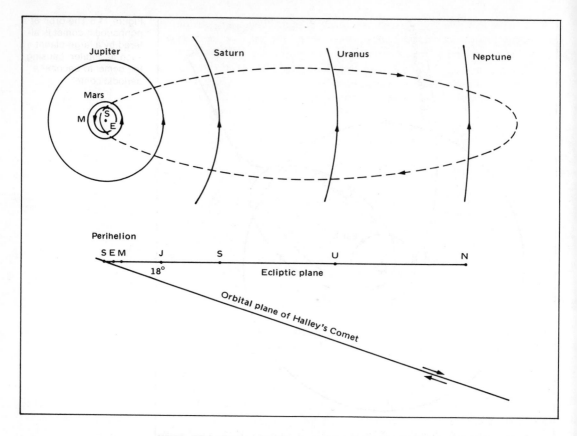

Figure 11.6 The orbit of Halley's comet takes it within 0.6 AU of the sun and out beyond Neptune.

ecliptic, in some instances at steep angles. When the inclination exceeds 90° the comet is described as moving in a retrograde direction (clockwise) in orbit around the sun. Almost all short-period comets, with the exception of Halley's comet and a few others, move in direct (counterclockwise) orbits. Of the entire known comet population, about half move in direct orbits, half in retrograde orbits.

Origin of Comets

With each passage around the sun some mass of a comet is lost, and we must recognize that eventually a comet will, in this manner, be "used up." It is estimated that comets make from 70 to 100 perihelions before disappearing, which means that a comet like Halley's comet, with a period of 75 years, will disintegrate in approximately 7500 years, if we take the optimistic view. Even comet Kohoutek, seen in early 1974, with a period of about 80,000 years, would last only 8 million years, a relatively short life span when compared to the 4.5 billion-year age of the Earth. This leads to the question of the source of comets.

Early philosophers like Aristotle thought comets to be the result of some meteorological event within our atmosphere, while Seneca thought they were similar to the planets but following different orbits. In the early nineteenth century, LaPlace considered the source of comets to be an interstellar cloud captured by the sun. At approximately the same time, J. L. Lagrange suggested that comets come from material violently expelled by volcanic activity from the large planets. Later an astronomer expressed the opinion that the Great Red Spot on Jupiter was the source of this expelled material. More recently the origin of the comets has been linked to the origin of the asteroids (see section 11.2). According to this hypothesis a planet, possibly between Mars and Jupiter, was destroyed. The heavier material supposedly formed the asteroids and the lighter materials formed the more fragile comets. Even at present some astronomers believe that those asteroids with highly elliptical orbits may be the inert nucleii of former comets from which the gases have been completely removed.

The modern astronomer Jan Oort has suggested a theory which is now considered seriously by many astronomers. The theory is that a great "comet cloud" of perhaps 100 billion comets exists at distances up to 150,000 AUs from the sun. These comets are the remnants of the nebula from which the sun and the planets were originally formed. In other words, the comets represent the primordial material of the solar system. The comets in this cloud have wide sweeping orbits which, according to the theory, are occasionally perturbed by nearby stars, sending the comet in toward the sun. The comet then becomes a long-period comet, spending hundreds of thousands to millions of years in this orbit before it disintegrates. On the rare occasions that such a comet passes through the inner solar system at perihelion, the gravitational force of Jupiter or one of the other large planets may alter the orbit of the comet, causing it to become a short-period comet that remains within the inner reaches of the solar system. It is felt that the supply of comets with short-period orbits are constantly replenished in this manner. Jupiter can also cause comets to achieve a hyperbolic orbit and propel it from the solar system back into interstellar space.

The real origin and structure of comets cannot be definitely established until it is possible to examine a comet at close range. Many questions need to be answered such as: Is the nucleus a "dirty snowball" or a "flying sandbank"? Does the nucleus rotate? What is the mechanism whereby materials are sloughed off? How large is the nucleus? These and other questions possibly could be answered by a probe to the vicinity of a large comet (Figure 11.7), although such a probe at present has a low priority.

11.2 ASTEROIDS

Ceres, the largest of the asteroids, is about 770 kilometres in diameter. It was discovered in 1801 by Giuseppe Piazzo in an orbit beyond Mars at 2.8 AU from the sun. Several more were discovered during the first decade of the nineteenth century in generally the same orbit. These include Pallas (490 km), Juno (250 km), and Vesta (490 km). No further discoveries were made until 1830 when the fifth asteroid, Astraea, was found. By 1890, about 300 more were discovered. At that time Max Wolf, a German astronomer, introduced the use of photography as a means of detecting these tiny objects. At the present time up to 2000 asteroids with definitely established orbits have been catalogued. Asteroids are identified by a number designated in order of discovery, and the discoverer has the right to name the new find.

The majority of the asteroids occur in orbit between Mars and Jupiter, and revolve around the sun in 3 to 6 years. The *asteroid belt* averages 2.8

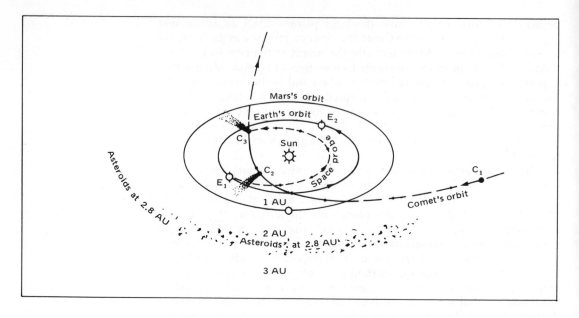

Figure 11.7 A proposed space probe is designed to intercept Halley's comet during its 1985–86 perihelion.

AU, but extends from 2.2 to 3.3 AU. The asteroids follow a direct orbit, and many have orbits that are significantly inclined to the ecliptic. The orbital inclinations average about 10° but may be as high as 52°.

It is estimated that approximately 50,000 of these small bodies, large enough to be seen with the 100-inch telescope, revolve around the sun in a direct orbit, and that close to half a million are larger than 1.6 kilometres (1 mile) in diameter. Doubtless as the size decreases down to sand-sized particles, the numbers increase tremendously. Because of the immensity of space these particles are not as densely grouped as one might think. Pioneer 10 and 11 traversed the asteroid belt with little difficulty. It was found that the distribution of the particles from the Earth through the belt varied with size. Medium-sized particles (1/1000 mm to 1/10 mm) were fairly uniformly distributed throughout the region. Small particles were more numerous closer to the sun, and larger particles were found in greater numbers toward the center of the belt. The total mass of all asteroid material is thought to be less than one-sixth of the moon's mass.

The asteroid belt has some distinctly vacant regions known as the *Kirkwood gaps*. These gaps occur at distances where the orbital period of the asteroid is a simple fraction(¼, ⅓, ½, etc.) of Jupiter's orbital period (Figure 11.8). Gravitational perturbation from Jupiter has caused the asteroids to be removed from these areas. For example, if we have an asteroid with a period of exactly half that of Jupiter's, we find that the asteroid makes two revolutions to Jupiter's one and they end up side by side at the same point in orbit on a regular basis. In this way Jupiter's huge gravitational force will have a perturbative effect upon the asteroid that is repeated on a regular basis. This is called *resonance* and results in a shift in the asteroid's orbit. Asteroids in orbits where the periods do not make simple fractions are also perturbed, but not at regularly recurring intervals, and therefore are not removed from their orbits. The effect is analogous to pushing a child on a swing. A push given at just the right moment at regular intervals will increase the amplitude of motion with relatively little effort because of

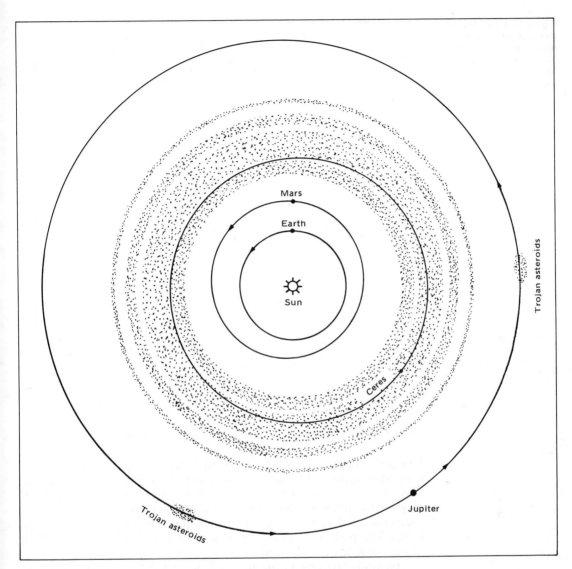

resonance. Pushes applied at random, regardless of effort, may increase or decrease the motion and have little effect.

In addition to the main body of asteroids orbiting between Mars and Jupiter, there are also several smaller recognized groups. One such group, the *Trojan group,* consists of about 15 catalogued asteroids plus others, all revolving around the sun in Jupiter's orbit. These are clustered at the *Lagrangian points,* sites where the gravitational force of Jupiter and the sun are in balance. Part of the group precedes 60° ahead of Jupiter in its orbit, forming an equilateral triangle with the sun, and part follows 60° behind Jupiter to form a similar triangle (Figure 11.9).

A second small group is the Mars-crossing asteroids, which experience perihelion between the orbits of Mars and Earth. There are approximately three dozen asteroids in this group and probably a great many smaller

Figure 11.8 The main group of asteroids occurs in orbit between the orbits of Mars and Jupiter. The spaces shown in the asteroid belt are the Kirkwood gaps.

Figure 11.9 The Trojan
asteroids precede 60°
ahead and behind Jupiter
in its orbit forming equilateral triangles with the
sun.

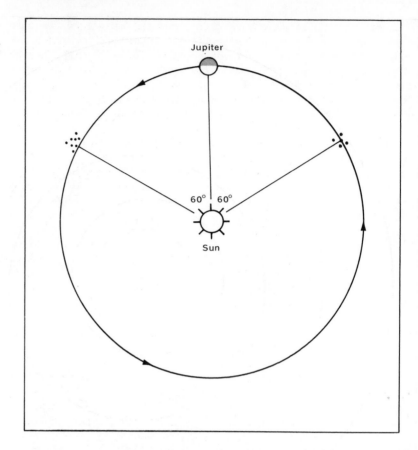

fragments not usually visible. The smaller fragments may be the source of meteorites striking the Earth (see section 11.4), and it is thought that the larger bodies occasionally crash into Mars. The source of this material may be the inner border of the main asteroid belt, where occasional collisions and the perturbing influence of Mars cause fragments to change orbits. The perturbing influence may cause some asteroids to establish orbits inside the Earth's orbit.

Asteroids crossing inside the Earth's orbit are known as the *Apollo asteroids*. These bodies have the most eccentric orbits of all the asteroids, reaching from the asteroid belt at aphelion to the orbit of Mercury at perihelion. One asteroid, Icarus, makes the closest approach to the sun of any known object in the solar system, coming within 0.19 AU at perihelion.

Some of the Apollo group occasionally make a close approach to Earth. Icarus, mentioned above, comes within 6 million kilometres. Apollo, discovered in 1932, has a perihelion inside the orbit of Venus and comes within 5 million kilometres of Earth. In 1937, Hermes was discovered, passing within 1 million kilometres of Earth. Most of the ten recognized asteroids in this group are about 1 kilometre in diameter, and a collision of one of these with Earth would be devastating. It is statistically estimated that such a collision could occur about once in 100 million years. The movements of Apollo asteroids in their orbits are watched with particular interest.

Several points of view have been expressed as to how asteroids originated. One theory proposes that there was planetary formation in the region between Mars and Jupiter which began in a normal fashion. Several planetismals formed and then the process stopped, caused possibly by the perturbing effect of the newly forming giant neighbor, Jupiter. Following this, collisions between the planetismals increased and the destruction of these bodies exceeded their growth. Collisions of this sort resulted in the fragmentation of asteroids. Many of the smaller bodies now detected are actually fragments of large, former asteroids.

Almost all of the asteroid fragments are too small to be seen, and knowledge of their characteristics is generally supplied by indirect methods. Periodic fluctuation in brightness may be due to the irregular shape of the body. Eros is one of the few asteroid fragments that can be seen telescopically. Its perihelion distance is 1.14 AU, which occasionally brings it within 0.2 AU of the Earth. Thus in 1931 Eros came within 23 million kilometres, permitting astronomers to view it directly, and it came this close again on January 23, 1975 (Figure 11.10). Eros is roughly brick-shaped, 36-by-15 kilometres, and can be seen to rotate in 5.27h. Another example is Hektor, found to be cylindrical in shape, 110 kilometres long and 40 kilometres in diameter.

Another manifestation of the collision process is groups of asteroids having similar orbital characteristics. These asteroids are thought to be fragments of a larger body that broke up as a result of a collision and as a group follow the same orbit as the parent body. The groupings are called *Hirayama families*, after the astronomer who first called attention to this phenomenon.

The collision theory of asteroid fragmentation has been challenged by some astronomers. Collisions, they say, would depend upon the rather random motion of the asteroids, but actually, asteroid motions are quite regular and organized; as a result collisions are rare.

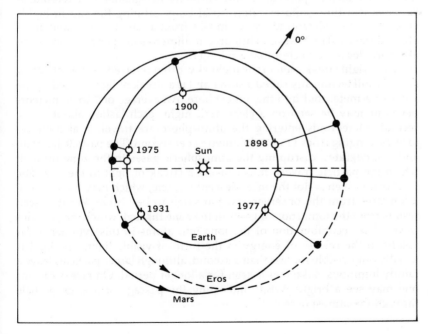

Figure 11.10 Some of the Apollo asteroids are shown in relation to Earth. The orbit of Eros, an asteroid that comes within 0.2 AU of Earth, is shown with various op-positions. The dashed portion of the orbit is below the ecliptic.

The discovery of asteroids in the early nineteenth century instead of the single large planet between the orbits of Mars and Jupiter predicted by Bode's "law," led astronomers to believe that the bodies represented the remnants of an exploded planet. The "exploded planet" hypothesis became generally accepted and is still considered seriously. In 1972, it was proposed that a planet with a mass approximately equal to that of Saturn occupied the site of the asteroid belt and blew up 15 to 20 million years ago. This would account for fragmentation, but the discrepancy in the mass between the hypothetical planet and the asteroids cannot be easily resolved. In addition, energy capable of accomplishing such an explosion cannot be supplied chemically, and the possibility of a nuclear explosion is highly unlikely. Tidal forces may have disrupted the planet if it approached too close to Jupiter, but there is no evidence of this having occurred.

When proposing his comet-cloud hypothesis, Jan Oort also suggested that some objects, now identified as asteroids, may actually be expired comets that have not dispersed. This is particularly true of some members of the Apollo group which, like some short-period comets, have highly eccentric orbits. A few asteroids have very tenuous comas, making them difficult to distinguish from comets.

11.3 METEOROIDS AND METEORS

The term meteor stems from the time of Aristotle, who considered the streaks of light seen in the sky as an atmospheric phenomenon. Meteors were ignored by astronomers until the end of the eighteenth century when two students at the University of Göttingen, using a crude triangulation technique, determined that the objects entered the atmosphere from a source beyond the moon. The term meteor was later applied to all cosmic particles, and this created some confusion. Through a recent redefinition the former meteor particle is now correctly designated a meteoroid. A *meteoroid* is a tiny, solid extraterrestrial body orbiting the sun in highly eccentric orbit. Meteoroids range in size from a few microns (micron = 10^{-3} millimetres) up to several hundred millimetres in diameter. Occasionally particles are detected that are larger.

The bright streak seen in the night sky, often referred to as a "shooting star," is still technically called a *meteor* and is a streak of light caused by the entry of a meteoroid into the atmosphere. An average of 5 to 10 meteors per hour may be seen on a clear dark night, each lasting about half a second. Meteoroids entering the atmosphere are traveling at high velocities, ranging from 11 to 72 kilometres per second. The particle heats up and disintegrates, disrupting the atmospheric gases in the zone through which the particle travels, thus causing a streak of light in the sky. The light that is seen is not the incandescent particle, which may be 80 to 120 kilometres from the observer and too small to be visible. What is seen results from the ionization of gases in the zone through which the particles travel. The recombination of the ions into atoms in this cylinder of air results in the release of energy in the form of visible light. The light is usually only visible for less than a second, although larger particles leave a faintly luminous wake that is seen for a longer period. On rare occasions one may see a bright *fireball*, indicating the passage of a large particle through the atmosphere.

The existence of meteors is visual evidence of the capture of meteoroid particles by the Earth's gravitational field. An estimated 100 million visible meteoroids fall toward Earth each day. Not all reach the surface, since the process of moving through the atmosphere results in a loss of mass by *ablation*, that is, by fragmentation, melting, and vaporization. Many hundreds of millions of micrometeoroids (0.5–200μ) also fall to the surface, so small that they remain relatively cool by radiating heat fast enough to prevent its accumulation. These particles drift down through the atmosphere to deposit on Earth amounts estimated to be 1000 tons per day.

Meteors, resulting from the entry of meteoroids into the atmosphere, are of two types. The *sporadics* originate from all directions and are random particles in orbit around the sun. An organized grouping of meteoroids will result in a *meteor shower*, a much more spectacular display of meteor activity. The particles in these displays enter the atmosphere along parallel paths and occur generally at regular intervals. Because of the perspective of distance, the meteor trails appear to come from a common point or *radiant* (Figure 11.11). For this reason a constellation may form a background for the radiant, and the shower will be named for it. For example, *Leonids* appear to come from the constellation Leo and occur annually each November 17 (see Appendix 2). The Leonid shower, although appearing each year, puts on a particularly unusual display approximately every 33 years. The last such demonstration was in 1966.

Meteor showers occur when the Earth encounters a meteor stream or meteor swarm. A *meteor swarm*, sometimes described as a "flying gravel pile," is a mass of meteoroid particles orbiting the sun as a group. The meteoroids in a *meteor stream* are distributed over an entire orbit. The diameters of the swarms and streams vary from 650,000 kilometres to over 80 million kilometres. These dimensions are generally calculated from the length of time required for Earth to pass through a stream. Despite the increase in meteor activity as the Earth encounters a swarm or stream, the

Figure 11.11 Particles entering Earth's atmosphere travel in parallel paths, but to the observer on Earth the resulting meteor trails appear to diverge from a common point or radiant.

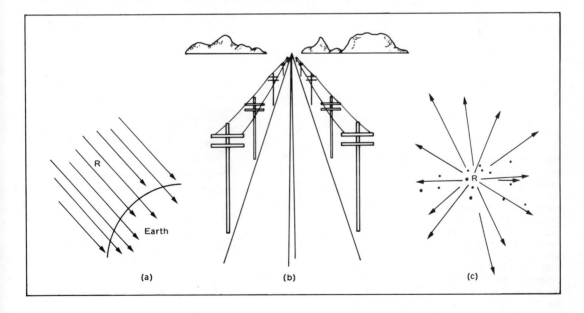

(a) (b) (c)

individual particles are widely spaced. For example, during the great Leonid shower in 1966 the distance separating each meteoroid particle was about 15 kilometres. In other shower displays the particles may be as much as 300 kilometres apart, and sporadics may be separated by as much as 650 kilometres. These figures help to illustrate the emptiness of space beyond Earth's atmosphere.

The ability to see meteors and the velocity with which they strike the Earth is dependent upon several factors, such as the time of day and the time of year. Since the Earth moves around the sun at a velocity of 30 km/sec, those meteoroids encountered head-on will show a higher velocity (up to 72 km/sec) than those overtaking the Earth. The opportunity for seeing meteors is greater after midnight, particularly in the early morning hours, than before midnight (Figure 11.12). This is so because after midnight that portion of the Earth experiencing morning hours is facing in the direction of orbital motion and overtakes most meteoroids, except those moving faster than Earth. Before midnight only the fastest meteoroids are able to overtake Earth. Ths position of Earth's *apex*, or that point on Earth's surface pointed toward the direction of motion, will bring about a seasonal variation in the frequency of meteors. More sporadics may be seen in the autumn when the apex is in the Northern Hemisphere and less in the spring. In the Southern Hemisphere the opposite will apply (Figure 11.13).

The source of meteoroid particles has been the subject of some study. From an orbital point of view there is evidence to indicate that meteor swarms, meteor streams, and comets are related. About 30 percent of the meteoroids appear to occur in orbits that coincide with cometary orbits, and the suggestion has been made that these particles represent debris left behind by a comet's passage in orbit around the sun. The remaining 70 percent are the sporadics; these particles may have originated from meteor swarms or streams that have become highly dispersed as a result of the perturbing effect of the planets. The planets may also influence the orbits of the entire swarm or stream as well. An older theory had proposed that meteoroids originated beyond the solar system. However, as a result of velocity studies on meteoroids, this theory has been disproved.

Figure 11.12 The meteoroids in Earth's orbit cannot move faster than 42 km/sec, which is escape velocity. Taking into account Earth's 30 km/sec orbital speed, meteoroids before midnight will impact the Earth at a maximum of 12 km/sec. After midnight the maximum inpact speed will be 72 km/sec.

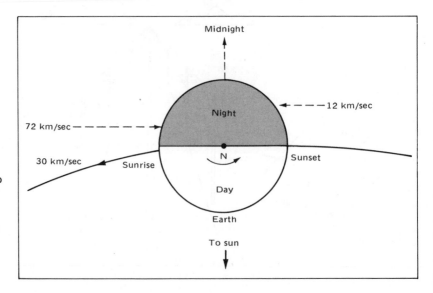

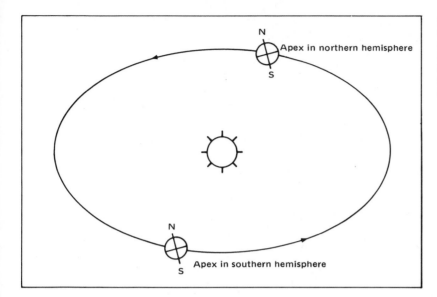

Figure 11.13 Seasonal
variation in frequency of
meteors results from the
position of Earth's apex
or that point on Earth's
surface pointed in the
direction of its motion.

11.4 METEORITES

In the previous section we discussed particles that are completely vapor-
ized in passing through Earth's atmosphere or that filter down as part of
the dust component in the air. Occasionally, a large particlé will survive
the plunge through the atmosphere and reach the surface in one piece or in
fragments. Such a particle is called a *meteorite.*

Are meteorites simply large meteoroid particles that have survived the
plunge through the atmosphere? Evidence gained from orbital studies
seemed to indicate not, but direct evidence did not become available until
1959. At that time a meteorite—the *Pribram meteorite*—seen as a fireball, fell
near Prague, Czech., and as luck would have it the entry through the
atmosphere was recorded from two sites about 40 kilometres apart. This
made it possible to determine with good accuracy the pre-entry orbit. The
data showed that prior to encountering the Earth the meteorite had an
orbit which at aphelion extended out near Jupiter and at perihelion neared
Venus. Such an orbit is characteristic of the Apollo asteroids, and the
Pribram meteorite may well have been one of many objects in an
asteroidlike orbit. A similar event occurred more recently near Lost City,
Oklahoma. Pictures taken by cameras tracking a fireball yielded data that
resulted in finding the Lost City meteorite within 600 metres of the
predicted site. This object also had an asteroidlike orbit that went out
beyond Mars at aphelion and at perihelion coincided approximately with
Earth's orbit (Figure 11.14). The use of various age-dating processes
revealed that the age of both objects was approximately 4 billion years.

Meteorites are recovered whenever possible, since they were until
recently the only extraterrestrial matter available for study. Meteorites are
classed as iron (siderite), stony-iron (siderolite), or stony (aerolite). The
iron meteorites are composed of about 90 percent iron, 5 percent nickel,
and smaller quantities of other metallic elements. They appear dark,
metallic, irregular in shape, have smooth thumbprintlike depressions, and
are of high density. Iron meteorites are easy to recognize but are the least

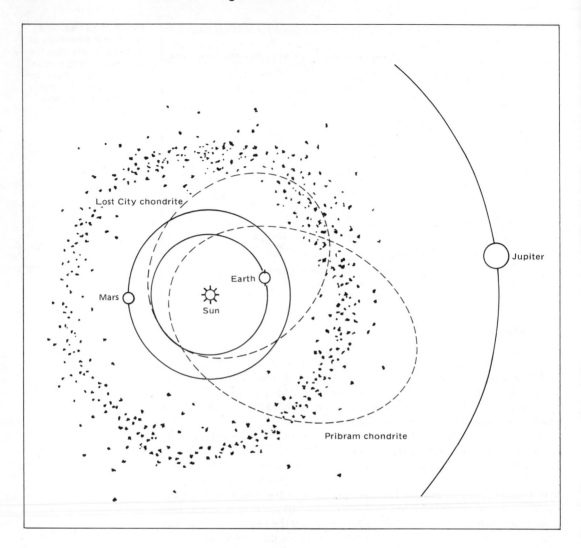

Figure 11.14 The trajectories of two meteorites, obtained from photographic data, made it possible to determine their orbits within the solar system.

common of the three types. They are believed to be similar in composition to the Earth's core.

Stony-iron and stony meteorites are more common than iron meteorites, but they are more difficult to identify because of their similarity to rock material on Earth. Their surfaces are usually brown to black, with a crust formed as a result of surface melting as the object passed through the atmosphere. The stony meteorites may contain 10 to 20 percent metal, whereas the stony-irons have a metal content intermediate between iron and stony meteorites.

Of interest, and also a matter of considerable controversy for over one hundred and fifty years, are the *carbonaceous chondrites*—a special form of stony meteorite. Based upon content, carbonaceous chondrites are subdivided into three types. Type I contains little mineral matter and a great amount of water and organic matter. Type III contains high amounts of mineral matter, some metallic components, and little water and organic matter. Type II falls between the other two. The controversy centers on the origin of the organic compounds. These compounds have been found, in part, to resemble amino acids normally occurring in living cells. Their origin, it is suggested, may be abiological—that is, not related to biological processes—or they may be of biological origin, which implies that life processes of some type may originate beyond the confines of the Earth. It has also been proposed that the organic components in these objects represent terrestrial biological contamination, which may have occurred after the chondrite landed on Earth's surface. The arguments have not yet been resolved. However, the finding of simple molecules in deep space (see chapter 19), which represent the precursors of more complex organic substances, lends some support to the idea of extraterrestrial origin of these materials.

Large meteorites are not common; there are only about three dozen catalogued as weighing more than one ton. The largest is the Hoba meteorite, an iron meteorite that lies partially buried in southwest Africa. This meteorite measures 2.7 × 3 × 1 metre and weighs approximately 55 metric tons.

The presence of large impact craters is evidence that much larger meteorites have crashed into the Earth's surface at very infrequent intervals during the Earth's 4.5 billion-year history. There is evidence that large meteorites like those that formed the huge impact craters on the moon have encountered the Earth during its history, and the impact craters would be visible were it not for the action of erosion and the effect of crustal deformation on Earth's surface. If such an encounter were to occur in modern times, the effect could be catastrophic. It has been suggested that the force of impact of a one-kilometre-diameter asteroid in the Atlantic ocean would generate a wave capable of washing over the Appalachian mountains.

There have been about 80 impact craters detected on Earth's surface, of which 14 are located in the United States. One such crater, the Barringer meteorite crater in Arizona, is 1280 metres in diameter and 175 metres deep (Figure 11.15). The meteorite that formed the crater, possibly 50,000 years ago, is estimated to have weighed about one million metric tons and to have been about 60 metres in diameter. The crater was formed by the explosive expansion of gases heated by the impact. Aerial and space

Figure 11.15 Barringer crater near Winslow, Arizona, formed by impact of a meteorite about 60 metres in diameter. (Bettmann Archive)

photography have recently revealed what appear to be "fossil craters," or *astroblemes,* of meteoritic origin in Canada. Other craters, where no meteoritic material has been found, have impact characteristics such as fractured rock patterns known as *shatter cones,* and a silica material, *coesite,* formed only under extreme high pressure.

One question that arises in a discussion of meteorite falls is, what is the chance of being struck by such an object? There has never been an authenticated account of a person being killed by a meteorite, although there have been stories to this effect. There are records of injuries to persons, as well as property damage, from meteorites. For example, in 1954 a woman in Alabama was slightly injured when a meteorite crashed through the roof of her home, bounced off a radio, and struck her on the arm. Such events are rare, however, and it has been estimated that in the United States a person may expect to be struck once in 9300 years, which makes the chances of being struck rather remote. Another question is, Who owns the meteorite once it lands? This has been the subject of extended legal action in several parts of the world. In the United States a court decided in favor of the owner of the property upon which the meteorite fell. In France cases have also been decided in favor of the landowner. In England a bill to claim all meteorites as property of the Crown was passed in Parliament. This was done to insure the availability of meteorites for scientific investigation rather than commercial exploitation.

11.5 SUMMARY

Comets, long considered omens of disaster, are now understood to be residents of the solar system just as are planets and their satellites. The comet is composed of a nucleus, coma, and a tail and surrounded by a hydrogen gas cloud. Composition of the nucleus is uncertain, but it may possibly be made of frozen gases in which are imbedded solid particles. The coma is a cloud of gases similar to those in the nucleus and will expand in size as the comet approaches the sun. The tail, a feature of large comets, may be nothing more than a very tenuous extension of the coma that points away from the sun. Comets have extremely elliptical orbits, typified by Halley's comet. This comet has an orbit extending from beyond Neptune to approximately 0.6 AU from the sun. Comets disappear after having made about 70 or more perihelions and are replaced by other comets, possibly from a vast comet-cloud in the outer reaches of the solar system. These comets make long-period sweeps past the sun and may be perturbed into short-period orbits around the planets.

Asteroids are the many thousands of bodies orbiting in a region between Mars and Jupiter. These are solid objects, generally quite small, that appear to be fragmented. There are several distinct groupings, including the main body of asteroids in the asteroid belt, the Trojan group in Jupiter's orbit, the Mars-crossing asteroids that cross inside the orbit of Mars, and the Apollo group which cross inside Earth's orbit. The origin of the asteroids is not known for certain. It has been suggested that their source is a planet in orbit between Mars and Jupiter that blew up, or that the asteroids are bodies that never developed into a planet at the time the solar system was forming.

Meteors are the bright streaks seen in the night sky caused by particles or meteoroids from space passing through Earth's atmosphere. Most of these are random particles, but some occur in organized groupings entering the Earth's atmosphere as a meteor shower. These groupings may exist as a mass of particles orbiting the sun as a meteor swarm, or may be particles distributed over an entire orbit as a meteor stream. Meteor streams and swarms are thought to be debris left behind by a comet's passage in orbit around the sun. The random particles, or sporadics, are thought to be particles dispersed from meteor streams or swarms by the perturbing effect of the planets.

Meteoroid particles that survive the plunge through the atmosphere are identified as meteorites. These, until recent moon landings, were the only source of extraterrestrial material available for close study. They are classed as iron, stony-iron, and stony, depending upon composition. Some meteorites, called carbonaceous chondrites, contain varying amounts of water and organic matter. They are of interest because of extraterrestrial biological processes implied by the presence of the organic matter.

When striking the Earth large meteorites form craters similar to craters seen on the moon. The craters are not as common on the Earth's surface as on the moon because of erosion and crustal deformation that occur on Earth.

QUESTIONS

1. Briefly describe the structure of a comet. What are the models suggested for the structure of the nucleus?
2. What is meant by parent molecules and daughter molecules in reference to a comet?
3. Describe the action of the sun on the formation and orientation of a comet's tail.
4. Some comets form two or more tails. What are the forces responsible for producing each type of tail?
5. Describe the orbital path followed by Halley's comet. Is Halley's comet a long-period or short-period comet?
6. Describe several suggested sources of comets. What is currently thought to be the best theory on the origin of comets?
7. What are the recognized groups of asteroids?
8. What is meant by the Kirkwood gaps?
9. How may the asteroids have originated?
10. What is the difference between meteors, meteoroids, and meteorites?
11. What is meant by a meteor swarm; a meteor stream?
12. What is the best time of year and best time of day for observing meteor activity?
13. Briefly describe the three types of meteorites.

EXERCISES AND PROJECTS

1. Make observations on meteor activity on several clear nights chosen at random. Then observe meteor activity during one of the periods of

expected meteor shower activity. Make a comparison between normal sporadic meteor activity and meteor shower activity. Several persons should be involved in the observations, each taking a portion of the sky to view.

FOR FURTHER READING

BROWN, P. L., *Comets, Meteorites and Men.* New York: Taplinger Publishing Co., 1974.

HAWKINS, G. S., *Meteors, Comets and Meteorites.* New York: McGraw-Hill, 1964.

HEIDE, F., *Meteorites.* Translated by E. Anders and E. DuFresne. Chicago: University of Chicago Press, 1964.

WOOD, J. A., *Meteorites and the Origin of Planets.* New York: McGraw-Hill, 1968.

Leo

CHAPTER 12   **THE SUN**

Give me the splendid silent sun, with all his beams full dazzling!

—*Walt Whitman*

The ancient Greeks thought the sun was a huge flaming rock about the size of Greece. In actuality, the sun is a unique body in the solar system, radiating tremendous amounts of energy, in contrast with the cold, dark planets orbiting around it. Its mass governs the movement of all other members of the solar system, and its radiation is the primary source of energy in the solar system. Solar radiation is emitted in all wavelengths of the electromagnetic spectrum (see section 4.1), but the visible and infrared portion of the spectrum account for about 99 percent of the total solar energy emitted. The sun is actually a star, appearing large because it is relatively close to the Earth; for this reason it yields much information on the nature of stars in general.

12.1 PHYSICAL CHARACTERISTICS

The sun is by no means one of the largest stars. Others are vastly greater in size, as, for instance, Betelgeuse, 490 light years from Earth, which has a volume 27 million times that of the sun. Actually, Betelgeuse is so large that the Earth's orbit could fit entirely within the body of the star, and Betelgeuse is over 10^5 times brighter than the sun. On the other hand, many stars are smaller than the sun, some being 10^{-4} as bright. The sun could, on this basis, be considered a typical average star.

The sun is a turbulent body of hot gas, emitting huge quantities of energy. It is about 1.4×10^6 kilometres in diameter (about 109 times the diameter of Earth) and includes over 99 percent of the total mass in the solar system. The sun (1.99×10^{33} gm) is 1000 times more massive than Jupiter and 332,000 times more massive than Earth. However, the sun's density and Jupiter's density are about the same, whereas the sun's density is only about one-quarter that of Earth's, 1.4 g/cm^3 compared to the Earth's 5.52 g/cm^3. The solar density is consistent with the belief that the sun is composed primarily of gaseous matter. The sun rotates on an axis, a fact first deduced by Galileo from the movement of sunspots across the face of the sun. By careful observation astronomers found that all parts of the sun do not rotate at the same speed. The sun's equator rotates once in about 25 days, while the period of rotation is 28 days at approximately 35° north and south of the equator, and 34 days, 75° from the equator. This variation in the speed of rotation is additional evidence that the sun is composed mainly of gaseous material. From sunspot movement it has also been determined that the solar equator is inclined 7° from the ecliptic.

The bulk of the gas comprising the sun is hydrogen, estimated to be about three-quarters of the solar mass, helium, about one-quarter of the mass; only one to two percent of the mass is composed of the heavier elements, mostly in an ionized form. Of the 92 naturally occurring elements, about 70 have been identified on the sun. None have been found on the sun which do not also occur on Earth, although helium, discovered on the sun in 1868, was not found on Earth until 27 years later.

Although we view the sun as a gaseous body, its structure can, for the purpose of discussion, be divided into several layers (Figure 12.1). The nature of the sun's interior can only be inferred; it is in the interior that the energy emitted by the sun is generated. The outer layers that may be seen from the Earth include the *photosphere*, which is the bright disk of the sun and is considered the surface; the *chromosphere*, a low-density region above the photosphere that may usually be seen during a total eclipse; and beyond this the *corona*, which together with the chromosphere make up the solar atmosphere. These various parts will be examined in more detail below.

12.2 THE SOLAR INTERIOR

The solar interior cannot be viewed directly, so knowledge of its characteristics is obtained through studies that apply the laws of physics to known data. Knowledge of the solar mass and the application of Newton's law of gravitation provide information about the internal pressure and temperature necessary to support the mass. These studies indicate that the pressure at the solar center is approximately one billion atmospheres and results in matter with a density of 150 g/cm^3. Pressure and density decrease rapidly as the surface of the sun is approached. The rapid falloff of density indicates that approximately 90 percent of the solar mass is concentrated within the inner half of the sun's radius. Temperatures at the core are about 1.5×10^{7}°K and decrease to an effective temperature of 5760°K (generally rounded off to a more convenient 5800°K) at the photosphere.

The tremendous energy radiated by the sun is generated by the transmutation of hydrogen to helium at the core. This energy, in the form of photons, diffuses outward through the *radiative zone*, requiring thousands of years to reach the photosphere before being radiated into space. In the deep interior, pressures and temperatures are such that collisions between atoms are violent and frequent, with the result that electrons are removed from their orbits. Photons are not readily absorbed by atoms under these conditions but can easily move outward from the solar center. Higher up, where temperatures are cooler, photons can readily be absorbed by an atom, displacing the outer, more weakly held electrons, and thereby slowing down the movement of the photons. This activity seems to occur in the *convective zone*, a region where hot gases rise to the surface to be cooled and returned again to the hotter interior by

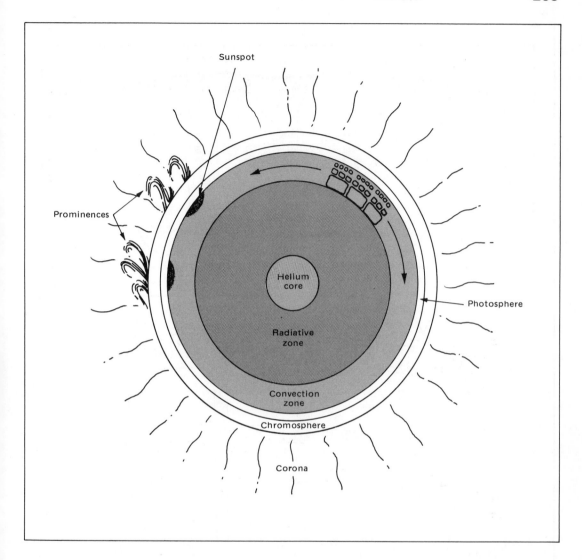

convection currents. The convective zone may extend to a depth of over 10^5 kilometres, and the convection cells may exist in several tiers (Figure 12.2). At the top of the convective zone is the photosphere or visible surface layer of the sun.

Figure 12.1 Cross-section of the sun showing internal structure. The various components are not to scale. (See detail of convection zone in Figure 12.2.)

12.3 THE PHOTOSPHERE

The photosphere is the bright, visible disk of the sun that, upon close examination, appears as a patchwork pattern of volatile granules. Normally the sun appears as a smooth, luminous disk without noticeable surface features. However, detailed high altitude photographs and satellite pictures show the solar surface to be quite mottled (Figure 12.3). This mottling has a fine-grained, cellular appearance called *solar granulation*. It is believed the granules result from the turbulent convective motion of hot gases rising to the surface from the interior. They are from 150 to 1000 kilometres in diameter.

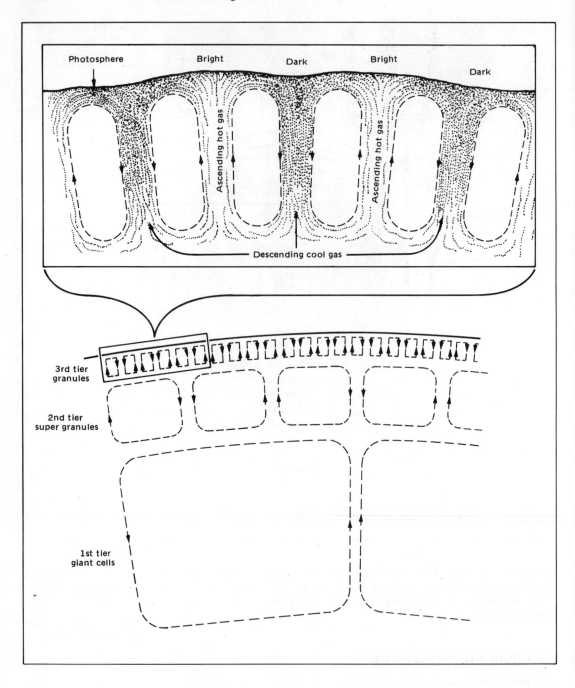

Figure 12.2 In the convection zone hot gases rise to the surface, cool, and descend to the interior of the sun. Convection cells may exist in several tiers of progressively larger cells with depth.

Doppler-shift measurements reveal that the center of each granule is an area of hot gases moving upward, and the darker edges are cooler gases from which energy has been dissipated and that are returning to the solar interior. The average lifetime of the granules is about 10 to 15 minutes. Doppler-shift measurements have also revealed larger-scale movements similar to that of the granules. This suggests larger-scale convection motion producing super granules, possibly generated by convection cells at a greater depth.

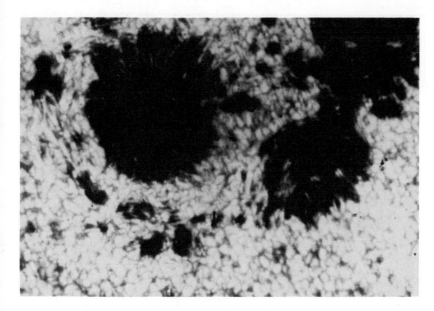

Figure 12.3 Granulation is the fine-grained cellular appearance in the photosphere shown in this illustration in association with a sunspot. (NASA)

Most of the light (photons) that we see coming from the sun emanates from a limited zone above the granular surface about 500 kilometres in depth. This entire zone is generally regarded as the photosphere. The photosphere is narrow compared to the 1.4 million-kilometre diameter of the sun, causing the sun's disk to appear to have a sharp, distinct edge when viewed from Earth. Density of the gases decreases very rapidly with elevation within the photosphere, which results in the photosphere ranging from complete opacity at the granular surface to transparency at the upper boundary. Photons emitted by the gases at the lower limits of the photosphere are allowed to escape into space because of decreasing gas density above.

Temperatures within the photospheric zone range from approximately 8000°K near the granular surface to about 4000°K at the photosphere-chromosphere boundary. Above this boundary temperatures rise again in the chromosphere. The 5800°K temperature given as an average temperature for the solar disk actually represents temperature at the center of the disk. Temperatures at the sun's edge or *limb* are about 300°K lower. The light from the sun, as seen by us on Earth, is noticeably brighter from the center than from the limb. Light from the center of the disk, coming from hotter, deeper gas layers, travels a minimum distance through the photosphere toward Earth. However, at the outer edges of the sun the light reaching us crosses obliquely through the photosphere, and we cannot penetrate as deep visually as at the center of the disk. We are viewing higher, cooler gases from which we receive less light and therefore experience *limb darkening* (Figure 12.4). In recent observations it has been reported that the limb at the sun's equator is slightly brighter than the limb at the poles. The variation in brightness has been attributed to the slight equatorial bulge, which amounts to approximately 65 kilometres. This is more than can be accounted for by the sun's rotation, and it has been suggested that the bulge is due to a rotation of the inner core once every four days instead of every 25 days as observed for the surface.

Although granulation is the dominant surface feature in the photosphere, occasional bright spots appear called *faculae*. These are areas

Figure 12.4 Limb darkening results from viewing higher, cooler gases in the photosphere, from which we receive less light.

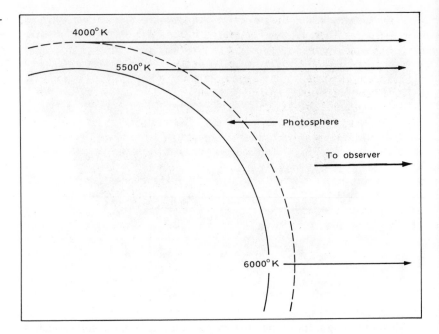

where higher-than-normal temperatures are detected and density of the gases is greater than in neighboring regions. Faculae are usually visible near the solar limb due to limb darkening and are not as evident near the center of the disk. The occurrence of faculae is in some way related to magnetic field activity.

In chapter 5 we discussed the nature of the continuous spectrum occurring in the sun. What is the source of this radiation continuum? Hot gases at the base of the photosphere emit energy in all wavelengths. This activity continues at a decreasing rate as higher levels are reached. At these higher levels the relatively cooler gases absorb energy at wavelengths characteristic of the elements present, and form the absorption spectrum that includes the well-known Fraunhofer lines. The dark lines in the absorption spectrum are useful in determining the identity of the elements present in the sun. If the temperatures and the pressures of the gases are known, it is possible to determine the relative abundance of the atoms of each element. Hydrogen, as previously stated, is by far the most abundant, with helium as the second most common element. Since temperatures in the photosphere are too low to produce helium absorption lines, the presence of helium is only inferred. Helium was originally discovered in the chromosphere and its abundance calculated from space probe measurements of the solar winds and from solar cosmic radiation. As yet we do not have an accurate figure for the amount of helium. All other elements detected in the sun, combined, make up approximately 1/1000 of the total number of atoms revealed in the absorption spectrum.

12.4 THE CHROMOSPHERE

The chromosphere, recognized as the lower portion of the solar atmosphere, is a region of nearly transparent gases extending about 10^4 kilometres above the photosphere. Just above the photosphere density

decreases rapidly, but as the upper chromosphere is approached the rate
of decreasing density lessens. The reverse is true for temperature. Initially,
the temperature increases slowly and toward the upper levels of the
chromosphere temperature increase accelerates, reaching 20,000°K.

The chromosphere is not visible against the bright disk of the sun and
in the past could only be seen during a total eclipse. At that time, just as the
solar disk was completely covered by the moon, the chromosphere would
appear for 15 to 20 seconds as a thin red crescent. Now the chromosphere
may be examined at will by means of a coronagraph, an instrument that
permits photographing the chromosphere and corona at a time other than
during an eclipse.

The spectrum of the chromosphere is known as the *flash spectrum*
because of the brief time during which it is visible during an eclipse.
During an oncoming eclipse the absorption spectrum will be visible until
the solar disk is covered. At that instant the absorption spectrum is
replaced by the emission spectrum or flash spectrum. Most of the dark
lines seen in the absorption spectrum now become bright lines, and a few
additional emission lines not seen in the absorption spectrum become
visible. Because of the increasing temperatures in the upper chromosphere, most of these lines disappear as the altitude increases.

The lines that first become visible in the flash spectrum are primarily
helium lines, which do not become visible except in the high temperature
of the chromosphere. Neutral helium will be revealed in the spectrum
when temperatures of 10,000°K are reached, and ionized helium at
temperatures at least 20,000°K. Since helium lines are present in the flash
spectrum, temperatures at these levels are assumed in the chromosphere.
Hydrogen, still the most abundant element, produces the strongest and
brightest lines in the flash spectrum, particularly at a wavelength of 6563
Å. The red color of this wavelength produces the characteristic red color of
the chromosphere.

Above the faculae in the photosphere (section 12.3) are the *plages* in the
chromosphere. These are bright, active zones of higher temperature and
density that emit more light than the gases in the surrounding region. Like
the faculae, the existence of plages is attributed to the presence of
magnetic fields.

Filaments are long, dark streaks which are visible in photographs of the
chromosphere. Unlike the plages, filaments are regions of cooler gas and
therefore absorb radiation, causing them to seem dark. When filaments
occur on the limb of the sun they appear as prominences. *Prominences* are
luminous gas streamers, extending from the chromosphere tens of thousands of kilometres into the corona. Temperatures of the gases making up
the prominences are far lower than coronal temperatures. This, along with
the higher densities that exist in the prominences, causes the atoms of gas
to capture electrons and emit photons, giving rise to the luminous
appearance of the prominences.

Two basic types of prominences are recognized, quiescent prominences and active prominences (Figure 12.5). Material in the quiescent
prominences is relatively stable and appears to stream down from the
corona into the chromosphere, an action which may continue for several
weeks. Active prominences, of which loop prominences are probably the
most familar, last for several hours and seem to be associated with solar
flares. Prominence activity is thought to be in some way influenced by
magnetic fields.

Figure 12.5 Large promi-
nence 212,000 kilometres
high. Photographed in
violet light of calcium
August 18, 1947.
(Courtesy of Hale
Observatories)

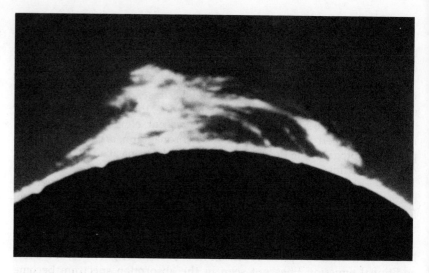

Figure 12.5 Large prominence 212,000 kilometres high. Photographed in violet light of calcium August 18, 1947. (Courtesy of Hale Observatories)

Solar flares are energetic eruptions that appear as bright spots in the chromosphere and apparently are related to sunspots as well as other solar activity. The duration of a flare ranges from a few minutes to several hours, during which it rises to maximum intensity very quickly and then fades relatively slowly. Flares liberate huge amounts of energy in a broad range of frequencies. These eruptions produce definite effects on the Earth, such as ionospheric disturbances, magnetic storms, interruptions in radio communication, unusual auroral displays, and a lowering of the average cosmic ray intensity. Giant solar flares are a hazard to manned space exploration in that, if unprotected, astronauts may be subject to lethal doses of radiation. Thus a reliable system of forecasting major solar flares is crucial to the safety of humans in space.

On the upper part of the chromosphere are many spikelike structures rising vertically or near-vertically through it. These features, called *spicules,* can best be seen when viewed near the limb of the sun in the light of hydrogen (Figure 12.6). The spicules, originating in the lower and middle chromosphere, may extend upward as much as 10^4 kilometres. The jets of gas move upward with speeds of 20 to 25 km/sec and last for only 10 to 15 minutes. The exact cause of the spicules is not known but their action indicates that they are in some way related to magnetic fields.

12.5 THE CORONA

The *corona,* the upper portion of the solar atmosphere, extends for an indefinite distance beyond the chromosphere. The pale, white light from the corona is only visible during a total eclipse or by means of a coronagraph, and the light intensity is only about half that of a full moon.

The strong emission lines seen in the spectrum of the chromosphere gradually disappear, to be replaced by the continuous spectrum of the inner corona. This portion of the corona, called the K corona (Ger.: Kontinuum), extends out about half the diameter of the sun from the photosphere and is the part most readily seen from the Earth during an eclipse. The K corona is made visible because light from the photosphere, colliding with electrons in the gas surrounding the sun, is redirected toward the Earth. Further from the sun the K corona is less dominant and here the sunlight is scattered by interplanetary dust. Examination of this outer portion of the corona reveals the dark absorption lines characteristic of the Fraunhofer lines and it is therefore called the F corona. Both these components of the corona extend far out into the solar system, even beyond the Earth. At these distances the solar gases are at extremely low densities. Near the Earth the corona may be detected as the *zodiacal light* seen on a dark, clear night just before sunrise or just after sunset. The zodiacal light is light reflected from dust concentrated in the ecliptic plane.

Emission lines have been detected in the solar corona, but these are relatively faint. When first discovered the lines could not be related to any known element, so the substance was thought to be a rare gas, which was called *coronium*. In 1942, it was discovered that the emission lines actually represented familiar elements in an unfamiliar form; that is, the atoms of the elements, particularly iron, were lacking a far greater than normal number of electrons and were in a highly ionized state. Iron had lost 13 out of the normal complement of 26 electrons, and this had resulted in *forbidden lines* in the spectrum—lines not normally seen in the laboratory.

The astronomer is able to deduce several things from the presence of forbidden lines in the sun's corona. First, the fact that the atoms of iron (also calcium and nickel) are so highly ionized indicates that temperatures in the corona must be at least $10^6°K$. This is far higher than temperatures in the photosphere, and the exact source of this energy is not known. One possible source may be shock waves produced by the turbulence of the gases at the sun's boiling surface. These shock waves are described as being similar to sound waves, and it is thought that the rapid succession of large numbers of waves would raise coronal temperatures. In addition to high temperatures, it is also possible to determine from the forbidden lines that gas density of the corona is extremely low.

Intense magnetic fields on the solar surface may also contribute to the high temperatures of the corona. The magnetic fields are thought to influence the shape of the corona and cause it to constantly change. This is particularly noticeable between periods of maximum and minimum

Figure 12.6 Spicules seen near the limb of the sun. (Courtesy of Hale Observatories)

Figure 12.7 Variation of solar corona with sunspot period showing solar corona during sunspot minimum (top) and sunspot maximum (bottom).

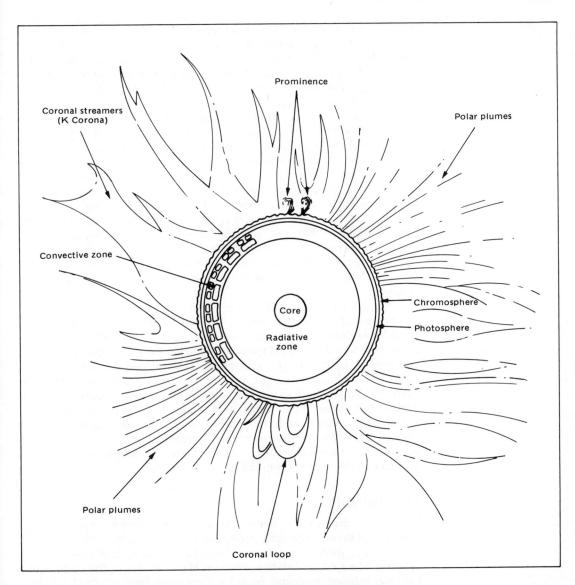

Figure 12.8 At sunspot minimum coronal streamers may be seen near the solar equator and polar plumes are seen at the poles.

sunspot activity (see section 12.7). During maximum sunspot activity the corona appears uniformly distributed around the sun (Figure 12.7), and the corona is very bright. At sunspot minimum the corona extends out quite a bit farther at the sun's equator than at the polar regions. At that time *coronal streamers* are very evident near the solar equator, and *polar plumes* become quite distinct at the poles (Figure 12.8).

12.6 SOLAR WINDS

In addition to all the forms of radiation emanating from the sun, there is also a constant flow of particles from the sun. These particles are a neutral aggregation of protons and electrons that make up what is called the *solar winds.* This flowing matter is sometimes called a *plasma* and is defined as a collection of charged particles in sufficient bulk so that its behavior is much like a fluid. The particles move outward radially from the sun, but because of the sun's rotation the movement of the plasma resembles the outward spiraling of water from a lawn sprinkler. The exact source of the plasma is not known, but the best guess is that it originates in the corona, and it is viewed as a continuous expansion of the solar atmosphere.

Satellite exploration in the vicinity of Earth has made it possible to determine characteristics of the solar winds at one AU from the sun. At this distance the wind has a velocity of approximately 400 to 500 km/sec., a flux (number of particles flowing through a given area in a given time) of about 5×10^8 protons per square centimetre per second, and a density of 10 protons/cm³. Calculations made on the basis of the strength of the solar winds at one AU indicate that it may extend at least to a distance of 50 AU. This would mean that the planets are orbiting within the physical body of the sun, if we include the solar winds as an integral part of the sun. These findings are restricted to the region close to the ecliptic. Little has yet been done to determine the properties of space well above and below the ecliptic.

Charged particles of the solar winds interact with the Earth's atmosphere but, fortunately for us, do not penetrate to the surface. The particles are influenced by the Earth's magnetic field, which causes the particles to spiral in toward the poles, ionizing the atoms and molecules of atmospheric gases. The recombination of the ionized atoms results in the emission of photons of light, causing the auroral displays seen in the northern and southern skies. The solar winds are also responsible for the disturbance of the upper atmosphere (ionosphere), especially during the eruption of a solar flare. In this event radio communication may fade out because radio waves that are normally reflected by the ionosphere pass through the ionosphere. This results from the interaction of ionized particles in the atmosphere and the solar winds.

12.7 SUNSPOTS

Sunspots on the sun have been of interest to the astronomer since Galileo first viewed them in the early seventeenth century. In the latter part of the eighteenth century, sunspots were thought to be holes in the sun's surface opening into a cooler interior. William Herschel was of the opinion that the solar interior, protected from the fiery photosphere by a cloud layer, may even be inhabited.

Sunspots appear as dark regions on the photosphere, composed of a dark center, the *umbra,* and a lighter edge, the *penumbra* (Figure 12.9). A sunspot is not truly black, but is dark by comparison to the surrounding area because the sunspot is cooler. Sunspot temperatures are about 1000°K to 1500°K lower than the normal temperature of the sun (Figures 12.9 and 12.10).

Sunspots vary in size and life span, most being relatively small—1000 kilometres in diameter and lasting only a few days—while a few may reach a diameter of many thousands of kilometres and will last for several weeks to several months. A record spot viewed in 1946 measured 96,000 by 145,000 kilometres, which is huge when compared with the Earth's diameter of slightly over 12,000 kilometres. Sunspots generally form in groups, with two spots growing larger than the rest. The major spots are

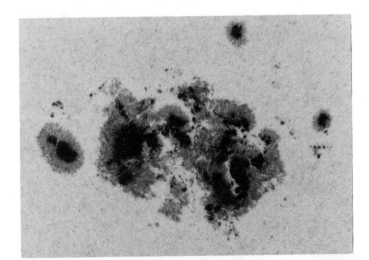

Figure 12.9 Large sun-spot group of May 17, 1951.(Courtesy of Hale Observatories)

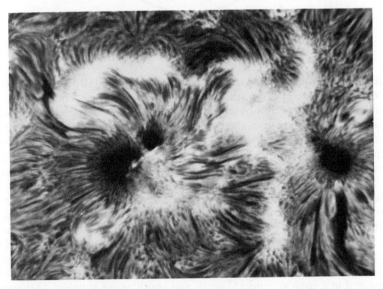

Figure 12.10 Bipolar sun-spot group photographed in red light of hydrogen alpha at Big Bear Solar Observatory, May 21, 1972. (Courtesy of Hale Observatories)

usually latitudinally oriented—that is, one following the other—and are associated with strong magnetic fields. One spot will have the opposite polarity of the following spot, much as if a horseshoe magnet were buried in the sun (Figure 12.11). The smaller sunspots will usually take on the polarity of the closest major spot. Although the exact cause of sunspot formation is not known, the reason for a sunspot remaining relatively cool during its existence is related to the magnetic fields. Strong magnetic fields tend to exclude charged particles, and it is the charged particles that carry heat from the sun's interior to the surface. The exclusion of heat as a result of the deflection of charged particles from a sunspot region causes the area to remain cool during the sunspot's life.

In the early part of the nineteenth century, Heinrich Schwabe, a German druggist with a great interest in astronomy, began keeping a systematic watch of the sun's surface. He was looking for the planet Vulcan, predicted by the French astronomer Urbain Leverrier to be in orbit between Mercury and the sun. The planet was never found, but

Figure 12.11 The sunspot
magnetic field appears as
if a horseshoe magnet
were buried in the sun.

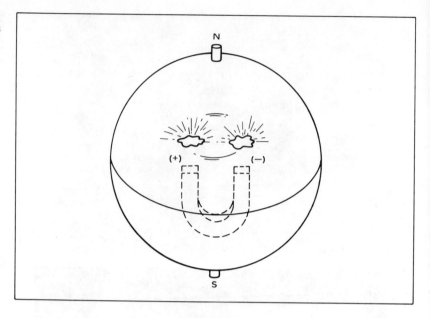

Schwabe's detailed record of the solar surface eventually led him to the discovery of the cyclic nature of sunspot occurrence. He found that sunspot activity varied from minimum to maximum and back to minimum in a cycle that averaged slightly over 11 years. This is an average period, since the length of the cycle has been found to vary from time to time by as much as several years either way. The first sunspot of the most recent cycle was observed in August, 1973. This means the next maximum could occur in 1978–79 and the cycle end in 1984. Schwabe announced his findings in 1843, but worked another 10 to 15 years before his efforts were recognized. Later it was found that sunspots varied in position on the solar surface as the cycle progressed. As the new cycle started at minimum, spots would appear 30° to 35° north and south of the equator. As the cycle progressed, new spots, larger and more numerous than the earlier ones, appeared closer to the equator, until at maximum activity most sunspots appeared in a zone around 15° north and south of the equator. At the end of the cycle, a few small spots appeared around 5° north and south of the equator, while at the same time new spots appeared 30° from the equator. The reappearance of sunspots at the high latitudes signals the start of a new cycle. There is no known reason for the cycle, nor is there an explanation for the distribution of sunspots during the cycle. In some way magnetic fields seem to influence sunspot activity.

   In the early part of the twentieth century, George Hale of the Mount Wilson Observatory investigated the properties of sunspots and discovered that the sunspots had a magnetic field. Hale made use of a discovery by Pieter Zeeman in 1896. Zeeman had found that the spectral lines of glowing sodium vapor broadened when a tube of the glowing vapor was placed between the poles of an electromagnet. The normally single spectral lines became double or triple under the influence of the magnetic field. This became known as the *Zeeman effect.* Hale noticed this effect in light from sunspots. By means of a polarizing attachment on the spectroscope, Hale was able to show that the leading spots of a sunspot group had one polarity, and the trailing spots had the opposite polarity. He was also able to observe that sunspot groups in the northern and southern hemispheres had reversed polarities (Figure 12.12). With the start of a new minimum cycle, the polarities of the two hemispheres were reversed, thus indicating that the point of minimum sunspot activity was the real start of each cycle. If the polarity of sunspots were considered along with sunspot activity, then the length of the cycle would be about 22 years.

12.8 SOLAR ENERGY

Energy released by the sun in the form of light and heat is the principal source of energy here on Earth. Ancient people probably thought that the heat and light from the sun was the same as that derived from their fire. Fire releases energy through the breakdown of organic compounds in the presence of oxygen, with carbon dioxide, some ash, and heat and light as byproducts of the process. Oxygen and carbon are detected on the sun in very small amounts, but solar temperatures are far too high to permit these substances to combine into molecules. Carbon is a factor in the release of energy in very large stars but not in the manner just described. Complex molecules cannot exist on the sun because of the high temperatures; therefore the sun is composed of atoms and subatomic particles.

The atom, or more precisely, the binding energy of the atomic nucleus, is the real source of energy from the sun. The release of this energy may be accomplished in two possible ways: *nuclear fission* or *nuclear fusion*. The nuclear fission process, which is the way energy is released in the atomic bomb, is not likely to be the means by which energy is released in the sun. In the fission process the atom of a heavy element, like uranium, is split into two lighter elements. The combined mass of the new products is slightly less than the mass of the original element, and the loss represents the mass that has been converted into energy. Heavy elements, such as uranium, do not exist in the sun in sufficient quantities for this type of

Figure 12.12 The polarity of sunspot magnetic fields are reversed from one cycle to the next.

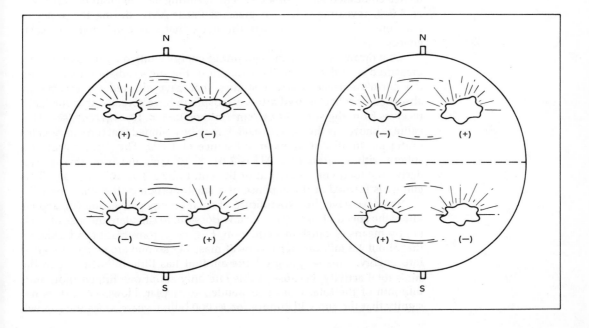

reaction to take place on a sustained basis. The fusion process requires the presence of a light element such as hydrogen, which is present in huge quantities on the sun. This makes nuclear fusion the likely mechanism responsible for the generation of the tremendous energy output of the sun.

The nuclear fusion process functions through a series of reactions called the *proton-proton reaction*, in which hydrogen nuclei are combined to form helium with an attendant release of energy. This is the same type of reaction that takes place in a hydrogen bomb. In the initial stage of the reaction two protons (hydrogen nuclei) are fused into deuterium, or heavy hydrogen. This is accompanied by the emission of a positron and a neutrino, as shown in the following reaction.

$$^1_1H + ^1_1H \longrightarrow ^2_1H + ^+_1e + \nu$$

Following this, the deuterium encounters and combines with a proton to form a light helium isotope. The positron reacts with an electron to form a gamma ray:

$$^2_1H + ^1_1H \longrightarrow ^3_2He + \alpha$$

A stable helium atom and two protons are formed by the reaction of two light helium isotopes.

$$^3_2He + ^3_2He \longrightarrow ^4_2He + 2^1_1H$$

In the entire series of reactions, four hydrogen atoms (or protons) were required to make one stable isotope of helium. Each of the hydrogen atoms has an atomic mass unit (AMU) of 1.008, so four hydrogen atoms have a combined AMU of 4.032. The resulting helium atom has an AMU of 4.003, revealing a loss in mass of 0.029 AMU during the series of reactions. This loss represents the mass that was converted into solar energy.

One means whereby the amount of energy emitted by the sun may be measured is to determine the strength of the sun's radiation received on the Earth's surface. Careful studies that have allowed for the energy absorbed or reflected by Earth's atmosphere have revealed that the Earth receives from the sun 1.94 cal/cm^2/min or 1.35×10^6 ergs/cm^2/sec. This value is known as the *solar constant*. It may be assumed that the sun radiates uniformly in all directions at a distance of 1 AU. The surface area of a sphere with a radius of 1 AU is 2.82×10^{27}cm^2, and from this we can derive the total energy output of the sun to be 3.8×10^{33} ergs/sec. This energy is formed at the expense of a small portion of the sun's mass.

Since we have the expenditure of solar energy per second (E), and we know the speed of light (c), it is now possible to calculate the loss of mass (m) by means of Einstein's equation, $E = mc^2$. From this it can be shown that about 4.3 million metric tons of mass per second are being converted into energy. This seemingly large amount has little influence upon the sun's total activity, because at this rate only about one fifteen-thousand-billionth of the total mass is expended each year. However, it does not signify that the sun will endure for 15,000 billion years, since the sun will not continue to lose energy in this manner to complete annihilation. The amount of mass lost in the reaction represents only about 0.7 percent of the total mass involved in the hydrogen-to-helium reaction. We could therefore expect the sun to have a total life expectancy of about 100 billion years if all the hydrogen were converted to helium. Even this figure is too great, for it is known that the sun and all stars (see chapter 14) become unstable when about 10 percent of the available hydrogen is used up. Thus a more realistic life expectancy for the sun would be 10 to 11 billion years. Since the age of the sun is now estimated to be 5 to 6 billion years, it is reasonable to expect, from present theory, that the sun has 4 to 5 billion years of life remaining.

12.9 USE OF SOLAR ENERGY

The sun is currently a very popular object for study, not only among astronomers, but also with scientists and environmentalists interested in tapping the sun as a source of energy. Only plants, through the photosynthetic process, make direct use of solar energy. There are a few practical uses to which people have put solar energy, for example, to evaporate sea water for the salt content or to obtain fresh water.

The increasing cost of fuel and the environmental costs of extracting oil from shale or coal from the ground make the use of continuously radiated energy from the sun an increasingly attractive alternative. Solar energy is attractive for several reasons. First, the source is constant and continuous and basically comes to us at no cost. A second advantage is that it is essentially nonpolluting. Since solar energy comes to us as heat, no heat is added to the total environment as is the case when fossil fuels are burned. Also solar energy does not add unwanted and undesirable substances to the environment. At the same time, we have to acknowledge that there are disadvantages. Energy from sunlight is diffuse and collecting it into usable quantities poses some difficult problems. In addition, not all parts of the world receive sufficient sunlight to make development of solar power practical. In the northwestern United States, for example, there may only be enough sunlight to operate auxiliary units that would provide heating and cooling during peak load periods. However, even this would be of benefit in large metropolitan areas.

Various schemes for utilizing solar energy have been proposed, but most rely upon some form of storage system because of the variation in the incidence of sunlight. The energy must be collected and stored for use at night or during cloudy periods. For this purpose, sunlight needs to be focused on heat-exchanging equipment containing a fluid, possibly molten salts, that can be heated up to 500° to 600°C. The heat thus stored would be used to generate steam to drive turbines. According to Aden and Marjorie Meinel of the University of Arizona, a system of solar collectors one square mile in area and a 300,000 gallon thermal storage tank would be required for a solar power plant capable of generating 1000 megawatts.

Another method that offers some promise is the large-scale generation of electric power by means of the photovoltaic effect. The *photovoltaic effect* depends on the ability of certain crystals to generate a voltage when light strikes them. The solar cells used in the space program to generate electricity for a variety of space satellites are photovoltaic cells. Unfortunately, the cells are at present extremely expensive to build. If the units can be inexpensively mass-produced, large-scale solar panels could be built to provide power in sunny areas such as the southwestern United States. Large storage batteries would be required to provide a steady flow of power at night and on cloudy days (Figure 12.13).

To overcome the need for storage batteries, a collecting array of photovoltaic cells could be placed above the Earth in synchronous orbit. The panels would be connected to a microwave relay antenna, which would convert power to microwaves and beam them to a receiving station on Earth. In this way areas not receiving a great deal of continuous sunlight could be supplied with uninterrupted power. Only during the equinoxes would the solar panels pass briefly into the Earth's shadow.

Despite the encouraging outlook for substituting solar energy for fossil fuel energy, there are many technical problems to be overcome. Even the most optimistic proponents of solar energy say that solar energy can only supply up to 20 percent of the world's energy needs, but this is still promising. In this case time, not money, is of the essence.

12.10 SUMMARY

The sun, our nearest star, is the primary source of energy for the solar system and controls the motions of all objects in the solar system. It is neither the largest nor the smallest observable star but appears to be average.

The sun is far more massive than any of the other components of the solar system; for example, it is 332,000 times more massive than the Earth. The composition of the sun is mainly hydrogen and helium, with the heavy elements only a few percent of the mass. The bulk of the sun's core is helium, formed by the transmutation of hydrogen to helium. Temperatures of $1.5 \times 10^{7}°K$ and pressures of 150 g/cm^3 occur at the core, decreasing toward the surface which is 5800°K.

Energy is transmitted by means of photons through the radiative zone up through the convective zone to the surface or photosphere. The upward movement of energy results in the surface appearing granular, much like the surface of a slowly simmering soup. The photosphere is a zone about 500 kilometres above the granular surface, where temperatures range

Figure 12.13 Methods whereby solar energy may be used to generate electric power. (a) Solar cell panels in orbit beam power to Earth. (b) An array of photovoltaic cells will generate electricity when light strikes them.

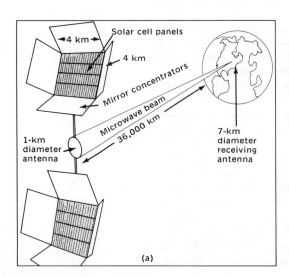

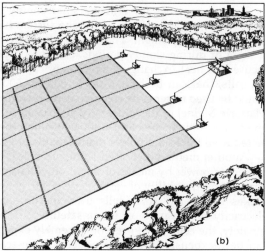

(a)

(b)

from 8000°K at the base to 4000°K at the top. Here, at the beginning of the chromosphere, temperatures increase again.

The chromosphere, recognized as the lower portion of the solar atmosphere, can only be observed during a total eclipse or by using a coronagraph. Helium was first discovered in the spectrum of the chromosphere, which is a bright line emission spectrum. Because of the preponderance of hydrogen emitting energy at a wavelength of 6563 Å, the chromosphere appears red. Dark streaks in the chromosphere are regions of cooler gases. These appear as prominences when viewed on the solar limb.

Beyond the chromosphere is the upper portion of the solar atmosphere or corona. The corona is composed of several parts. The K corona is generally seen during total eclipses, and beyond that is the F corona. Coronal temperatures reach 1,000,000°K, a fact that has been revealed by the presence of forbidden lines in the spectrum. The forbidden lines represent spectral lines not normally seen in the laboratory and are of known elements that are highly ionized. This can only occur at extremely high temperatures.

There is also a continuous flow of particles emanating from the sun. This is known as the solar wind, which may extend out to at least 50 AU in the solar system.

Sunspots, first viewed by Galileo, are still intensely studied. Sunspots occur in cycles of about 11 years from minimum through maximum to minimum activity again. Sunspots vary in size from about 1000 kilometres to many tens of thousands of kilometres in diameter. They last from a few days to several months. Sunspots are associated with strong magnetic fields. The polarity of the sunspots reverses with the beginning of each new cycle.

Energy is released by the sun as a result of the conversion of a small amount of mass during the formation of helium from hydrogen. The reaction, known as the proton-proton reaction, is similar to that which takes place in the nuclear fusion process of a hydrogen bomb. About 4.6 million tons of mass per second is being converted to energy by this process. Despite this the sun will continue its life for another 4 to 5 billion years.

Means to convert some of this energy are being sought to reduce our reliance on dwindling supplies of fossil fuels. One method is to collect the solar heat to generate steam to turn turbines. Another method generates electricity directly by means of the photovoltaic effect. This method, used in the space program, depends upon the ability of certain crystals to generate a voltage. Many problems still need to be solved before either of these methods becomes practical.

QUESTIONS

1. How does the sun compare in size and density with the Earth; with Jupiter?
2. What would a cubic centimetre of the sun's core weigh on Earth? How does this compare with material at the core of the Earth?
3. Why does the sun's photosphere have a granular appearance?

4. Why does the center of the photosphere appear brighter than the edge?
5. Explain how the Fraunhofer lines are formed in the spectrum of the photosphere.
6. Why does helium appear in the spectrum of the chromosphere but not in the spectrum of the photosphere?
7. What is meant by the "flash spectrum?"
8. What type of spectrum do we receive from the photosphere; the chromosphere; the corona?
9. Briefly describe the properties of the solar winds.
10. Describe the changes that occur with regard to sunspots during the course of a sunspot cycle. How does the polarity of the sunspots vary during the cycle?
11. Through what steps is hydrogen converted into helium in the sun?

EXERCISES AND PROJECTS

1. Make a report on how solar energy may be collected for use in heating and cooling an average-sized house. Pay particular attention to the construction of apparatus required for collecting the energy.
2. If a small telescope is available, plot the movement of a sunspot group moving across the sun's surface. *Warning:* Never look at the sun directly through a telescope or any other means. Allow the sun's image to fall on a white surface on which the image of the sunspots may be traced from day to day.

FOR FURTHER READING

BRANDT, J. C., *The Sun and the Stars.* New York: McGraw Hill, 1968.

GAMOW, G., *The Birth and Death of the Sun.* New York: New American Library, 1952.

GINGERICH, O., *New Frontiers in Astronomy.* San Francisco: W. H. Freeman and Company, 1975.

McCREA, W. H., *Physics of the Sun and Stars.* London: Hutchinson, 1950.

MENZEL, D. H., *Our Sun.* Cambridge, Mass.: Harvard University Press, 1959.

Canis Major

CHAPTER 13 **THE STARS**

Scintillate, scintillate, globule vivific,
Fain would I fathom thy nature specific,
Loftily poised in ether capacious,
Strongly resembling a gem carbanaceous.

—Anonymous

People have been viewing the sky for thousands of years, and all of the
ways we have explained the universe to ourselves have been based on
physical observations. Ancient people thought of the sky as an inverted
bowl and the stars as lights placed on it; others thought the bowl was
pierced with holes and the stars were light shining through from heaven.
Very little was known about the real physical properties of the stars prior
to 1838 because the true distances to stars was not known. When an
astronomer looked at Polaris, the North Star, he was unable to tell if it was
a relatively dim star nearby or a very bright star that happened to be far
away. Prior to the nineteenth century scientists had been able to separate
light into its component colors, but the true significance of the spectrum
was not fully understood. In 1838, the first stellar distance was measured.
Later in the nineteenth century the secrets of the spectrum were discovered and the true nature of starlight began to be revealed. Astronomers
soon recognized that the light coming from stars was their only source of
information on stellar properties.

13.1 STELLAR DISTANCE

The ancient Greeks understood the principle of parallax, which has as a
basis the process of triangulation used by surveyors in determining the
distance to an inaccessible point. The Greeks were unable to demonstrate
parallactic displacement of stars and used this as an argument against the
heliocentric concept. Triangulation is a relatively simple procedure. The
distance between two points is measured to form a baseline. A transit is set
over one point and sighted toward the inaccessible point. The angle
formed between the baseline and the line of sight is measured. By
repeating this procedure at the other end of the baseline, a triangle is
established in which the baseline and two including angles are known.
The distance to the inaccessible point may then be calculated. Essentially

285

Figure 13.1 The parallactic ellipse of a star is the projection of the Earth's orbital ellipse through the star's position onto the celestial sphere. Points *A* and *B* on the Earth's orbit will correspond to Points *A* and *B* on the celestial sphere. The size of the projected ellipse will be inversely proportional to the star's distance from Earth. The ellipse will suffer progressive foreshortening from the ecliptic pole to the plane of the ecliptic. The parallactic ellipse as seen projected from the plane of the ecliptic (not shown in diagram) would be essentially flat.

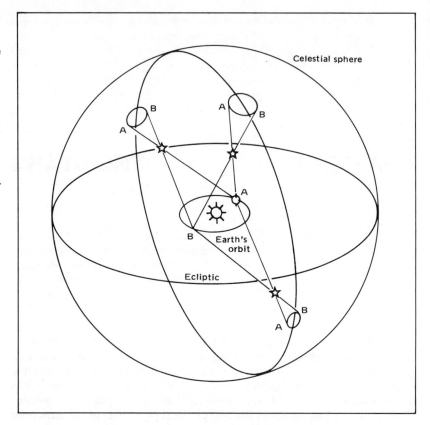

the same procedure is used to measure the distance to a star. A baseline is established, which in this case corresponds to the radius of the Earth's orbit. The displacement of the star is noted against a background of more distant stars over a six-month interval when the Earth is at opposite points of the orbit (Figure 13.1). The parallactic angles measured by this technique are extremely small, even for the nearest stars, and are less than 1.0 second of arc. Thus we are dealing with what might be termed an extremely "skinny triangle"—a right triangle with one of the angles at less than 1."0 (one second) of arc and the other angle at almost 90°.

There are some problems related to measuring such small angles, since the recordings are made six months apart. During this interval the solar system as well as the star is moving through space, and it is therefore necessary that a number of measurements be made over several years to reduce error introduced by this motion. Because of the very small size of the angle, the refraction of light as it passes through the atmosphere is a factor that needs to be considered, as well as the peculiarities of the equipment being used. In 1838, the German astronomer F. W. Bessel made very accurate measurements of the parallactic displacement of the star 61 Cygni. He chose this star because it moved through space at an angular distance of 5" of arc per year, which indicated to him that 61 Cygni was quite close to the Earth. Bessel took into consideration the possible sources of error cited above and was able to measure the angular parallax at 0."292 of arc, which translated into a distance of 696,000 AU or about 11 light years.

Another unit of measure used to indicate stellar distance is the *parsec* (pc). The parsec, an abbreviation for parallax at one second of arc, represents the distance to a point in space where the angle of parallax equals 1.ʺ0 of arc. The distance of one parsec equals 206,265 AU or 3.26 light years. This would make the distance to Alpha Centauri, with an angle of parallax 0.ʺ761 of arc, 1.31 pc and the distance to 61 Cygni, 3.4 pc. The relationship between distance and parallax may be given by the following equation:

$$d_{pc} = \frac{1}{p(\text{seconds of arc})} \qquad (13.1)$$

where d is the distance in parsecs and p is parallax in seconds of arc.

The use of the Earth's orbital radius as a means of measuring angular parallax is tedious and time-consuming. Because of this, in the years between Bessel's work and the end of the century, stellar distances were determined for fewer than 100 stars. In the early 1900s, photography was introduced as an astronomical technique and greatly speeded up the process. Up to the present this technique has permitted the direct measurement of stellar distance of several thousand stars. It must be recognized that with increasing distance the reliability of the angular parallax measurement decreases. The angle, at best, is extremely small, thereby limiting the practical use of the method to stars that are relatively close to the solar system. The error introduced as the distances become greater limits the parallax technique to about 100 pc. Beyond this distance indirect techniques for measuring distance are used.

In section 4.3 we discussed the relationship of stellar brightness to distance and the application of the inverse-squares law: the brightness of the light from a source is inversely proportional to the square of the distance from the source. In other words, if we double the distance to a star, we decrease the brightness of the star to one-fourth. Making use of this technique requires some knowledge of stellar properties, and later in this chapter we will show that stars may be grouped according to similar characteristics. If we have two stars with the same general physical characteristics (size, temperature), they will also be the same brightness. Thus the brightness of a nearby star, whose distance may be determined by parallax method, may be compared with the brightness of the more distant star. For example, a distant star found to be 1/16 as bright as the similar nearby star will be four times the distance from the observer.

13.2 STELLAR LUMINOSITY AND MAGNITUDE

While the relative brightnesses of stars is used to determine distance, knowledge of distance is necessary to determine a star's absolute brightness or luminosity. *Luminosity* is the total energy radiated by a star into space each second. Because the total luminosity of the sun is known (3.8 × 10^{33} ergs/sec), stellar luminosity is measured in submultiples and multiples of the sun's luminosity. Luminosity of a star may be found by comparing brightness of any star with the brightness of the sun if placed side by side at some standard distance. Since we know the luminosity of the sun, we may place it at a convenient standard distance of 1 pc and compare it with the apparent brightness of any star. Sirius, the brightest star seen in the winter sky, at its usual distance of 2.65 pc would have an apparent brightness (b) of 3.6 times that of the sun at the specified standard distance. If we decrease Sirius' distance to the standard distance of 1 pc, its apparent brightness would increase d^2 in keeping with the inverse-squares law. Its luminosity (L) would then be expressed as the following relationship:

$$L = b \times d^2$$

where L is the star's luminosity in terms of solar luminosity, b is in units of solar brightness and d is the distance in parsecs. In the previous section, d was stated as $1/p$, and by substituting for d^2 we have

$$L = \frac{b}{p^2} \tag{13.2}$$

If we apply the above equation to Sirius, whose apparent brightness is 3.6 times that of the sun and whose parallax is 0.″371, we find that

$$L = \frac{3.6}{(0.371)^2} = 26.2$$

This means that Sirius is radiating 26.2 times more energy per second than is the sun.

Stars vary enormously in absolute luminosity, ranging from 10^6 times as luminous as the sun to as little as 2×10^{-6} the luminosity of the sun. The latter would be about as bright as the full moon. Barnard's star (so named after E. E. Barnard, its discoverer) has a luminosity 1/100 that of the sun, and a planet 1 AU from that star would experience a temperature of about $-130°C$.

Astronomers make use of a unit called *magnitude* to express the relative brightnesses of stars. Hipparchus, in the second century B.C., developed a catalogue of stars in which the brightest stars were assigned to the first magnitude and those stars just visible to the naked eye to the sixth magnitude. Other stars were assigned to intermediate values for magnitude. The invention of the telescope revealed the presence of many stars not formerly seen and new numbers were added to the system, with the progressively larger numbers indicating fainter and fainter stars.

TABLE 13.1   RELATIONSHIP OF DIFFERENCE IN MAGNITUDE TO BRIGHTNESS

Magnitude Difference	Brightness Ratio
0	1
1	2.512
2	6.3
3	16
4	40
5	100
7.5	10^3
10	10^4
15	10^6
20	10^8
25	10^{10}
30	10^{12}

The system is still useful, but astronomers now use a *photometer* to more accurately measure brightness. In this way it was discovered that a star of the first magnitude was about 100 times brighter than a star of the sixth magnitude; this relationship has now been adopted as a standard. The brightness ratio of two stars with a difference of five magnitudes will be 100:1. This means that a star of the fourth magnitude is 100 times brighter than a star of the ninth magnitude. The brightness ratio between two stars differing by one magnitude is considered the fifth root of 100, or 2.512. In other words, a star is 2.512 times brighter than a star with the next higher numerical magnitude (Table 13.1). The foregoing tells us that an increase in the brightness of a star is not proportional to an increase in magnitude. In other words, doubling the brightness does not result in a doubling of the magnitude. If the magnitude of two stars is known, then the difference in brightness may be found by raising 2.512 to a power equal to the difference in their magnitudes. For example, if star A has a magnitude of 3 and star B has a magnitude of 6, then star A would be $(2.512)^3$ times brighter than star B.

What we have been considering here is the *apparent magnitude* of the star, which depends upon its luminosity and distance. A star may be very bright but because it is very distant it will appear faint to the observer. The sun, which is close to us, appears very bright and, based on apparent magnitude, is the brightest object in the sky. It appears much brighter than first-magnitude stars, therefore bringing about the need for negative magnitude values. The sun has an apparent magnitude of –26.7 and the full moon, seen as the second brightest object in the sky, has an apparent magnitude of –12.7. Next in this scale is Venus which at maximum brightness has an apparent magnitude of –4.2. The brightest star in the sky, Sirius, has an apparent magnitude of –1.4.

In discussing luminosity, we stated that Sirius was about 3.6 times brighter than the sun if the sun were 1 pc from the observer. At 1 pc the sun would have an apparent magnitude of about 0.0. Sirius, with an apparent magnitude of –1.4, would be $(2.512)^{1.4}$ or 3.6 times brighter than the sun under those circumstances.

It is often desirable or necessary to compare brightnesses of several stars on an equivalent basis. For example, the sun appears much brighter than Sirius, although if both were compared at the same distance Sirius ($26.2L_\odot$) would be far brighter. For this reason the concept of absolute magnitude has been introduced. *Absolute magnitude* represents the magnitude of a star if it were located at a standard distance of 10 pc. This means that stars at a greater distance than 10 pc would have a lower numerical value for absolute magnitude than for apparent magnitude, and stars closer than 10 pc would have a higher value for absolute than for apparent magnitude (Table 13.2). The sun, which has an apparent magnitude of –26.7, has an absolute magnitude of 4.9 and would therefore appear quite faint at 10 pc.

If the distance and the apparent magnitude are known, it is possible to determine absolute magnitude using the following equations:

$$M = m + 5 + 5\log p \qquad\qquad (13.3)$$

or

$$M = m + 5 - 5\log d \qquad\qquad (13.4)$$

where M and m are absolute and apparent magnitude, respectively, p is parallax in seconds of arc, and d is distance in parsecs.

TABLE 13.2　　Some Properties of Nearby Prominent Stars

Star	Distance Parsecs	Apparent Magnitude (mv)	Absolute Magnitude (Mv)	Spectral Class
Sun	—	−26.7	4.9	G2V
*Alpha Centauri	1.3	− 0.01	4.4	G2V
†Barnard's Star	1.8	9.54	13.2	M5V
*Sirius	2.7	− 1.46	1.4	A1V
†Epsilon Eridani	3.3	3.73	6.1	K2V
†61 Cygni	3.4	5.22	7.6	K5V
Epsilon Indi	3.4	4.68	7.0	K5V
*Procyon	3.5	0.37	2.6	F5IV
Tau Ceti	3.6	3.5	5.7	G8V
*40 Eridani	4.9	4.43	6.0	K1V
Altair	5.1	0.8	2.2	A7IV
Fomalhaut	6.9	1.2	2.0	A3V
Vega	8.0	0.04	0.5	A0V
Arcturus	11.0	− 0.06	−0.3	K2III
Pollux	12.0	1.2	1.0	K0III
*Castor	14.0	1.6	0.9	A1V
*Capella	14.0	0.05	−0.6	G2III
Aldebaran	16.0	0.9	−0.2	K5III
*Regulus	26.0	1.4	−0.6	B7V
Achernar	30.0	0.5	−1.0	B5V
Canopus	30.0	− 0.7	−3.1	F0I
*Spica	70.0	1.0	−3.6	B1V
Bellatrix	145.0	1.6	−3.6	B2III
Betelgeuse	150.0	0.4	−5.5	M2I
*Rigel	250.0	0.1	−6.8	B8I
Deneb	430.0	1.3	−6.9	A2I

*Multiple star system.
†Accompanied by dark companions.
The abbreviations mv and Mv stand for "magnitude visual." It is customary to use a lower case m for apparent magnitude and an upper case m for absolute magnitude.

TABLE 13.3　　Relationship of Color Index to Stellar Class, Color, and Temperature*

Color Index	Spectral Class	Color	Surface Temperature, °K
−0.6	O	Blue white	50,000
−0.3	B0	Blue	27,000
−0.16	B5	Blue	16,000
0.0	A0	White	10,400
0.15	A5	White	8200
0.30	F0	White	7200
0.45	F5	White	6700
0.60	G0	Yellow	6000
0.65	G5	Yellow	5500
0.80	K0	Orange	5100
1.15	K5	Orange	4300
1.40	M0	Red	3700
1.70	M5	Red	3000

*Values given are for stars on main sequence only.

Photographic magnitude (B) may be determined by recording a star's brightness on a blue-sensitive photographic plate. If a yellow-sensitive plate is used *visual magnitude* (V) is obtained which most closely resembles magnitude as seen with the human eye. The difference between photographic magnitude and visual magnitude, or B – V, provides a value called the *color index* (C.I.) of a star. The color index is a numerical expression of stellar color and temperature (Table 13.3).

13.3 STELLAR TEMPERATURES

Another intrinsic property of stars is temperature. The inaccessibility of the sun and stars requires that measurement of such properties as temperature be accomplished indirectly. In this case it is possible to determine temperature from the intensity of total radiation. An assumption is made that the sun or star is a perfect radiator—that is, it radiates energy with maximum theoretical efficiency. For such an object, called a black body, the temperature may be calculated using the *Stefan-Boltzmann law*. This law states that the energy emitted per square centimetre of surface area per second is proportional to the fourth power of the absolute temperature. This may be stated as

$$E = 5.67 \times 10^{-5} \, T^4 \tag{13.5}$$

where E is the energy in ergs/cm^2/sec and T is the temperature °K. We know that the sun radiates 6.23×10^{10} ergs/cm^2/sec and by rearranging the above equation we can obtain the temperature as follows:

$$T = \sqrt[4]{\frac{6.23 \times 10^{10}}{5.67 \times 10^{-5}}} = 5760°K \tag{13.6}$$

This is the effective temperature of the sun's surface.

Temperature may also be calculated by determining the amount of energy radiated at each wavelength. In section 4.2 we discussed the relationship between temperature and the energy emitted at each wavelength, and now we may view this relationship graphically. In Figure 13.2 the energy radiated at different wavelengths by the sun is shown to reach a peak at about 4700 Å. Also shown is the curve for a red star where radiation reaches a peak at approximately 9000 Å in the infrared region, and a star where peak radiation occurs at 3000 Å in the ultraviolet region.

In the late nineteenth century, William Wien was able to derive an equation for calculating temperatures of stars from spectroscopic data showing the wavelength at which radiation was most intense (see section 5.5, equation 5.5). Based on Wien's equation, the sun's temperature is about 6100°K. This temperature does not coincide exactly with the effective temperature of the sun, but the difference is within tolerable limits. The assumption that the sun or star is a perfect radiator is only approximately true, and this leads to some difference in the results between methods.

Temperature is also related to color. We recognize that iron when heated becomes a dull red and as it is heated further becomes white hot. The same principle applies to stars. Red stars are relatively cool compared to yellow stars like the sun or white stars such as Sirius or blue-white stars like Rigel. The colors of many stars have been determined, and therefore it

Figure 13.2 Energy radiated by Barnard's star (a) reaches a peak at 9600 Å; for the sun (b) 4700 Å; and for Sirius (c) at 2900 Å.

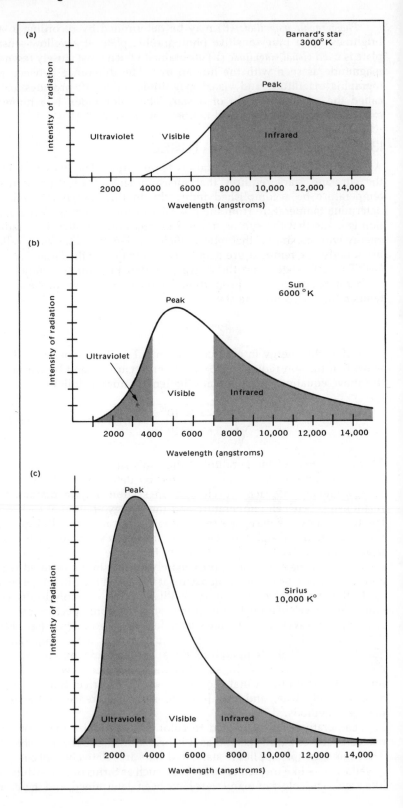

is possible to approximate the surface temperatures of these stars. A more precise indication may be obtained from the color index calculated from the difference between the photographic magnitude and the visual magnitude (see section 13.2). Each color index number is related to a specific spectral class of stars, which in turn is associated with a particular temperature. The color index is the logarithm of the ratio of photographic to visual magnitude, which has been adjusted to be equal for stars identified as type AO (see section 13.7 and Table 13.3).

13.4 STELLAR DIAMETER

Measuring the diameter of planets in the solar system or the sun is possible because these objects are sufficiently close to be observed as disks. Stars, for the most part, are too far away and appear as pinpoints of light, even in large telescopes. Astronomers have developed several ingenious methods to overcome this problem, and therefore sizes of many stars are now known. The first direct determination of stellar diameter was made in 1920 by A. A. Michaelson with a device called an interferometer mounted on the 100-inch telescope on Mount Wilson. With this instrument he and F. G. Pease found the angular diameter of Betelgeuse to vary from 0.″034 to 0.″042. Betelgeuse is an irregular variable star (see section 14.4) that varies 1 magnitude in 5 to 6 years. Its diameter varies from 500 to 800 solar diameters, and if centered at the sun, Betelgeuse's surface would extend out beyond the orbit of Mars. Only a few stars are large enough to permit direct measurement by this technique. Table 13.4 lists the diameters of stars measured by the interferometer techniques described above and in the following paragraph.

An interferometer technique utilizing two telescopes separated by 200 metres was devised recently in Narrabri, Australia. It was successful in measuring angular diameters as small as 0.″005, and has enabled astronomers to determine the diameters of a few more stars. By using this

DIAMETERS OF STARS MEASURED WITH INTERFEROMETER   TABLE 13.4

Name	Measured Diameter (Sec. of Arc)	Diameter (Sun = 1.0)	Distance Parsec
Betelgeuse	0.054	800	150
Antares	0.040	400	120
Rigel	0.0025	77	250
Aldebaran	0.020	45	16
Epsilon Orionis	0.0007	40	490
Canopus	0.0066	24	30
Arcturus	0.020	23	11
Epsilon Canis Majoris	0.0008	16	210
Bellatrix	0.0007	14	145
Beta Crucis	0.0007	13	150
Achernar	0.0019	8	30
Spica	0.0087	7	70
Beta Carinae	0.0016	5	29
Regulus	0.0014	4.3	28
Vega	0.0032	3	8
Procyon	0.0055	2.1	3.5
Altair	0.003	1.9	5.1
Sirius	0.0059	1.7	2.7
Fomalhaut	0.0021	1.8	6.9

technique, the angular diameter of Sirius was measured at 0."00585, and its diameter in terms of solar diameters may be obtained by the following equation:

$$D_\odot = 107d\beta \tag{13.7}$$

where $D_\odot$ is solar diameters, d is distance in pc and β is the angular diameter. Sirius is 2.7 pc from the Earth, and by substituting we find that Sirius is 1.7 solar diameters.

If the luminosity and the temperature of a star are known, it is possible to determine its diameter. The luminosity is the total energy emitted by a star, and this in turn is a function of the size and temperature of the star. This relationship may be shown as follows:

$$D_\odot = \left(\frac{5760}{T_e}\right)^2 \sqrt{L} \tag{13.8}$$

where $D_\odot$ is solar diameters, T_e is effective temperature °K of the star, and $L_\odot$ is luminosity in terms of solar luminosity. Again we may use Sirius to illustrate the diameter determination. Its temperature is 10,000°K and luminosity $L_\odot$ 26.2. By substituting in the equation, we find that $D_\odot$ equals 1.7 solar diameters. This is in good agreement with the direct method where angular diameter was measured, and therefore the above relationship is satisfactory as an indirect measurement for stellar diameters. It must be realized that the reliability of this or any other technique for gathering data upon stellar properties is dependent upon the accuracy of the components used in the equations.

13.5 STELLAR MASS AND DENSITY

Mass of a single stellar object is difficult to obtain except when based on indirect data. Planetary masses of those planets associated with satellites may be readily found, but the problem becomes more complex for planets like Mercury and Venus, which have no satellites. Stellar masses are determined for the most part from binary stars (see section 15.1), from which the gravitational effects of the components upon each other may be studied. Stars vary greatly in brightness and diameter, but the range in mass is generally quite narrow. When compared with the sun, the majority of the stars range in mass from 0.1 to 10 times that of solar mass. Stars such as HD 698, which is over 100 solar masses, or Luyten 726-8, which is 0.04 solar masses, are exceptions.

Binary stars are systems that orbit around a common center of gravity similar to the Earth-moon system. Visual binaries are those in which the components are visible to the observer on Earth, and are therefore useful in determining stellar masses (see Figure 15.2). Mass for such systems may be determined by the use of Newton's modification of Kepler's third law, expressed as follows:

$$M_1 + M_2 = \frac{a^3}{p^2} \tag{13.9}$$

where M_1 and M_2 represent masses of the separate components of the binary system, a is the mean distance between the components in AU, and p is the period in years required for the components to complete one

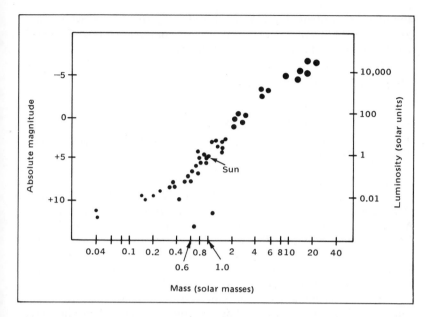

revolution. Once again we may use Sirius as an example, for it was discovered to be a visual binary in the mid-nineteenth century. The approximate period for this system is 50 years, and the 2 components are 20.5 AU apart. Thus:

$$M_1 + M_2 = \frac{(20.5)^3}{50^2} = 3.44 \text{ solar masses}$$

This represents the combined mass of Sirius A, the bright component, and Sirius B, the dark companion.

It is possible to calculate the masses of the individual components if their relative distances to the barycenter are known. Sirius B is approximately twice as far from the barycenter as Sirius A. The masses of Sirius A and Sirius B are approximately 2.2 and 1 solar mass, respectively.

A very useful relationship between luminosity and mass is recognized and has been developed by charting (diagramming stars of known luminosity and mass) (Figure 13.3). The two stellar characteristics may be determined for many stars and their position located on the chart. Luminosity and mass of stars have been found to be related in that the more massive stars are also more luminous. This relationship is sufficiently exact that masses of stars may be estimated if their luminosities are known. The technique is useful in determining the mass of single stars where mass cannot be calculated directly.

Since both stellar mass and volume (computed from diameter) data of stars are available, it is possible to determine density. *Density* is the mass of a substance for a given volume. Generally, density is stated as grams per cubic centimetre, and from mass and volume data of the sun the mean solar density is found to be 1.4 g/cm³. Previously it was stated that stellar mass is generally limited to a fairly narrow range. Density, on the other hand, varies widely because of the wide range in stellar sizes. Many stars have densities far below that of the sun, and in fact are sometimes referred to as "hot vacuums." One very large star, Epsilon Aurigae, has a density about 10^{-8} solar densities. On the other hand, stars such as white dwarfs (see section 14.5), with a great deal of mass compressed into a far smaller volume than the sun, have densities of many tons/cm^3.

13.6 STELLAR MOTION

The unaided eye is unable to detect stellar motion because of the tremendous distances between the stars and the observer. Even over long periods of time the motion that does occur appears slight; constellations described by the Greeks several thousand years ago are essentially unchanged today. A few nearby stars exhibit sufficient motion, so that Edmund Halley in 1718 was able to detect changes in position when he compared star charts of his day with those made by Hipparchus almost 2000 years earlier. Halley was able to see that stars such as Aldebaran, Arcturus, and Sirius had changed positions, and he felt that the differences were too great to be attributable to observational error. He therefore concluded that the stars did move. The motion of stars was in keeping with Newton's law of gravitation, since motion of all objects in space is necessary to prevent objects from falling into each other.

The motions of many stars have been studied since Halley's discovery. The process is complex and tedious, requiring great patience and precision. The Earth's revolution around the sun must be considered, as well as precession and the motion of the entire solar system through space. In recent years photography has facilitated the examination of stellar motion. Photographs are taken of sections of the sky, covering a field of view 10 × 10 degrees and including many thousands of stars. When these photographs are compared with those taken 10, 20, and up to 50 years previously, astronomers are able to detect even very small displacement resulting from motion.

The type of motion just described is the star's proper motion. *Proper motion* is the angular rate of change of position of a star on the stellar plane

Figure 13.4 Barnard's star in the constellation Ophiuchus has the greatest known proper motion. The illustration shows the displacement in position from August 24, 1894 (left), to May 30, 1916 (right) (22 years). (Yerkes Observatory photograph)

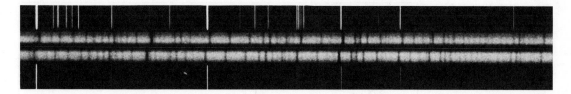

in seconds of arc per year. The annual rate is small, but over a number of years the displacement becomes noticeable. Stars at very great distance exhibit very small change and can therefore serve as reference stars for measuring the rate of change of nearby stars. Most proper motion is small, with Barnard's star in the constellation Ophiuchus exhibiting an exceptionally large proper motion of 10."25 per year. Barnard's star is second closest (after Alpha Centauri) to the sun at a distance of 1.81 pc. Although displacement of Barnard's star is relatively large compared to the proper motion of most stars, it still requires about 175 years for the star to move a distance equivalent to the diameter of the moon. W. J. Luyten, an astronomer who has measured the proper motion of many stars, has stated that less than 350 stars exhibit a proper motion of greater than 1" per year, and that the average annual displacement for naked-eye stars is about 0."1 per year (Figure 13.4).

Figure 13.5 Two spectrograms show radial velocity shifts in the spectroscopic binary Alpha Geminorum. Radial velocity shifts at two different times are shown. (Lick Observatory photograph)

Proper motion provides us with information on the apparent direction a star is traveling in relation to the sun, but doesn't tell us tangential velocity. The *tangential velocity* is the linear velocity in a direction perpendicular to the line of sight. Once proper motion and the distance to a star is known, tangential velocity may be determined by the following equation:

$$v_t = 4.74 \frac{\mu}{p} \, \text{km/sec} \qquad (13.10)$$

where μ is proper motion in seconds of arc per year, p is distance in pc, and v_t is tangential velocity in kilometres per second.

The proper motion is not an indication of the true motion. To determine true motion it is also necessary to obtain the radial velocity. *Radial velocity*, expressed in kilometres per second, is a measure of the velocity of a star's approach or recession from the observer on Earth. From the Doppler effect in the spectrum of the star, the velocity and direction may be determined (see section 5.5, Figure 13.5). Radial velocities have been determined spectroscopically for more than 10,000 stars, but it must be pointed out that this does not represent the absolute velocity of the star. What is being measured is the velocity with which the star and our sun are approaching or receding from each other. The velocity expressed for the star is in reference to the sun, and to obtain an accurate velocity it is necessary to correct for the Earth's orbital velocity of 30 km/sec along the line of sight of the star. Stars vary in radial velocity, but most stars in the vicinity of the sun average about 35 km/sec. None exceed 60 km/sec.

When tangential velocity (v_t) (Figure 13.6) and radial velocity (v_r) have been determined, it is possible to establish the true space velocity (V) of a star with respect to the sun. The *space velocity* represents the hypotenuse of a right triangle whose sides are the tangential and radial velocities. The magnitude of V may be determined by applying the Pythagorean theorem:

$$V = \sqrt{v_r^2 + v_t^2} \qquad (13.11)$$

Figure 13.6 Knowledge
of tangential velocity (v_t)
from A to B and radial
velocity (v_r) from A to C
makes it possible to de-
termine the true space
velocity (V) from A to D.

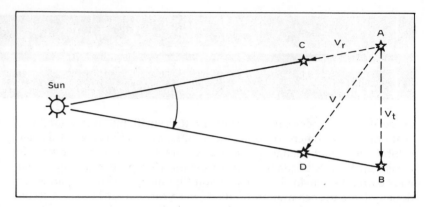

Up to this point we have been discussing stellar motion as if the sun were standing motionless, which is not the case. The sun does move with respect to local stars, a fact first realized by William Herschel in 1783. Motion of the sun or the stars can only be discussed in relative terms—that is, the motion of any one of these objects can only be measured relative to other objects also in motion. If we use the galactic center as a point of reference, we find that the stars in the neighborhood of the solar system all have about the same velocity as the sun. This velocity is about 240 km/sec with respect to the galactic center. The observed motion of stars within a distance of about 50 pc from the sun (our local neighborhood) is due to the slight difference in the eccentricities and inclinations of the orbits of these stars revolving around the galactic center. The stellar motions are analogous to skaters in a rink. From the center the skaters all appear to be moving at about the same speed. However, some may be moving modestly faster and some slower and in slightly different direction from the point of view of a single skater in motion around the center. To describe this type of motion among the stars, astronomers have established a useful frame of reference called the *Local Standard of Rest*, or LSR. Stars within the neighborhood of the sun would appear at rest with respect to the LSR— that is, their velocities would average zero if the sun were at rest. But the sun is moving with respect to the LSR, and each star will reflect this motion.

The motion of the sun relative to the local stars can be ascertained by studying the proper motion and radial velocity of these stars. We would find that in one area in the sky most of the stars have a negative radial velocity of –19.4 km/sec. This indicates that the stars are for the most part moving toward us from what is the *apex* of the sun's direction of motion. The –19.4 km/sec indicates that the sun is moving toward the apex at 19.4 km/sec. The average proper motion at the apex averages to zero, since there are as many stars moving in one direction as there are stars moving in the opposite direction. At a point 180° from the apex is the *antapex*, where stars have a positive radial velocity of 19.4 km/sec. This reflects the sun's motion at this velocity away from the antapex. Proper motion will average zero at the antapex as at the apex. Proper motion at the apex and antapex would be analogous to walking through a forest. Stars at the apex would appear to be moving away from the apex as the sun approaches, much as trees in a forest appear to separate as they are approached (Figure 13.7). At the same time the stars at the antapex would seem to be closing in behind as the sun moves away from them. The maximum effect of proper

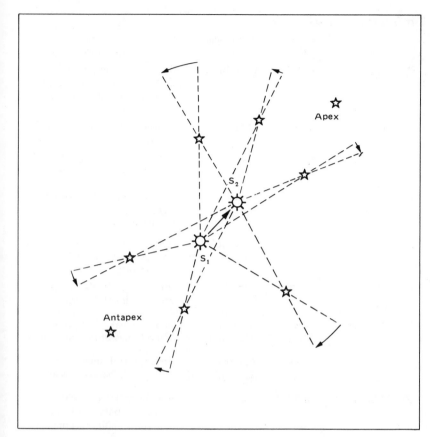

Figure 13.7 The motion of the sun may be determined by studying the proper motion and radial motion of nearby stars. Stars in the line of the sun's motion will have little proper motion and large radial velocity relative to stars at right angles to the line of the sun's motion.

motion can be measured at points at right angles to the sun's direction of motion, whereas radial velocity averages out to zero at these points.

An analysis of local stellar motion reveals that the sun is approaching Vega, a star in the constellation Lyra slightly to the east of Hercules, at a velocity of 19.4 km/sec. This means that the solar system is traveling through space a distance of 4.09 AU per year. When the motions of more distant stars than those within a radius of 50 pc are included, the apex shifts toward the constellation Cygnus. This more closely reflects the motion of the sun in orbit around the galactic center.

13.7 STELLAR CLASSIFICATION

In the previous sections we discussed various properties of stars from which astronomers are able to obtain data. The data by itself, although interesting, is of little value unless some conclusions may be reached or inferences drawn from the relationships of measurements made. Thus it was inevitable that astronomers would categorize stars into groups with similar properties. The systematic classification of rocks, minerals, plants, and animal life has been found very useful and led astronomers to the question of whether or not stars could be similarly classified. Was each star a unique object in space different from all others or did the stars have common characteristics which permitted logical grouping?

In the mid-nineteenth century, Angelo Secchi grouped stars into five categories, based on the arrangement of absorption lines seen in their

TABLE 13.5   THE SPECTRAL CLASSES

Class	Color	Temperature, °K	Representative Stars	Physical Characteristics
W		100,000	Velorum	Broad emission lines of He, O, N, Si, or C resulting from material ejected at 2000 km/sec from stellar surface.
O	Intense blue	40,000	10 Lacertae	Lines of ionized He, O, N, Si. H lines weak.
B	Blue	20,000	Spica, Achernar	He neutral. Stronger H and some ionized O, N, and Si.
A	Blue white	10,000	Sirius, Fomalhaut	Strong H lines. Ionized Ca, Fe, Mg, Ti. Lines of neutral metals appear.
F	White	7500	Canopus, Procyon	Weak H lines. Neutral metals stronger than in Type A. Ca lines strong.
G	Yellow	5500	Capella, Sun	Strong Ca. Metals such as Fe are prominent. H weak.
K	Orange	4500	Arcturus, Aldebaran	H lines weak. TiO present. Strong lines of neutral metals.
M	Red	3000	Antares, Barnard's Star	Strong metal lines. TiO strong. VO bands appear.
R	Orange	4000	Rare	Similar to Type K with molecular bands of C and cyanogen (CN) present.
N	Red	3000	Rare	Similar to Type M with strong bands of C, CN and CH present instead of TiO.
S	Red	3000	Rare	Similar to Type M with ZrO instead of TiO.

spectra. Later Henry Draper modified this system by taking into account variations in the stellar spectra not included by Secchi. With improvements in the techniques of spectrographic analysis, E. C. Pickering was able to refine classification of stars by including and accounting for greater detail in the stellar spectrum. This led to the Harvard system of spectral classification in the early twentieth century, based on the examination of over 200,000 stellar spectra by A. J. Cannon, A. C. Maury, and W. P. Fleming. Initially, the letters of the alphabet from A to Q (J was omitted) were assigned to stellar spectra in a purely empirical manner, determined by the appearance of the spectrum of each star. The stage of evolution and differences in temperature of the stars were not considered. The letter A was assigned to spectra showing the strong broad hydrogen lines as they occurred in the spectrum of Sirius. When other lines were present, such as those at wavelengths 4026 Å and 4471 Å, the spectra were placed in group B. In this manner categories were developed, with the letter Q used for spectra so peculiar that they did not fit any of the other categories.

After many years of work it was found that approximately 99 percent of the stellar types fit into seven main spectral classes, which constituted a temperature scale when properly rearranged. On such a scale the stars ranged in color from blue, indicating high temperature, to red at the low end of the temperature scale. The letter designation, as originally ascribed to spectral classes when rearranged according to temperature, read in the following order: O, B, A, F, G, K, M. After the adoption of the system it was found that the single letter was not sufficient for the variation in intensity of the spectral class represented by the letter, nor for the difference in the number of lines present. As a result each spectral class was divided into ten subclasses and identified by a number 0 to 9 to indicate placement within each class. In this system the sun is a G2 star, placing it near G between G and K. A star designated B5 would indicate spectral characteristics resulting from temperature differences of stars between B0 and B9.

There are several minor classes of stars prominent enough to be mentioned. These stars are similar to stars in the main spectral classes, but certain characteristics place them in a separate grouping. Type W (Wolf-Rayet) are very hot stars with broad emission lines and are generally placed above type O stars (Table 13.5). Types R and N are cool red stars similar to types K and M, respectively, except that absorption bands of carbon compounds are visible. Type S stars show conspicuous bands of zirconium oxide and other metals characteristic of low-temperature stars.

Spectral classification has proved to be a valuable device. By identifying the color of a star, it is possible to infer other characteristics, such as temperature and the prominent lines that will occur in the spectrum of that star. Table 13.5 shows a listing of spectral types, their color, average temperature, and prominent lines usually found in stellar spectra at the indicated temperature. An examination of the data in the table may lead to the erroneous conclusion that stars differ greatly in composition. Stars are composed primarily of hydrogen, with smaller amounts of helium, and very minor amounts of other elements, as in the sun. Differing temperatures result in variations in intensity of spectral lines, causing helium, for example, to be prominent in very hot stars and hydrogen appear to be nearly absent. The fact that hydrogen does not appear in the spectrum does not preclude its presence in the star.

13.8 HERTZSPRUNG-RUSSELL DIAGRAM

With the accumulation of data based on spectral classification, it became inevitable that correlations of some sort would be discovered. In 1905, the Danish astronomer Einar Hertzsprung speculated on the possibility of a relationship between a star's spectral class and its luminosity. By making such a comparison he discovered that white and blue stars were generally highly luminous, whereas red stars had a low luminosity. He did find some exceptions to this: a few red stars were highly luminous, indicating great size. A few years later an American astronomer, Henry Russell, made a similar discovery and the combination of his work with Hertzsprung's led to the Hertzsprung-Russell or H-R diagram (Figure 13.8).

The H-R diagram is composed of the absolute magnitude on the ordinate (vertical line) and spectral class or color on the abscissa (horizontal line). Each star, represented by a dot, is plotted on the diagram according to its absolute magnitude and spectral class. The color index is now frequently used instead of spectral class, since star colors can be precisely measured. In looking at the distribution of stars in the diagram, it

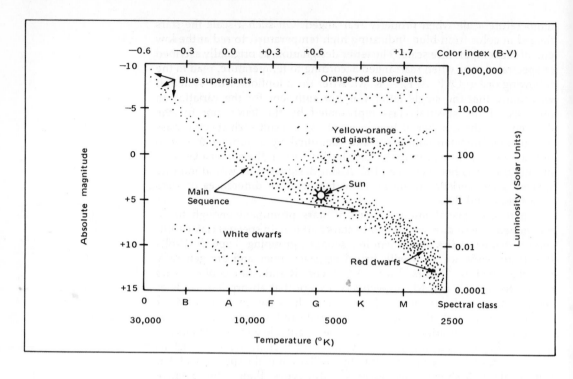

Figure 13.8 The Hertzsprung-Russell diagram.

becomes evident that the majority of stars within a radius of 50 pc fall into a rather narrow band. The band extends from the upper left portion of the diagram, where hot luminous stars are found, to the lower right, populated by cool red stars. This band is called the *main sequence*. Stars in the main sequence are considered to be normal stars, and this is the group of stars of which the sun is a member. These stars differ from each other only in terms of mass, which is the factor determining temperature and brightness. The bright luminous stars in the upper lefthand corner are very massive, although the upper limit is about 60 solar masses. There are a few that are more massive. The stars become progressively less massive in descending to the cool red stars in the lower righthand corner. The lower limit is at least 0.1 solar masses and may extend down to 0.01 solar masses (Jupiter is 0.001 solar masses). The sun lies approximately midway between the most massive and the least massive stars in the main sequence.

In the upper righthand side of the diagram are several distinct groups of stars designated as *giants* and *supergiants*, depending upon their luminosity. These are the relatively cool red stars with a high luminosity due to their very large size. The distinction between giants and supergiants is purely arbitrary, and represents a difference in degree of luminosity. In 1925, another group of stars, the *white dwarfs*, were discovered. These stars were of low luminosity but white in color, indicating very high temperature. Therefore these stars fitted in the lower lefthand corner of the H-R diagram. The low luminosity was evidence of their small size. Knowledge of stars is reasonably complete within a range of only a few parsecs (Figure 13.9), and there are no giants or supergiants within this radius. It is necessary to reach out at least 50 pc to include these very large stars, as is shown in Figure 13.8. The very massive stars are relatively rare compared to the red stars, which are the most common in the main sequence. Within this radius approximately 89 percent of the stars are on the main sequence, 9 percent are white dwarfs, and 1 percent are giants and supergiants. A few variable stars are included in this grouping and will be discussed in chapter 14.

The designation of spectral class provides us with information of the star's color, temperature, and spectral characteristics but nothing about its absolute magnitude. For example, a G2 star may have an absolute magnitude of approximately 5 (sun = 4.9 Mv), 0, or –5, but this is not indicated by the G2 designation. The Morgan-Keenan Luminosity Classes (Table 13.6 and Figure 13.10) show that a G2I star, which is a supergiant, is quite different from a G2V star, like the sun, found on the main sequence.

Is the H-R diagram simply an interesting array of points derived from stellar distance, luminosity, and temperature? Or is it a means whereby astronomers can make valid comparisons of stellar properties and thereby

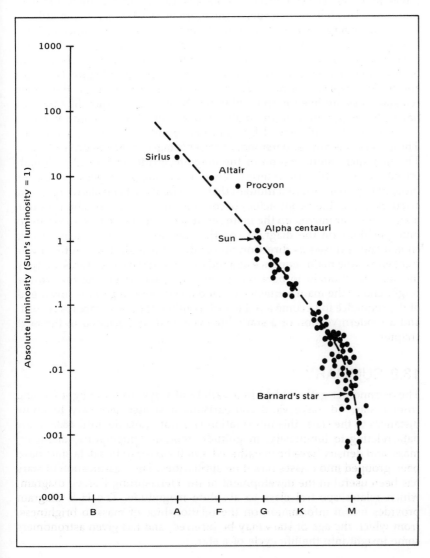

Figure 13.9 The Hertzsprung-Russell diagram of stars within 6 parsecs of the sun reveals no giants or supergiants.

TABLE 13.6   Morgan-Keenan Luminosity Classes

Luminosity Class		Example
I	Supergiant	Betelgeuse, Canopus
II	Bright giant	Epsilon Canis Majoris
III	Giant	Aldebaran
IV	Subgiant	Altair
V	Main sequence	Sun, Sirius
VI	Subdwarf Sd	
VII	White dwarf Wd	Sirius B

learn something about stellar relationships? Actually, a great deal of information has been obtained from the H-R diagram. We have already indicated that size and luminosity are related, and that highly luminous stars are also very massive stars and vice versa. The relationship is not direct, however, as a star 5 times as massive as the sun may be 50 times as luminous. This tells us that fuel in the more massive stars is being utilized to generate energy at a far greater rate than their size would indicate. From this it is possible to conclude that bright massive stars are relatively short-lived, compared to smaller cooler stars on the main sequence. Those stars on the upper lefthand portion of the main sequence in Figure 13.9 may endure for a few tens or hundreds of millions of years, whereas the cooler red stars may survive for 15 to 20 billion years. Yellow stars like the sun have a life expectancy on the main sequence of 10 to 12 billion years.

What is the part played by stars not found on the main sequence? These stars, it is now understood, represent stages in the life cycle of stars. The first such interpretation of the diagram, made by Russell, although erroneous is of historic interest, since it did suggest a possible line of thought to pursue. Russell thought that a cloud of dust and gas first formed a red giant and then contracted, becoming progressively brighter. In this manner the star moved up the main sequence from the yellow, to white, to blue positions on the diagram. As the star used up its nuclear fuel, it cooled and reversed its direction on the diagram, ending up on the lower right end of the main sequence as a red dwarf. On the basis of this analysis the sun was thought to be a relatively young star ascending toward the bright end of the main sequence or an old star returning to the lower end of the sequence to become a red dwarf. Neither of these choices is correct, and a modern version of a star's life cycle will be discussed in the next chapter.

13.9 SUMMARY

The astronomer gathers data in a variety of ways from the light coming from a star. To make valid comparisons it is first necessary to know distances to the stars; this information was not available until 1838. Now data relating to luminosity, magnitude, motion, temperature, diameter, mass, and density have been gathered. On the basis of this data stars have been grouped into classes based on similarities. The classification of stars has been useful in the development of the Hertzsprung-Russell diagram, which relates spectral class to absolute magnitude. The H-R diagram provides useful information on the relationships of mass to brightness, from which the age of stars may be inferred, and has given astronomers some insight into the life cycle of a star.

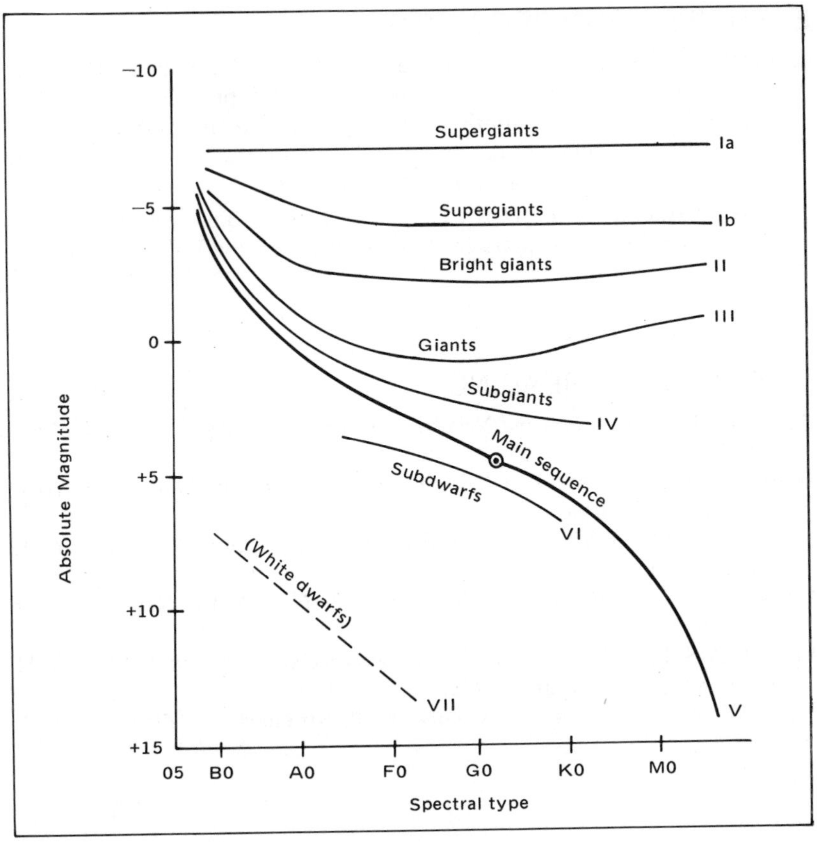

Figure 13.10 The Morgan-Keehan luminosity classes.

QUESTIONS

1. Briefly describe the triangulation process which is the basis for measuring distances to stars. To what distance is this method useful?

2. What are some of the problems in measuring distances to stars through triangulation?

3. Star A has an apparent magnitude of 4; star B has an apparent magnitude of 8. Which star is brighter and by how much?

4. What is the difference between apparent magnitude and absolute magnitude?

5. Briefly describe a method for determining a star's temperature. What data is needed?

6. Define briefly proper motion and radial velocity and describe how they are determined.

7. How is true space velocity determined?

8. Describe the system of stellar classification. What is the relationship between spectral class and temperature?

9. What type of stars are found on the main sequence of the H-R diagram? How do the stars differ from the lower righthand portion of the main sequence to the upper lefthand portion?

10. What does the position of a star on the main sequence tell us about its mass and possible longevity?

EXERCISES AND PROJECTS

1. We now know that stars move, resulting in modification of the configuration of constellations over extended periods of time (see Figure 15.12). Choose another prominently known constellation and determine the relative movements of the stars, showing how the constellation may appear in the distant future.

2. Make a tracing of the H-R diagram in Figure 13.8. From the data in Table 13.2, locate the stars in their position on the H-R diagram.

FOR FURTHER READING

ALLER, L. H., *Atoms, Stars and Nebulae*. Cambridge, Mass.: Harvard University Press, 1971.

BRANDT, J. C., *The Physics and Astronomy of the Sun and Stars*. New York: McGraw-Hill, 1966.

CLAYTON, D. D., *Principles of Stellar Evolution and Nucleosynthesis*. New York: McGraw-Hill, 1968.

GAMOW, G., *The Birth and Death of the Sun*. New York: New American Library, 1952.

GINGERICH, O., ed., *New Frontiers in Astronomy*. San Francisco: W. H. Freeman and Company, 1975.

ROSE, W. K., *Astrophysics*. New York: Holt, Rinehart and Winston, 1973.

Orion

CHAPTER 14   BIRTH AND DEATH OF A STAR

Bright star, would I were stedfast as thou art...

—*John Keats*

The eternal and everlasting stars were for the ancient astronomers an indication of the unchanging nature of the universe. The constellations appeared year after year on a schedule little altered and were therefore a reliable signpost of the passage of time. Gradually, small changes were noticed. Over the centuries stars shifted their positions ever so slightly with respect to each other, which in time led astronomers to realize that they were, in fact, moving in space. Novae, representing another form of change, were witnessed, although the exact nature of the phenomenon was not understood until relatively modern times. Along with this understanding came the knowledge that stars, like animals and plants, have a birth, a main life, and an ultimate death.

In pursuing this subject there are several questions that might be asked: What is the path followed by a star from birth to death? Do all stars follow the same sequence of events during their life cycle? We have discussed the variation in mass of stars found on the main sequence and can speculate on whether the very massive stars like Rigel evolve in the same manner as the sun or a star one-quarter solar mass.

14.1 BIRTH OF A STAR

The birth process of a star is probably the least understood phase of stellar evolution, but it is now recognized that new stars form in the huge interstellar clouds of dust and gas called *nebulae*.

The Nebulae

The nebulae are composed of a number of gases, principal among which is hydrogen. Neutral hydrogen, emitting radiation at 21-centimetre wavelength, was first detected in 1951 by Harold Ewen and E. Purcell of Harvard University. The hydrogen atom consists of an electron and a proton spinning in reverse, or *antiparallel*, direction and therefore is in its lowest energy state (Figure 14.1). The direction of spin may be reversed to

Figure 14.1 A hydrogen atom consists of a proton and an electron which when spinning in reverse, or antiparallel, results in the atom's lowest energy level. If the direction of spin of the electron is reversed, the atom is at its highest energy level. Returning to low energy level results in the release of energy at 21-centimetre wavelength.

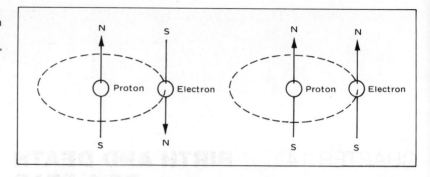

parallel spin, where the proton and electron are spinning in the same direction. This is brought about by the absorption of an appropriate amount of energy which raises the energy level of the atom, an action accomplished by the collision of the hydrogen atom with a photon. Returning to its lowest energy level results in the release by the atom of a small amount of energy at 21-cm wavelength. This event for any given hydrogen atom occurs once every 11 million years, but the enormous amount of neutral hydrogen in interstellar space makes it possible for this radio emission line to be readily detected. Areas in interstellar space where neutral hydrogen occurs are identified as H I regions.

Hydrogen also exists in the ionized state, generally in regions of the galaxy where stars have recently been formed. The source of energy responsible for ionizing the hydrogen gas is ultraviolet radiation from the newly formed hot stars, which have surface temperatures of more than $2.5 \times 10^4 °K$. Photons from the ultraviolet radiation have sufficient energy to cause the electrons in the hydrogen atom to be ejected. This process is known as *photoionization* and is reversible. Ionized hydrogen may recapture an electron and emit a photon in a *recombination* process. Regions in space where this is occurring are known as H II regions.

Hydrogen is the most abundant gas in interstellar space, comprising about 90 percent of the total of all gases. Helium is the next most abundant, making up approximately 10 percent of the total, with all other elements adding up to 0.1 percent. Although elements other than hydrogen and helium comprise only a very small fraction of the total, their importance in the interstellar media should not be overlooked. It is from these elements that the planets and life upon the Earth were formed. In descending order of quantity are oxygen, nitrogen, carbon, magnesium, silicon, iron, and sulfur. The balance of the elements make up less than 0.04 percent of the total.

In recent years it has been discovered that molecules formed of these elements have slowly been evolving in the interstellar spaces. Most of the molecules are combinations of carbon, oxygen, hydrogen, and nitrogen, the same materials that are basic to the life process on Earth. It appears that the processes of organic (carbon) chemistry that occur on Earth also occur among the stars. The implications of this with respect to life in the universe will be discussed in chapter 19.

Formation of Stars in Nebulae

Because stars are formed in nebulae they are not uniformly scattered in space but instead appear to be grouped in clusters (see chapter 15). The

apparent clustering of stars suggests that stars are not formed singly but that tens, hundreds, or even thousands of stars are, in a sense, "hatched" in the same nebula. The Pleiades (Seven Sisters), a group of seven stars when seen with the naked eye in the winter sky, represents such a cluster (Figure 14.2). Through the telescope several hundred stars may be seen in this group, and these stars are thought to have formed about 100 million years ago. A hazy nebulosity—not visible to the unaided eye—can be photographically detected and represents dust and gas remaining from the recent birth of these stars.

Figure 14.2 The Pleiades, a star cluster, is thought to be a center of star formation. (Lick Observatory photograph)

The Great Nebula in Orion is of distinct importance to astronomers studying stellar formation because of the presence of highly luminous stars within the cloud. The high luminosity indicates that energy is being expended very quickly. Their life span is relatively short, and therefore they are relatively young stars. Four of the stars form a group called the Trapezium (Figure 14.3). These are mainly responsible for the energy causing the Orion nebula to glow. The glowing portion of the nebula is only a small portion of an immense cloud of dust and gas that is visible as a dark area in and around the glowing gas. Recent developments in infrared astronomy and molecular radio astronomy have enabled astronomers to learn something about what is occurring in this immense cloud. Temperature changes taking place within the cloud can be detected with techniques developed in infrared astronomy. Observers can "see" the embryo stars in

 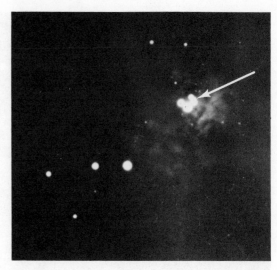

Figure 14.3 The Orium nebula and the Trapezium. (Left) Shows constellation. (Right) Trapezium with exposure of 180 minutes. (Yerkes Observatory photograph)

the infrared portion of the spectrum long before portions of the cloud where stars are forming become visible. The molecular radio astronomer can identify certain molecules in the dark clouds. Hydrogen and helium are known to be the major components of the clouds, but radio astronomers are interested in the presence of carbon monoxide, which represents only a fraction of 1 percent of the mass of the cloud. The carbon monoxide yields information on temperature and density of the cloud. An increase in temperature and density of portions of the cloud is a signal that star formation may have been initiated.

The Protostar

In the initial stages of star formation a huge dust and gas cloud begins to condense, but not uniformly. Rather, a huge cloud many light years in diameter will develop regions of higher density, and these regions will condense more rapidly into protostars. The *protostars* are clouds of much denser material than the typical interstellar clouds, and may be observed as dark "holes" in the otherwise illuminated regions of a cloud like the Orion nebula. What causes a cloud to increase in density? No one is certain, but it has been suggested that galactic shock waves initiate compression waves which pass through interstellar clouds, causing variation in cloud density. When sufficiently dense, mutual gravitational attraction of the dust and gas particles within the cloud cause further contraction at a more rapid pace. As a result segments of a huge cloud, perhaps one-quarter light year in diameter, collapse to one billion kilometres in diameter in about 10^5 years. Collapse of this sort, which is very rapid, may be detected by studying the spectrum of molecules like carbon monoxide or formaldehyde in the cloud. Motion of the molecules will indicate the rate of collapse and whether or not the cloud is beginning to rotate.

During the contracting process an increase in temperature will occur in the protostar. Initially, the cloud is incredibly cold, somewhere in the vicinity of $5°K$ to $10°K$, and movement of the atoms will be quite sluggish. As the cloud collapses dust and gas particles will collide at an increasing rate, resulting in rising temperatures. When the density of the cloud has increased 10 to 100 times, temperatures will have risen to 100°K to 150°K, and the speed of the hydrogen atoms in the cloud will be about one to two kilometres per second. In 1973, an object called *IRS-5* (infrared source-5) was detected, which was larger than the solar system and emitted 3×10^4 times more energy than the sun. Its temperature was about 350°K.

Gravitational energy released by the continued collapse of the protostar further heats the interior of the newly forming star. The energy thus radiated may be detected in the invisible infrared portion of the spectrum, but much of the heat being emitted is contained in a cocoon of dust and gas, which may eventually form a planetary system. In 1974, a group of such infrared objects was discovered near the center of the Orion nebula. The infrared objects—five have been detected—are thought to be less than 10^5 years old and are associated with a large dust and gas cloud producing radiation from excited carbon monoxide molecules.

The Star's First Appearance

Temperatures continue to rise and the density of the dust and gas continues to increase. At temperatures of 10^4°K, electrons begin to be displaced from hydrogen and helium atoms at the center of the protostar because of high speeds atoms are now experiencing—up to 30 km/sec. The protostar, a mixture of ionized gas (atomic nuclei and electrons), has now contracted down to about 150 million kilometres and is beginning to appear luminous. In a very short time, perhaps a year or less, the temperatures will have tripled to 1.5×10^5°K in the protostar's interior, and the surface temperatures will be sufficiently high to cause the protostar to appear as a huge, red, luminous object 50 to 75 million kilometres in diameter. Although the amount of energy emitted per unit of area by the new star is far less than is emitted by the sun, the protostar's huge size will cause it to appear far more luminous than the sun. Therefore the protostar could now be placed in the upper righthand quadrant of the H-R diagram in the vicinity of the giant red stars (Figure 14.4). It must be recognized that the protostars are not red giants, for red giants are known to be aging stars. At this time the surface temperature of the newly forming star would be about 3500°K, and would be visible to astronomers on Earth. The actual appearance of a new star may have been observed in 1936 in the Orion nebula. Seeing the star, FU Orionus, appear in this way was unusual, for such an event is statistically expected in our vicinity of the galaxy only once in 500 to 1000 years. Since then several other suspected "births" have been sighted in other parts of the galaxy—one is V 1057 Cygni in the constellation of Cygnus and another is located in the Orion nebula. Also included with these may be the T Tauri stars which are very young, bright, type 0 and B stars. They are variable and have thick chromospheres from which materials are being ejected at high speeds. The T Tauri stars are associated with clouds of dust and gas from which they appear to have been formed.

With the internal generation of energy the protostar is no longer subject to the force of gravity alone. As temperature increases, collisions between atoms and subatomic particles within the protostar become more violent, thereby increasing internal pressure. The internal pressure, countering the force of gravity, causes a slowdown in the rate of collapse of the protostar. This decrease in the rate of collapse is such that it requires about 10 million years for a protostar with a mass equal to that of the sun

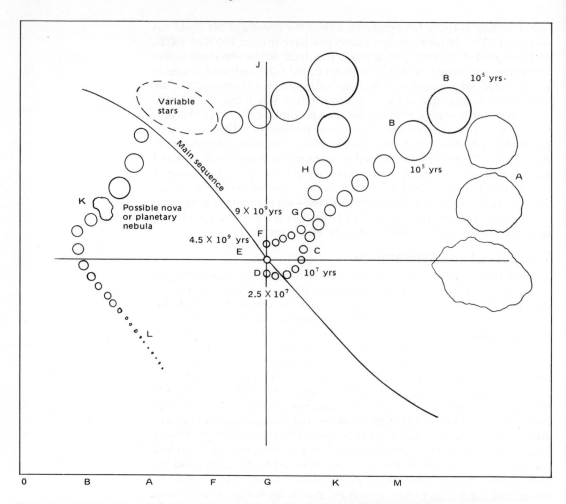

Figure 14.4 Sequence of events in the life cycle of a star of approximately one solar mass. (A) Cloud begins to contract and temperature increases. (B) Star becomes visible in infrared about 10^5 years. (C) Star 10^7 years old and about twice the diameter of the sun. (D) Star 2.5×10^7 years old arrives at ZAMS. (E) Present position of sun about 4.5×10^9 years old. At the turn-off point, star is 9×10^9 years old. From (F) to (G) is the period of constant luminosity. Star is 10^{10} years old. (H) Rapid increase in size and temperature to red giant stage. At the end of this stage temperature rise is very rapid in what is known as helium flash. (J) Star shrinks in size and becomes unstable leading to possible variable stage. (K) Star possibly a nova or ejects a planetary nebula. (L) Star becomes a white dwarf, loses residual energy by radiation and becomes a black dwarf.

to shrink down to a diameter of about 2 million kilometres, which is almost twice the diameter of the sun. At this stage the internal temperature of the protostar will have reached about 10 million °K.

Gravity and Temperature

Up to this point two forces have influenced the development of the newly forming star: gravity, responsible for the collapse of the dust and gas cloud, and the opposing internal pressures generated by the increasing internal temperatures. The internal pressures caused a slowdown in the rate of collapse, particularly in the later stages of the star-forming process. During this period temperatures increased, reaching a level of 10 million

°K. At this temperature protons (hydrogen without electron) have sufficient energy to permit a thermonuclear reaction to occur by means of the proton-proton reaction (see section 12.8), and the protostar now becomes a star.

Nuclear fusion generates considerable heat in the star's interior, and although contraction continues it is at an extremely slow rate, requiring 15 to 20 million years for the star to reach the approximate diameter of the sun. At that point the star is in a state of *hydrostatic equilibrium*—gravitational collapse is in equilibrium with internal pressures and the star takes up a position on the main sequence, which is called the *zero age main sequence* (ZAMS).

The size and mass that a star will ultimately achieve is dependent upon the size of the dust and gas cloud that formed the protostar. This mass is important, as it is the determining factor in the length of time required for the transition from a dust and gas cloud to a star on the main sequence. In the case described above, it required about 20 to 25 million years for a star of one solar mass to accomplish the transition. Stars of much higher mass require a few million years or less. For example, a star of 30 solar masses may require only 3×10^4 years for the entire process. On the other hand, a star of 1/4 to 1/10 solar mass may need hundreds of millions of years. Dust and gas clouds of about 1/12 solar mass or less have insufficient mass to sustain a thermonuclear reaction and do not become stars on the main sequence.

14.2 ON THE MAIN SEQUENCE

How long does a star remain on the main sequence? This again is dependent upon its initial mass (Table 14.1). A star of one solar mass remains reasonably stable for an extended period—perhaps 10 to 12 billion years. This represents about 85 to 90 percent of its life expectancy. The sun has now progressed through approximately half of its main-sequence residency. During the main-sequence tenure, a star very slowly moves up the H-R diagram as a result of a gradual increase in size (about 30 percent) and an approximately one-magnitude increase in brightness. Temperatures at the core of the star are about 12 to 15 million °K, and the proton-proton reaction is converting hydrogen to helium and, in the process, generating tremendous energy. The helium produced as a result of hydrogen "burning" accumulates at the center of the star to form a helium core. Hydrogen, it must be pointed out, does not "burn" in the normal sense, but the expression is a convenient way of denoting the use of an element in a nuclear reaction.

LIFE EXPECTANCY OF STARS ON THE MAIN SEQUENCE   TABLE 14.1

Solar Mass	Spectral Type	Absolute Magnitude	Life Expectancy (Years)
30	O7	−5	2×10^6
10	B3	−3	3×10^7
3	A2	1	6×10^8
1	G2	5	1×10^{10}
1/3	M5	9	2×10^{11}
1/10	M7	14	3×10^{12}

The same process occurs in a more massive star. However, temperatures are much higher and the rate at which energy is being generated and radiated is far greater than in a star of one solar mass. As a consequence, the life expectancy of such a star may only be tens or hundreds of millions of years. High internal temperatures in excess of 15 million °K will also permit hydrogen burning to form helium by means of the carbon-nitrogen cycle, if carbon and nitrogen are present in the star.

In the *carbon-nitrogen cycle*, carbon and a proton (hydrogen nucleus) form a light nitrogen isotope, accompanied by the emission of a gamma ray. This is followed by the light nitrogen isotope, forming a heavy carbon isotope, a positron, and a neutrino, as shown in the following reactions:

$$^{12}_{6}C + ^{1}_{1}H \longrightarrow ^{13}_{7}N + \gamma$$

$$^{13}_{7}N \longrightarrow ^{13}_{6}C + _{1}e^{\circ} + \nu$$

The heavy carbon atom reacts with a proton to yield nitrogen and a gamma ray, and the nitrogen combines with a proton to form light oxygen isotope and a gamma ray.

$$^{13}_{6}C + ^{1}_{1}H \longrightarrow ^{14}_{7}N + \gamma$$

$$^{14}_{7}N + ^{1}_{1}H \longrightarrow ^{15}_{8}O + \gamma$$

The light oxygen isotope yields a heavy nitrogen isotope, a positron, and a neutrino, and the heavy nitrogen in turn reacts with a proton to form carbon and a helium atom.

$$^{15}_{8}O \longrightarrow ^{15}_{7}N + _{1}e^{\circ} + \nu$$

$$^{15}_{7}N + ^{1}_{1}H \longrightarrow ^{12}_{6}C + ^{4}_{2}He$$

This cycle contributes a small amount (about 10 percent) of the energy generated by the sun and other stars of approximately one solar mass, but is the dominant energy-producing reaction in more massive stars.

Small stars with only a fractional solar mass remain on the main sequence for periods up to 100 billion years. Such stars as, for example, Barnard's star are stable, small red dwarfs, less than half the sun's diameter and only a few percent as bright. During the end of the small star's tenure on the main sequence, it becomes a little brighter, signaling the beginning of its aging period.

14.3 THE AGING STAR

The burning of hydrogen and the accumulation of helium at the core do not continue unabated for an indefinite time. When the hydrogen at the core is exhausted, the core, no longer supported by internal pressure from the nuclear reactions, begins to collapse under gravitational influence.

This results in rising temperatures and densities at the core, causing the hydrogen in the shell surrounding the helium core to heat. Hydrogen burning commences in the shell, and the conversion of hydrogen to helium is resumed with an accelerated release in energy. The heat energy generated by the collapse of the core and the hydrogen burning in the shell is greater than the energy generated during the star's main-sequence period. As a result the gas comprising the star expands outward, overcoming the inward force exerted by gravity, and the star begins to expand at what is called the *turnoff point* on the H-R diagram. At this time the star leaves the main sequence and proceeds toward the *red giant stage.*

Much of the energy generated by the star shortly after leaving the main sequence at the turnoff point is expended in the expansion process, so the luminosity of the star is not immediately changed to an appreciable extent. This is a period of *constant luminosity,* where the surface of the star cools as it expands and the energy radiated per unit area is reduced. The star therefore moves horizontally to the right on the H-R diagram from its position on the main sequence (Figure 14.4). The expansion process continues for about a billion years for a star of one solar mass, during which the expansion process doubles the size of the star.

At the end of this stage hydrogen burning increases dramatically and the star's size and luminosity increase very rapidly, while at the same time surface temperatures decrease. The star is now a red giant of very low density with its diameter about 50 times that of the present sun. The surface of the sun at this stage would reach out to the orbit of Mercury. Toward the end of the red giant stage internal temperatures continue to rise and very rapidly reach a level of 100 million °K. Such a rise in temperature should result in expansion. However, the helium core is essentially a solid, expanding only slightly and thus causing temperatures to rise further. At this point, helium burning is initiated. The higher the temperature becomes, the faster is the rate of helium burning, in a continuous cycle, until the rate increase is explosive. The process just described occurs in only a few hours and is called the *helium flash.*

At temperatures of 100 million °K, helium burning of the core begins, and the *triple alpha process* is initiated in which three atoms of helium are converted to one of carbon. At the onset of this reaction two helium atoms (alpha particles) fuse to form beryllium, followed by a beryllium and helium atom combining to form carbon.

$$_2^4\text{He} \ + \ _2^4\text{He} \ \longrightarrow \ _4^8\text{Be}$$

$$_4^8\text{Be} \ + \ _2^4\text{He} \ \longrightarrow \ _6^{12}\text{C}$$

Some astronomers believe that following the triple alpha process the fusion of carbon and helium nuclei is responsible for the formation of the heavier elements.

The energy released during the helium core burning process continues to heat the star as the star shrinks in size. This causes the position of the star to shift to the left on the H-R diagram. The evolution of a star like the sun is not too well understood beyond the stage just described. There is some evidence that the star becomes unstable and may eject a *planetary*

Figure 14.5 NGC 7293.
Planetary nebula in
Aquarius. Photographed
in red light with 200-inch
telescope. (Courtesy of
Hale Observatories)

nebula, which expands outward at 20 to 30 kilometres per second and represents about 10 percent of the star's mass (Figure 14.5). The nebula diffuses rather rapidly and is thought to last only a few tens of thousands of years. The relatively short life span of planetary nebula and the fact that about a thousand are visible lead astronomers to conclude that they are a fairly common occurrence.

When the helium at the core of more massive stars is exhausted and replaced by a carbon core, helium burning occurs in a shell surrounding the core, and this shell in turn is surrounded by an outer shell of burning hydrogen. The star now moves to the right on the H-R diagram to reoccupy a position in the red giant region, and it is thought that this horizontal movement on the diagram is repeated several times. How the red giant stage of massive stars ends is not known exactly, but as with the smaller stars it is thought that the star experiences a period of instability, pulsating as a variable star.

Stars less massive than the sun enter a red giant stage, but temperatures do not become sufficiently high to result in helium burning at the core. Stars of fractional solar mass enter the red giant stage twice before expiring. Between the two red giant stages the small star may exhibit some instability, corresponding in action to RR Lyrae variables (see page 320). Because of the long life expectancy of low-mass stars, only very old stars may have had time to reach this stage.

14.4 VARIABLE STARS

During the latter part of the red giant stage, it is thought that stars become unstable and result in a group of stars collectively called variable stars. These stars are sufficiently numerous for astronomers to consider them as a stage in the normal life cycle of at least some stars. Variable stars may be placed in two broad classes— *pulsating stars,* which are intrinsically variable as a result of a change in the diameter and temperature of the star, and the *eruptive variables,* which change explosively during a nova or supernova episode. Binary stars that vary in brightness as a result of one member of a pair eclipsing the other are not considered variable stars and will be discussed in chapter 15.

Pulsating Stars

When the variation in the brightness of a star is plotted against time, a light curve is obtained which indicates the period of variability (Figure 14.6). At the same time if the star is examined spectroscopically, a Doppler shift of the spectral lines indicates an expansion and contraction of the star's surface with respect to Earth. The expansion and contraction process is in phase with the light curve (Figure 14.7). The maximum rate of contraction (*A*) indicated by a positive velocity in km/sec coincides with the point of minimum light, because the stellar surface is at its coolest. The rate of contraction then slows down (*B*) as the star heats up and increases the internal pressure against the contracting force of gravity. At the same time the star begins to brighten. The contraction does not stop when internal pressure equals the force of gravity, but rather overshoots that equilibrium point. Contraction continues until the smallest diameter is reached (*C*), and the star begins to expand again (*D*) with a continuing increase in brightness. When the maximum rate of expansion is reached (*E*), the star is at its brightest. The rate of expansion decreases (*F*), and by the time the star has reached its largest diameter (*G*) it has decreased markedly in brightness. The expansion process again overshoots the point where internal pressure and the force of gravity are in equilibrium.

There are essentially four important types of pulsating variable stars, based upon the period obtained by plotting the light curve.

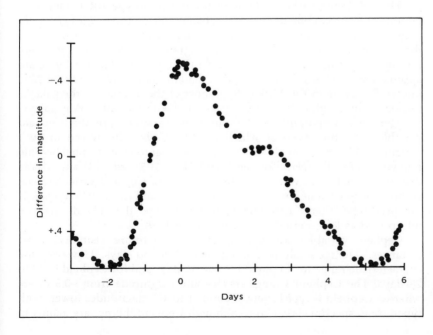

Figure 14.6 The light curve of a type I Cepheid variable. The star is Eta Aquilae.

Figure 14.7 The expan-
sion and contraction of a
Cepheid variable is in
phase with the light
curve.

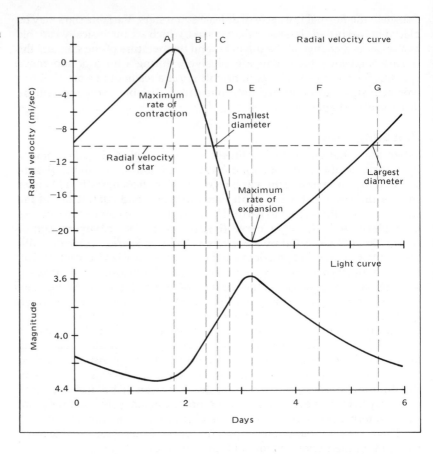

Figure 14.7 The expansion and contraction of a Cepheid variable is in phase with the light curve.

The RR Lyrae variables (named for the prototype RR Lyrae) have periods of less than one day, and the absolute magnitude varies generally between +0.5 to +1.0. This variation in magnitude is not very great, and therefore their positioning on the H-R diagram is restricted to a relatively thin horizontal line between A5 and F5 (Figure 14.8). The RR Lyrae are approximately 50 to 100 times brighter than the sun. There are about 2500 known RR Lyrae in the Milky Way, most of them found in the galactic nucleus or in the globular clusters associated with the Milky Way galaxy.

The most important of the pulsating variables are the Cepheid variables, the prototype of which is *Delta Cephei*, discovered to be a variable in 1784 by John Goodricke. Stars similar to Delta Cephei became known as *classical Cepheids* or *type I Cepheids* in 1952, when Walter Baade discovered that there were actually two types of Cepheid variables. The second group became known as *type II Cepheids* and the star *W Virginis* is the prototype. Some astronomers feel that type II Cepheids may be interpreted as long-period RR Lyrae (Figure 14.9).

Cepheid variables are supergiants and therefore visible at great distances. Their overall period range is 1 to 50 days. However, the characteristic range for Cepheid I is 5 to 10 days and for Cepheid II, 12 to 28 days. The Cepheid I range in absolute magnitude from –0.5 to –6, whereas Cepheid II types average about 1 to 1½ magnitudes lower. Both types are in spectral class F to K, although Cepheid II types are mainly in

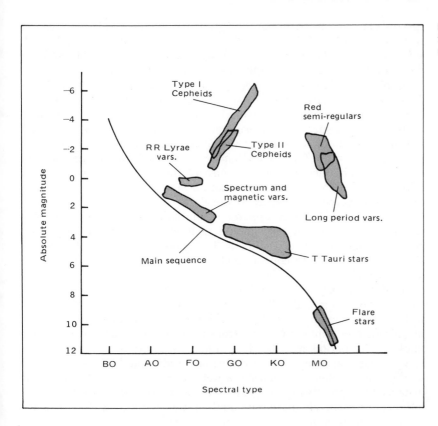

Figure 14.8 Position of the pulsating variable stars on the H-R diagram.

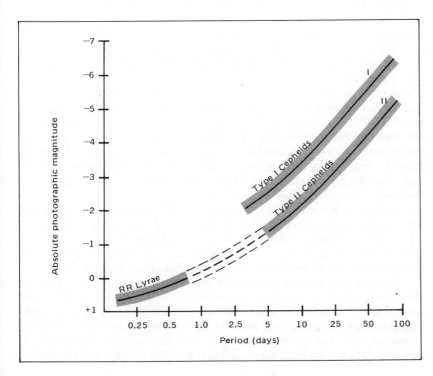

Figure 14.9 Period-luminosity curves for Cepheid variables and RR Lyrae stars.

spectral class G. The nature of the light curve differs slightly between the two types. Cepheid I variables brighten rapidly and dim gradually, with some slight irregularity during dimming (Figure 14.6). Cepheid II variables dim more slowly than the Cepheid I types.

Cepheid variables have become a powerful tool in determining distances to external galaxies as a result of a discovery by Henrietta Leavitt in 1912. She found the relationship between the period and the apparent magnitude to be such that the brighter the average luminosity, the longer was the period. This discovery was made while Leavitt was studying variable stars in the *Small Magellanic Cloud,* a small external galaxy seen near the south celestial pole. Because of the great distance (not then known) to the SMC, it was felt that all stars in that galaxy could be regarded as being equidistant to the Earth. Therefore the relationship between the period and the apparent magnitude of the Cepheid variables could be considered similar to the relationship of the period to the absolute magnitude. This resulted in the Period-Luminosity (P-L) diagram shown in Figure 14.9. By convention RR Lyrae variables are included in the same diagram.

Harlow Shapley recognized the value of the P-L diagram of Cepheid variables and attempted to find the zero point for the absolute magnitude. If this were found it would be possible to determine the absolute magnitude of a Cepheid variable from the diagram once the period were known. The period may be determined by direct observation, and a comparison of apparent and absolute magnitude would yield the distance to the star. Cepheid variables are not close enough to measure their distance by triangulation, so Shapley made use of statistical parallax, an indirect and relatively inaccurate method. By determining absolute magnitude in this fashion, it was possible to measure the distances to nearby external galaxies. This technique when first used resulted in erroneous data until it was recognized by Baade that two types of Cepheids existed. His corrections in the early 1950s led to an increase in the distance scale of the universe by a factor of two (see section 17.2).

Another common type of pulsating variable is the *Mira-type* or *long-period* variable that has periods ranging from 100 to 800 days. Mira (Omicron Ceti), the prototype for this group, has a period that averages 331 days, but the intervals may vary by as much as 3 weeks, indicating some instability in the cycle. Those with the longest periods appear to be the least regular in their cycle. The Mira-type variables are red giants or supergiants of spectral class M, having a surface temperature of 2000 to 2500°K. From minimum to maximum they typically range through approximately 5 magnitudes, which means they are about 100 times brighter at maximum than at minimum.

Similar to but not included with the Mira-type variables are the semiregular or irregular variables, of which *Alpha Herculis* is the prototype. These stars exhibit no set period but oscillate between 100 and 200 days. They are also giant and supergiant M-type stars, of which Betelgeuse, seen in the winter sky in the constellation of Orion, is an example.

Several minor groups of variable stars are recognized, but their place in the scheme of stellar evolution is not known. One of these, the *flare stars,* are small red dwarfs that brighten several magnitudes in a few seconds and gradually dim to normal. Some of these stars have been found to emit bursts of radio energy simultaneously with the flare. Another group of spectral class A stars has been found that exhibits variable spectra, that is,

the intensity of certain lines in the spectrum vary periodically. Some of these A-type stars have magnetic fields that fluctuate in strength, and some undergo polarity reversal. The magnetic and spectral variability may be linked.

Figure 14.10 Nova Herculis 1934, showing large change in brightness between March 10 (left) and May 6, 1935 (right). (Lick Observatory photograph)

Eruptive Variables

Stars in this group include the *novae* and *supernovae*, which are characterized by an enormous variability in brightness, up to 20 magnitudes in the case of supernovae. *Nova* (novae, *pl.*), meaning "new," refers to a "new star," although in reality it represents an existing star that has literally exploded. Approximately 200 novae have been observed in the Milky Way, and most of these have been photographically recorded in the twentieth century (Figure 14.10). Telescopes and photographic techniques have enabled astronomers to discover about 25 novae each year in the neighboring galaxy of Andromeda—a spiral galaxy much like the Milky Way. From this it is assumed that a similar number occur in the Milky Way each year. Because we are a part of the Milky Way and are unable to continuously monitor all parts of the sky, and because of the dust and gas in the galaxy that obscure portions of it, the astronomer on Earth is able to detect only a few novae each year.

The supernova that resulted in the *Crab nebula* was observed by the Chinese in 1054. The Chinese called these types of stars "guest stars," because they suddenly appeared and then slowly disappeared. Tycho's star seen in 1572 in the constellation Cassiopeia was brighter than Venus and visible even in daytime. Kepler's star was observed in the constellation Ophiuchus in 1604 and rivalled Jupiter in brightness.

Novae characteristically increase in brightness very suddenly, in two phases. The first phase requires two to three days, bringing the star to within 2 magnitudes of its maximum. Then, after a brief pause, an additional day is required for the so-called "fast" novae, and several weeks for the "slow" novae, to reach the final phase. Stars of from one to five solar masses are considered fast novae and those of less than one solar mass are slow novae. After several days at maximum the decline in brightness begins, rapidly at first and then more gradually (Figure 14.11).

Figure 14.11 Light
curves of a fast nova (a)
and a slow nova (b).

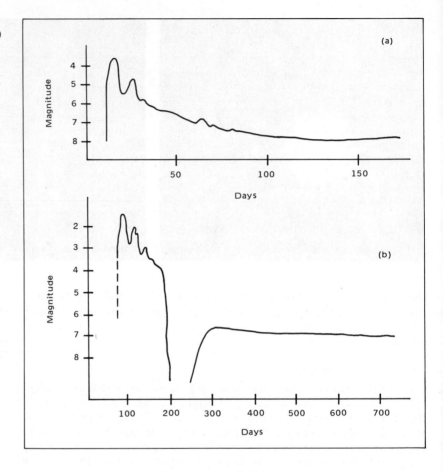

The brightness fluctuates during the decline, varying several magnitudes during the six months to several years required for the star to return to normal. The slow novae require the longer period. The total change in brightness may be from 10 to 12 magnitudes for the fast novae and a magnitude or so less for the slow novae. During the eruptive stage gases are ejected from the stellar surface at velocities of several thousand km/sec for fast novae, and up to 1500 km/sec for slow novae.

Some novae have been observed to erupt several times and are known as *recurrent novae*. Eruptions of this kind are not cataclysmic because the star is not destroyed. Only a small amount of the total stellar mass, perhaps as little as 0.01 percent, is shed during an eruption.

It is believed by many astronomers that all novae are members of binary star systems. One model of such a system assumes that material from a giant red star escapes to its smaller companion, a white dwarf, which is a star that is approaching the end of its life cycle (see section 14.5). The hydrogen-enriched material, after accumulating for some years, may stimulate the nuclear reactions on the surface of the aging star. The resulting burst of energy represents the nova eruption. Intervals between such eruptions may vary from 25 to 100 years.

One interesting result of the study of novae is that knowledge of the rate at which material is ejected has enabled astronomers to determine the distance to the nova. The shell formed from the expanding stellar material may, after several years of observation, be resolved as a disk. From this it is possible to determine the angular displacement of the shell's surface from the star. Spectroscopic examination provides the velocity with which the surface of the shell is approaching and receding from Earth. It may be assumed that the transverse velocity is equal to the velocity of approach and recession, because the erupting shell would be approximately spherical and moving equally in all directions. Using these elements—the annual displacement in seconds of arc per year (μ) and the transverse velocity in km/sec (v_t)—we can solve for the distance to the nova in parsecs (p).

$$p = 4.74 \, \frac{\mu}{v_t} \, \text{km/sec}$$

Supernovae are eruptive variables that are far more spectacular than novae, attaining magnitudes of –18 to –20. There are two types of supernovae, recognized partly from differences in their spectra and partly on the basis of their respective light curves. Type I supernovae are perhaps 2 magnitudes brighter at their maximum than type II, and the light curve for type I reaches a well-defined peak followed by an initial rapid drop and a slower fadeout. In addition, type I supernovae have broad-line spectra, but it has not yet been established whether these are emission or absorption lines. The type II supernovae have a rather irregular light curve and at maximum exhibit continuous spectra.

The amount of mass ejected during a supernova is great—estimated to be equivalent to 1 solar mass or more, which represents a large fraction of the star. This material is ejected at velocities of up to 10^4 kilometres per second. The tremendous energy required for these velocities is thought to be initially generated by a rapid collapse of the star with implosive force during what is called the *nucleosynthesis stage.* During this stage, elements in the iron group (iron, cobalt, nickel) are formed from carbon nuclei. This reaction is an exothermic reaction in which heat is given off, causing the core to reach temperatures of $3 \times 10^{9}\,°\text{K}$. Since the reaction occurs in a very short period of time (less than two minutes), the sudden release of the energy results in shock waves being set up internally and transmitted rapidly to the stellar surface. The surface of the star is not sufficiently stable to withstand the shock, and the energy is released in a supernova. It is thought that during the explosive phase the very heavy elements above iron on the periodic scale are formed through neutron capture by heavy nuclei. In this manner the interstellar medium is enriched with heavy elements, which may subsequently become part of newly forming stars.

14.5 THE DYING STAR

We know that there is a degree of uncertainty as to what occurs following the red giant stage in stellar evolution. The gap in the H-R diagram between the red giant stage and the white dwarf or final stage in the evolution of stars like the sun is thought by some astronomers to be occupied by the pulsating variable stars (Figure 14.12). The stars are contracting from the red giant stage, but are generally unstable and therefore exhibit the pulsating characteristic. This stage is thought to be relatively brief, lasting only 10^4 to 10^5 years.

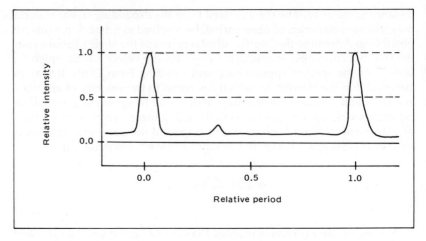

Figure 14.12 The periods of pulsars range from three seconds for the longest to 0.33 seconds for the shortest. The intensity of radiation averaged over many periods of a typical pulsar shows a main pulse and a smaller interpulse.

Beyond the red giant stage and the possible pulsating stage, astronomers recognize several types of stars that typify the final state in stellar evolution. These include the white dwarfs, the neutron stars, and the intriguing, recent addition to the group, the black holes. The path to stellar finality is determined by the star's mass.

White Dwarfs

In stars like the sun the temperatures at the helium core may not reach a level high enough to initiate carbon burning. However, the core may contract and temperatures increase sufficiently to permit helium shell burning, which causes the outer surface layer to expand and separate as a planetary nebula. The star from which the planetary nebula originates is now a very dense, hot white dwarf. Stars less than 1/3 solar mass generally become white dwarfs without helium burning of the core and the ejection of a planetary nebula.

Many white dwarfs are companions to more massive stars and are detected by the movement of the larger component. Such a star is Sirius B, which along with other white dwarfs occur on the H-R diagram below the main sequence and in the lefthand quadrant. Their surface temperature is high, ranging from 5000 to 50,000°K, and their size is small, perhaps 0.02 solar radii (2 Earth radii) or smaller, down to the size of Earth. If a star of one solar mass shrinks down to the size of the Earth, there must be an accompanying rise in density, and white dwarfs are exceedingly dense. At these densities electrons can no longer move at random, but have their motion restricted by close packing with other electrons. Because of the motion permitted to electrons in these circumstances, there is a limit beyond which electrons cannot be compressed. Eventually the white dwarf will contract to the point where electrons occupy their smallest possible volume, and they exert an outward pressure preventing further contraction of the star. The atomic nuclei become arranged much like close-packed marbles, and material at the center of the white dwarf begins to resemble more the properties of a solid than a gas. Such a material is called a *degenerate electron gas* and may achieve densities up to 10^8 g/cm^3, compared with 150 g/cm^3 for the present density at the center of the sun.

Under the conditions just described, all nuclear reactions have ceased and the star has contracted gravitationally until reaching the limits of contraction imposed by the degenerate gas. No internal energy is being generated, so the star gradually cools, eventually becoming a "black dwarf." The upper limit for white dwarfs was theoretically calculated to be 1.4 solar masses by S. Chandrasekhar, and is called the *Chandrasekhar limit.*

Neutron Stars

Stars with greater than 1.4 solar mass will shed mass through the supernova mechanism. The collapse or implosion of the remaining fraction of the star not lost in the supernova explosion results in matter being compressed to a density of over 10^{14} g/cm^3. As this density is approached electrons accelerate to velocities near the speed of light and are forced into the atomic nuclei to combine with protons to form neutrons. The neutrons will react at high pressures in the same way that electrons do—that is, when occupying their smallest possible volume they exert an outward pressure as a *degenerate neutron gas,* thus halting further collapse. The star is now a neutron star with a diameter of 10 to 30 kilometres.

Can we detect a neutron star? The existence of these objects was first suggested in the 1930s but not confirmed until after the discovery of pulsars in 1967. Two young British astronomers at Cambridge were studying radio waves from a distant galaxy in the constellation Vulpecula, using a radio telescope. They detected radiation that was not a steady signal but rather a series of precisely timed, very short-period pulses. Because the source could not be identified, several theories were suggested to account for the signals. One theory proposed that the source was one of the many artificial satellites orbiting the Earth, but this was immediately rejected. Another imaginative suggestion was the LGM theory, according to which an advanced civilization on a planet orbiting a distant star had placed navigational beacons in space to aid them in space travel. The signal came from one of the beacons oriented so that the signal was inadvertantly beamed in the direction of the Earth. The first discoveries of these radio sources (about 100 are now recognized) were identified as LGM I, LGM II, etc., but later changed to a more appropriate designation, CP for Cambridge Pulsar. Needless to say, the LGM stands for "little green men."

The idea that the *pulsar* (pulsating radio star) was actually a neutron star was not long in coming. The first clue was the discovery that a pulsar with the shortest period of any thus far found was located close to the center of the Crab nebula, a supernova remnant. This star, the only pulsar thus far detected in the visual portion of the spectrum, had been observed since 1942 and was suspected to be the remnant of that supernova. The period of that pulsar (0.033 sec) is the same in the visible, radio, and X-ray portions of the spectrum, and the pulse lasts only a millisecond. Since light travels only about 300 kilometres during a millisecond, the object cannot be more than 300 kilometres in diameter. The periods of pulsars range from 3 seconds for the longest to 0.033 seconds for the shortest (Figure 14.12 and Figure 14.13).

The pulses have been intensely studied and can either be related to the star's pulsating characteristics or to the rotation of the star. We have previously discussed the characteristics of various types of pulsating stars

Figure 14.13 Photograph of Crab nebula pulsar taken at maximum (top) and minimum (bottom) light. (Lick Observatory photograph)

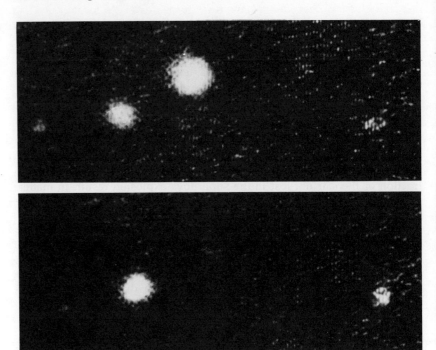

whose periods were measured in hours, days, or even years. These were generally very huge stars of low average density. Small stars such as white dwarfs with high densities, if unstable and capable of pulsating (which they are not), would have a period of several seconds. If the pulses were in some way related to the star's rotation, the period of the pulses would be approximately the same as if the star pulsated, for it has been calculated that the pulsating period of a star is close to the fastest period of rotation of that star. The rotating star is the most acceptable model for a pulsar, which rotates at high speed as a mechanism for preserving angular momentum. For example, a star on the main sequence rotates, and the angular momentum of this rotation is dependent upon the star's mass and surface speed. As the star collapses its diameter decreases, but its mass remains essentially unchanged or at most is reduced only slightly compared to the reduction in the size of the star. Because the mass of the star is unchanged, the period of rotation is drastically decreased to conserve angular momentum. Now we have a pulsar rotating at high speed, but we can ask the question, What is the energy source of the pulses?

The source of energy of the pulses may be a very strong magnetic field. On the main sequence a star may have a magnetic field of average strength. During the collapse to the neutron star stage, the magnetic field also contracts but without a loss of energy, resulting in a field of very high intensity. Although there is not general agreement on the final model for a pulsar, it is felt that the high speed of rotation and the strong magnetic field are the source of the pulses. As the dying star rotates, electrons and atomic nuclei are thrown out into the magnetic field from the magnetic poles. The interaction of the charged particles with the magnetic field accelerates the particles to near the speed of light. This generates a strong *synchrotron radiation*, which is detected as pulses. If the magnetic poles, which do not necessarily coincide with the rotational axis, are oriented toward the Earth, the radiation pulses will be detected as the star rapidly spins. This action is analogous to a lighthouse beacon sweeping across the sea and with each rotation briefly illuminating a ship.

Let us briefly summarize the relationship of a neutron star and a pulsar. A neutron star was theorized to exist as the remnant of the supernova in the Crab nebula. A pulsar was discovered in the Crab nebula in the position thought to be occupied by the neutron star. The pulsar revealed traits which were consistent with what was theorized about a neutron star, and the conclusion was drawn that pulsars and neutron stars were the same. If the interpretation of the source of the pulses is correct, then only neutron stars with their magnetic fields properly oriented toward the Earth can be detected. From this we can infer that there may be many more neutron stars in existence whose presence we cannot detect.

The continuous monitoring of the pulsars has revealed that there is a very gradual slowdown in their rate of rotation. Typically, the rate of increase in the pulsars' period is about a microsecond (10^{-6} sec) per month. The gradual slowing down is consistent with the theory that neutron stars are of different ages. The Crab nebula neutron star, having the fastest rotation of any of the observed neutron stars, would also be the youngest, according to this theory. The loss of energy reflected in the slowdown can be estimated from the energy lost in synchrotron radiation by the star to the surrounding nebula. This loss can be equated with the decrease in the rotational speed of the neutron star, and that in turn is related to the increase in the pulse period. Another variation in rotational speed has been noted in both the Crab nebula and the Vulpecula (Vela) neutron stars. In both cases there was a one-time increase in period, and this has been interpreted as a starquake. It was suggested by M. Rudermann that neutron stars have a solid crust which is disrupted periodically by the gradual slowing down of the star's rotation. This interpretation is the subject of some controversy.

Black Holes

We now come to the final and dramatic end in the life cycle of a massive star. Neutron stars will contract to a point where the outward pressure of the degenerate neutron gas prevents further collapse. There is however a maximum-mass limit beyond which neutron stars are unable to sustain themselves against further contraction by gravity. This limit, originally set at 0.7 solar mass by J. R. Oppenheimer in 1939, has more recently been revised upward to as much as 2.5 to 3 solar masses. This would include the very large stars with masses up to 50 or more solar masses. In many cases not enough mass is shed during the nova and supernova stage to permit these stars to end up as stable white dwarfs or neutron stars. In this event gravity becomes the dominant force, because of the large mass causing the star to collapse to a smaller and smaller object. As the star contracts, the material of which the star is composed increases in density and the gravitational field increases in strength.

Let us digress for a moment to recall that to escape from the Earth objects are required to reach a certain velocity— approximately 12 km/sec. Now if we were to shrink the Earth to one-half its present radius without a loss of mass, the escape velocity would increase to about 17 km/sec. An object that is more massive than the Earth but the same size would also

Figure 14.14 A diagram
of a black hole showing
the photon sphere, event
horizon, and singularity.
(By permission of Harper
& Row, Publishers)

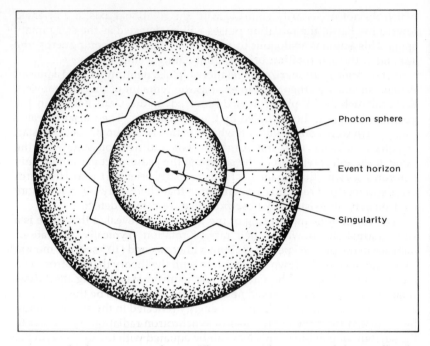

require a higher escape velocity. The greater the mass and the smaller the
size, the higher will be the escape velocity. From this it should not be too
difficult to conceive of a massive star that has collapsed down to a very
small radius and has a gravitational field so great that the escape velocity is
equal to the speed of light. A star that contracts beyond this radius, the
Schwartzschild radius, is a black hole. A *black hole* is a star whose gravitational
field becomes so strong that the escape velocity exceeds the speed of light.
What is the size of such an object? This is dependent upon the mass of the
original star. For a star of one solar mass the Schwartzschild radius would
be 3×10^5 centimetres or 3 kilometres. Actually, stars less than 2.5 solar
masses would not be capable of collapsing to the Schwartzschild radius.
For stars more massive than 2.5 solar masses, the radius is determined by
multiplying 3×10^5 cm by the mass of the star in solar masses. Thus a star
of 3 solar masses would have a radius of 9 kilometres at the time it reached
the Schwartzschild radius.

Figure 14.15 A beam of
light approaching a black
hole may be deflected,
go into circular orbit of
the photon sphere, or be
sucked into the black
hole. (By permission
of Harper & Row, Pub-
lishers)

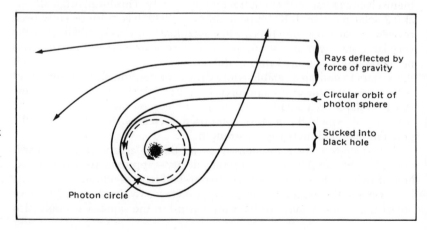

In theory the star could continue to collapse, because there is no force to counter the force of gravity. Having contracted beyond the Schwartzschild radius, it is no longer possible to see or learn about events on the star. An indestructible person on its surface could no longer communicate with the outside universe, nor could anyone communicate with that person. In a very real sense he has passed over a horizon—the *event horizon*, or the point at which an object disappears from view as it approaches a black hole because the escape velocity equals the speed of light. Continued collapse would result in the star contracting to a very tiny volume, ultimately reaching zero volume, which is known as the *singularity*. At this point the intensity of the gravitational field would be infinite, and the matter of which the star was composed would be subject to infinite pressure and would be at infinite density (Figure 14.14).

Although many of the concepts just discussed are not new, they are so strange that scientists have been reluctant to consider them seriously until the last decade. With the discovery of pulsars and their relationship to neutron stars, black holes no longer were considered so unlikely, and the hunt for such objects was on.

But how does one look for an object from which no light nor any form of radiation can escape? One method would be to observe how a beam of light would react when passing a suspected black hole. Recall that Einstein predicted light would be deflected when passing an object with a strong gravitational field (section 3.6). What would be the gravitational effect on a light beam of an object with the properties of a black hole? The closer the beam of light is to the black hole the more the light beam would be bent. Finally, the light would be sufficiently close to cause it to go into a circular orbit around the black hole in what is called the *photon sphere*. If the beam were to come closer it would be sucked into the black hole. However, this

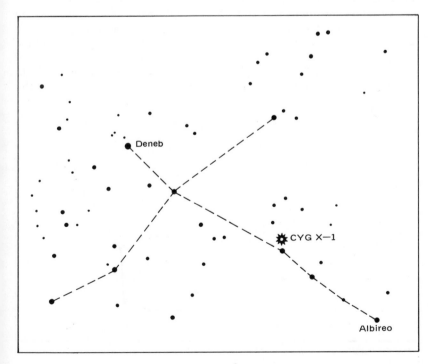

Figure 14.16 An X-ray source, Cygnus X-1, appears to be a black hole. The object has been identified as HDE 226868, a binary star about 2450 pc from Earth.

Figure 14.17 Artist's conception of the black hole of Cygnus X-1. The painting shows gas being drawn from the large star into the companion black hole. (Courtesy of Lois Cohen, Griffith Observatory)

is not a practical approach to detecting suspected black holes, since the orbit of the light beam would be extremely unstable (Figure 14.15).

A more practical approach to finding a black hole has been to attempt to detect its effect on matter or, more particularly, its effect on another star. It is well known (see chapter 15) that many of the stars we see are double-star systems, and there is a good probability that a black hole may be detected in conjunction with a normal star. Such a possible relationship has been detected in the constellation Cygnus. In 1962, an X-ray source was discovered in the constellation of Cygnus, but the location was not pinpointed until 1973 with the aid of the Uhuru satellite launched in 1970. Information obtained from the satellite in conjunction with data from the

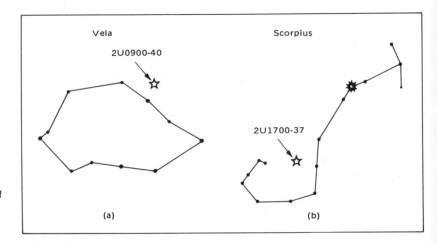

Figure 14.18 Two additional stars suspected of being black holes are 2U0900-40 in Vela and 2U1700-37 in Scorpius.

National Radio Astronomy Observatory identified the object as HDE 226868 (Figure 14.16), a binary star system about 2450 pc from Earth. This system is composed of a highly luminous type BO star of about 30 solar masses and an unseen companion of about 6 solar masses, orbiting each other every 5.6 days. The small, unseen companion, now known as Cygnus X-1, appeared to be the X-ray source, and was identified as a possible black hole. The X rays are generated, it is thought, by gas being pulled from the surface of the large star into the black hole (Figure 14.17). In so doing, the speed at which the gas flows will increase as it approaches the black hole. Temperatures of the gas will rise in excess of $10^{6\circ}$K, and as a result X rays will be emitted as the gas nears the event horizon. The X-ray emissions appear to be of a pulsing nature, indicating that the source is rotating at high speed. Two additional X-ray sources with black hole characteristics similar to Cygnus X-1 are being studied. These are identified as 2U0900-40 in the constellation Vela and 2U1700-37 in the constellation Scorpius (Figure 14.18).

14.6 SOME BIZARRE ASPECTS OF BLACK HOLE ASTRONOMY

Theoretical aspects of the black hole have led astronomers to speculate upon the ultimate fate of matter in a black hole. One mathematical model of space around a black hole, developed by Einstein and Rosen in the 1930s, indicated that the event horizon connected our universe with another universe. This connection became known as the *Einstein-Rosen Bridge*, or the *wormhole*. Through the wormhole, it was suggested, matter from a black hole is funneled to emerge as a *white hole* in another universe (Figure 14.19). Robert Hjellming, an astrophysicist at the National Radio Astronomy Observatory, speculates on the possibility that quasars (see chapter 17) may be white holes from another universe and that the flow of matter between the two universes is maintained in equilibrium in this manner. Could we travel from one universe to another through the black hole? Not likely, but a rotating black hole, it has been theorized, would have two event horizons. An intrepid astronaut crew could avoid the singularity by entering the black hole between the event horizons and emerge in the other universe or at another place or time in our own universe.

It has been suggested by Stephen Hawking, astrophysicist at Cambridge, that at the time the universe was formed numerous mini-black holes were created during the initial "big bang." Many of these might be the size of a dust particle and have the mass of the Earth. According to some Atomic Energy Commission researchers, these mini-black holes could be useful as an inexpensive and nonpolluting source of energy. They suggest that if such a mini-black hole could be found and relocated in Earth orbit, tremendous amounts of energy could be generated by firing hydrogen pellets into the black hole. The tremendous gravitational field would cause a nuclear reaction with the resulting radiation of hot gases. If the gases were passed through a magnetic field produced by a space station near the mini-black hole, an electric current would be generated and by means of microwave technology beamed to antennae on Earth where the microwaves would be converted to electric power.

Figure 14.19 The Einstein-Rosen bridge, or wormhole, through which matter is funneled from a black hole in one universe to emerge as a white hole in another universe (a) or in some different part of our universe (b). (By permission of Harper & Row, Publishers)

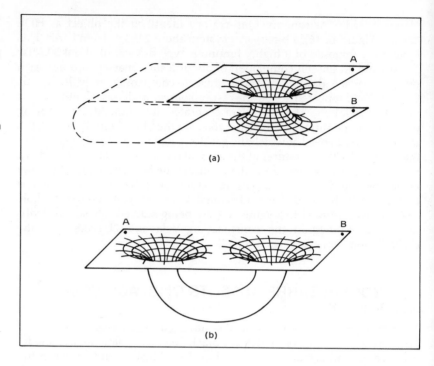

It must be recognized that the kinds of things just discussed are pure speculation and, while theoretically feasible, do not have any practical application in the foreseeable future. However, study of black holes is one way to understand the universe and might prove to open paths to as yet unthought-of concepts.

14.7 SUMMARY

Stars are formed from huge clouds of dust and gas called nebula. More than one star will form in a nebula, resulting in stars appearing in distinct groups. The stars form when the nebula begins to condense. This tends to concentrate mass, which in turn hastens the contraction further. As the material forming the protostars contracts, it heats up until the temperature is sufficiently high to initiate a thermonuclear reaction. Hydrogen, the main constituent of the protostar, is converted to helium through the proton-proton reaction, and the protostar becomes a star on the main sequence. At this point the star has achieved a state of equilibrium, where the internal pressure generated by the nuclear reactions is in balance with the collapsing force of gravity.

The place on the main sequence occupied by a star is dependent upon its mass. Massive stars remain there a relatively short time compared to the sun, whereas very small stars occupy their position on the main sequence longer than the sun. Ultimately, the hydrogen at the core will be exhausted. Hydrogen around the helium core will begin burning and the energy generated will cause the star to expand into the red giant stage.

Some variation in the aging pattern followed by a star will occur, depending upon the star's mass. A small star of less than one solar mass may enter the red giant stage twice before contracting down to a white dwarf without helium burning at the core. Stars of approximately one solar mass may become somewhat unstable after the red giant stage, and shed mass by ejecting a planetary nebula before collapsing to the white dwarf stage.

Stars larger than 1.4 solar masses may become Cepheid variables after the red giant stage or may shed mass through the supernova mechanism. If they shed sufficient mass to bring them below 2½ to 3 solar masses, they will collapse to form neutron stars. Neutron stars were first detected as pulsars. Large stars of 3 solar masses or more will contract beyond the neutron star stage to become black holes.

QUESTIONS

1. Neutral hydrogen emits radiation at a wavelength of 21 centimetres. How is this energy generated?
2. What is the difference between an H I and an H II region?
3. Where are stars most generally formed and why?
4. Trace the development of a star of one solar mass from the initiation of formation as a dark cloud until it reaches zero age main sequence.
5. How is helium formed if carbon and nitrogen are present in a massive star?
6. What is meant by the triple alpha process? Where does this reaction occur?
7. After a star has exhausted a part of its hydrogen fuel, it begins to age and leaves the main sequence. Trace the process from main sequence to white dwarf stage.
8. Briefly describe the main types of pulsating variable stars.
9. In what way may the Cepheid variables be used to determine the distance to a star?
10. What is the difference between a fast nova and a slow nova?
11. How do novae and supernovae differ?
12. Make a table showing the mass, radius, and density of white dwarfs, neutron stars, and black holes.
13. What are the characteristics of a neutron star?
14. What is the source of the pulses from a pulsar?
15. Define the following: singularity, photon sphere, event horizon, wormhole.

EXERCISES AND PROJECTS

1. By making a number of sample star counts, it is possible to obtain an estimate of the total number of stars visible to the unaided eye. Some device such as a cardboard mailing tube may be used to regulate the area of the sky to be sampled. The entire celestial sphere will then be

limited to the length of the tube, which will represent the radius of the celestial sphere. From this a measure of the total area of the celestial sphere may be determined. The area covered by each sample star count will be equal to the aperture of the tube.

$$\text{Area of sphere} = 4\pi r^2 \qquad \text{Area of aperture} = \pi r^2$$

Make at least two dozen sample counts under as ideal conditions as possible. A dark clear moonless night far from any source of artificial light is best. Take samples from all parts of the sky from horizon to zenith. Notice that stars are not uniformly distributed over the entire sky, so sampling must be representative. Once the sample counts have been made, determine the average count. Determine the total number of stars visible as follows:

$$\frac{\text{Area of sphere}}{\text{Area of aperture}} \quad \times \quad \begin{array}{c}\text{average number of} \\ \text{stars per sample}\end{array} \quad = \quad \begin{array}{c}\text{Total number of} \\ \text{stars visible}\end{array}$$

Remember that this provides an estimate of the stars visible to the unaided eye in the entire sphere. Under ideal conditions this should be about 6000 stars. We can only see half of the celestial sphere at any one time, so total stars visible will only be half of the amount calculated above. Make a similar star count in a metropolitan area if possible and compare the two counts. By what percentage is visibility reduced by light pollution and other atmospheric conditions indigenous to metropolitan areas?

FOR FURTHER READING

Aller, L. H., *Atoms, Stars and Nebulae.* Cambridge, Mass.: Harvard University Press, 1971.

Brandt, J. C., *The Physics and Astronomy of the Sun and Stars.* New York: McGraw-Hill, 1966.

Clayton, D. D., *Principles of Stellar Evolution and Nucleosynthesis.* New York: McGraw-Hill, 1968.

Gamow, G., *The Birth and Death of the Sun.* New York: New American Library, 1952.

Gingerich, O., ed., *New Frontiers in Astronomy.* San Francisco: W. H. Freeman and Company, 1975.

Rose, W. K., *Astrophysics.* New York: Holt, Rinehart and Winston, 1973.

Cepheus

CHAPTER 15   **THE MULTIPLES**

> Great fleas have little fleas upon their backs to bite 'em,
> And little fleas have lesser fleas, and so ad infinitum.
> And the great fleas themselves, in turn, have greater fleas to go on;
> While these again have greater still, and greater still, and so on.
>
> —*De Morgan*, A Budget of Paradoxes

The fact that stars are formed in close association in dust and gas clouds has led to the specific grouping of stars. The simplest of the groupings is the binary, composed of two stars in close relationship. The number of components in a cluster may be as many as 10^6 stars or more.

15.1 BINARY STARS

The sun is a single star, but this is not the status of the majority of the stars. Close to 50 percent of the stars seen as singles without the aid of a telescope are actually multiple systems with two or more components. Two stars related in this manner are known as *binaries,* and the individual stars in the system are the *components*. Although the components in a binary system are generally close together, this alone does not identify them as such. Two stars may appear close because of the coincidence of alignment but actually may be a considerable distance apart along the line of sight (Figure 15.1). These are known as *optical binaries* and are rather rare. Stars in a binary system orbit around a common center of mass or barycenter similar to the Earth and moon.

    The first binary star was discovered in 1650 by an Italian astronomer, Jean Baptiste Riccioli, who observed that Mizar, the middle star in the handle of the Big Dipper, contained two components. The binary may readily be seen by anyone with a pair of good binoculars.

    Many more binaries were discovered after the initial Mizar find, and in the early nineteenth century William Herschel published a catalogue of more than 800 two-component binaries. In his study of binaries Herschel found one component to be brighter than the other, and he therefore assumed that the smaller one was more distant. His intention was to find the distance to the smaller star by measuring the parallax of the nearer star. However, by 1803 he had determined that the two stars in Castor (seen with the naked eye as a single star in the constellation Gemini) were

Figure 15.1 Two stars that may appear to an observer on Earth to be a binary pair may actually be a considerable distance apart along the line of sight. These are optical binaries.

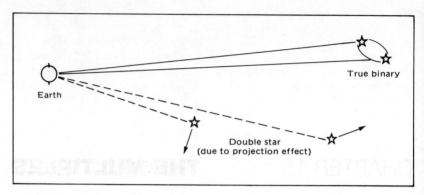

Figure 15.2 The paths across the sky of Sirius *A* and *B*. The small inset shows the possible real motion of the two components around the center of mass *C*.

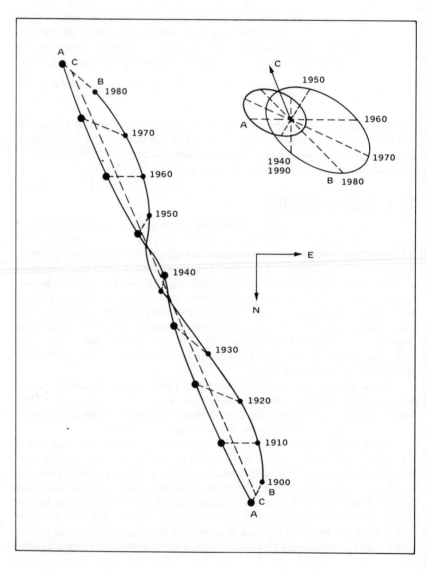

actually revolving around a common center of gravity. He recognized the distinction between optical and real or *visual binaries* as they are now known, and realized that he had found something of great interest.

Toward the end of the nineteenth century, E. C. Pickering discovered that Mizar A, the brighter of the two, was a spectroscopic binary, detected by Doppler shifts in its spectrum. After the turn of the twentieth century, E. B. Frost found that Mizar B, the fainter component, was also a spectroscopic binary. At present about 65,000 binaries are recognized and listed in the Double Star Index and Observation Catalogue by the U.S. Naval Observatory.

Visual Binaries

Visual binaries are those that can be resolved into two stars telescopically. These stars are observed revolving around a barycenter and, as a result, move across the sky on a wavy course (Figure 15.2). The serpentine path of the larger component is less pronounced than that of the smaller. However, of all the recognized binaries only about 300 have been observed for a long enough period to permit the calculation of orbits. The calculation of an orbit is accomplished by observing the motion of the fainter component with respect to its brighter companion. The angular separation or distance between the components and the position angle measured from north through east (Figure 15.3) are determined at regular

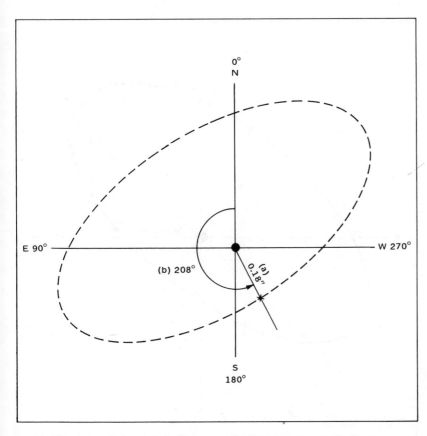

Figure 15.3 The angular separation between two components is measured (a) in seconds of arc, and the position angle is measured from north through east (b).

intervals over an extended period. This will provide the *apparent relative orbit* of the smaller component with respect to the larger component of the system (Figure 15.4). It is recognized that in an apparent relative orbit the fainter component follows an elliptical orbit and obeys Kepler's second law (law of areas), but the brighter companion does not fall on a focus of the ellipse. This is due to the fact that the apparent relative orbit is generally not perpendicular to the line of sight, but is inclined at some random angle to what is called the *plane of the sky*. By careful study of the properties of the apparent relative orbit, it is possible through geometric methods to obtain the angle and thereby determine the *true relative orbit* of the system (Figure 15.5). The true relative orbit is still measured relative to the major component in the system. In order to determine the true motion of both components through space, it is necessary to measure their position relative to nearby stars. When the orbital period and distance between a binary pair are known, their respective masses may be determined (see equation 13.9).

Astrometric Binaries

Both components need not be visible in order for an astronomer to realize that the observed object is a binary star system. In the early half of the nineteenth century, F. W. Bessel, while studying the proper motion of Sirius and Procyon, detected the fact that their paths were wavy and not straight. From this he deduced the presence of unseen companions and in

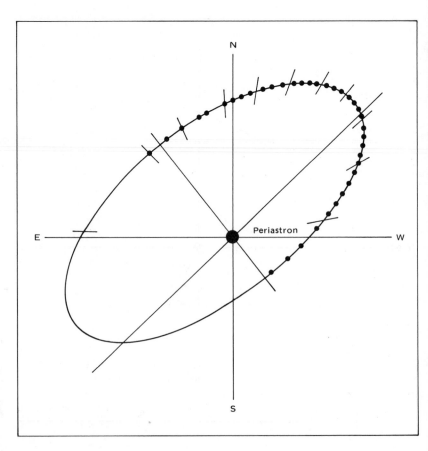

Figure 15.4 The method diagramed in Figure 15.3 will, over an extended period, result in an apparent relative orbit of the smaller component with respect to the larger as if the larger component were stationary. Periastron is the smallest distance between components.

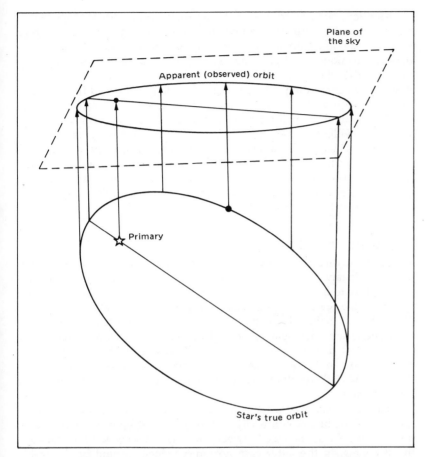

Plane of
the sky

Apparent (observed) orbit

Primary

Star's true orbit

Figure 15.5 By a careful study of the inclination of the orbital plane of the apparent relative orbit, it is possible to determine the true relative orbit of the smaller component with respect to the larger.

1844 suggested the possibility that both stars were binaries. This turned out to be true, for in 1862 Sirius B was detected telescopically by Alvin Clark who was testing a new 18.5-inch lens. Procyon B was seen in 1896.

Binaries of the type just described can only be detected by virtue of the influence of the invisible component on the visible component. They are known as *astrometric binaries*. When the invisible component is finally seen, the system takes its place with the visual binaries.

Other examples exist where the fainter component has not yet been seen. Probably the best-known example is Barnard's star, 1.83 pc from Earth. Peter van de Kamp has analyzed over forty years of data on the motion of this star. From his studies he concluded that Barnard's star is orbited by a dark companion of 0.0018 solar masses; that is, the companion has a mass about 1.8 times that of Jupiter. Since this mass is considered insufficient to initiate a thermonuclear reaction, the companion is thought to be a large planet about 4.5 AU from the star with an orbital period of 25 years. Another interpretation of the data indicated the presence of two planets, one with 1.1 times Jupiter's mass with a period of 26 years and the other 0.8 Jupiter's mass with a period of 12 years. The presence of any planet associated with Barnard's star has been challenged by G. Gatewood and H. Eichhorn. In an independent study, upon which they reported in 1973, they were unable to observe the periodic wavy motion ascribed by van de Kamp to Barnard's star. Their conclusion was that no planet existed.

Another star, Laland 21185, 2.5 pc from Earth, also is suspected of having a dark companion about 10 times the mass of Jupiter. Both Barnard's star and Laland 21185 are quite close to the solar system, and the dark companions are large compared to planets in the solar system. Thus far only large planets have been found to be associated with stars, but this does not preclude the presence of smaller planets. Viewed from another planetary system the combined mass of the gas giants may be detected, but it is unlikely that the presence of the nine planets around the sun would be evident. Other nearby stars suspected of having dark (planet) companions include Epsilon Eridani, 3.3 pc; 61 Cygni, 3.4 pc; and Luyten's star, 3.8 pc. Sufficient data for making a positive statement about the mass of the dark companions is not yet available. On a statistical basis, however, the data, if correctly interpreted, would indicate that planets are definitely not unique to the solar system.

Spectroscopic Binaries

The ability to distinguish the components in a binary system is dependent upon the distance between the components. In many instances the stars are too close to be visually separated, but it is possible to resolve them by means of the spectroscope. Mizar, the first visual binary, also has the distinction of being the first *spectroscopic binary*. In such a system the components are not visually distinguishable, but may be observed by the Doppler shift in their spectral lines. E. Pickering in 1889 observed this phenomenon in the spectrum of the large component of Mizar, and thought it was the result of a variation in the radial velocity of two closely orbiting stars.

The most ideal circumstances for detecting such variations occur when the orbit of the binary pair is edge-on to our line of sight. More generally, the orbit would be inclined between the edge-on view and the plane-of-the-sky view. In the latter case no radial velocity would be detectable except that of the entire system. Figure 15.6 represents a hypothetical binary system, with two stars in circular orbit around a barycenter. At position A the orbital motion of the two stars is transverse to the line of sight from the Earth, and their radial motion is the radial motion of the entire system through space. The radial motion of the two components is equal. At position B component 1 is moving away from the Earth, displaying a positive radial velocity with respect to the system. The spectrum of component 1 would be red-shifted. Component 2, moving toward Earth, displays a negative radial velocity and its spectrum is blue-shifted. Now the two components are moving in opposite directions with respect to the barycenter of the binary system. The radial velocity of component 1 with respect to component 2 is the sum of the radial velocities of the two components with respect to the barycenter. This represents the maximum radial velocity relative to the system as it appears in its inclined position with respect to the observer on Earth. If the binary system were edge-on to the Earth, the radial velocity would be greater. In position C the components have returned to a net zero radial velocity with respect to the binary system. At position D component 1 is moving toward the Earth, displaying a negative velocity with respect to the binary system, and has a blue-shifted spectrum. Component 2, moving away from the Earth, has a positive radial velocity and its spectrum is red-shifted. The spectrum for each position would vary as shown in Figure 15.6.

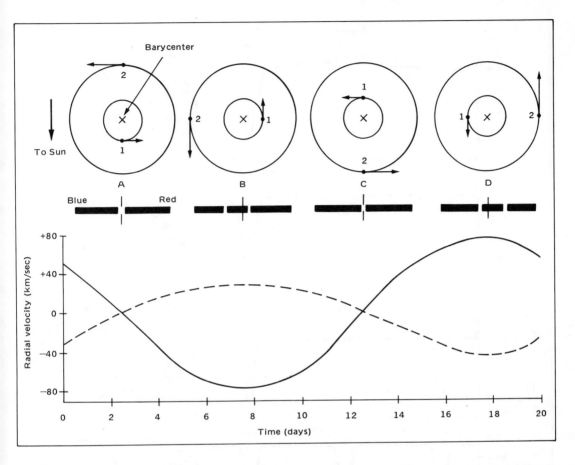

Figure 15.6 A hypothetical spectroscopic binary showing relative velocities over the period of the orbit and the spectral shift resulting from radial motion.

In section 13.5 we discussed the procedure for determining stellar mass using Newton's modification of Kepler's third law. In this case a, the distance between the two components, is dependent upon the maximum radial velocity and may be calculated in the following manner:

$$a = \frac{vP}{2\pi}$$

where v is the relative velocity for the components and P is the period of their relative orbit. Recall that the maximum radial velocity was less than the actual radial velocity because of the inclined orbit. As a result, the determination of mass would yield only a lower limit for the mass of the components. An accurate measurement of mass would be dependent upon a determination of the inclination of the orbit, and this measurement is accurate only if the orbit is in an edge-on position with respect to Earth, so that one component can partially or completely eclipse the other.

Eclipsing Binaries

If the orbit of a binary system is close to 90° with respect to the plane of the sky or, in other words, approximately edge-on to us on Earth, one star will periodically eclipse the other. This is known as an *eclipsing binary*. Algol (Beta Persei) or the "demon star," as it was called by ancient astronomers, was observed by them to be "winking" in the sky. In 1889, H. Vogel, using

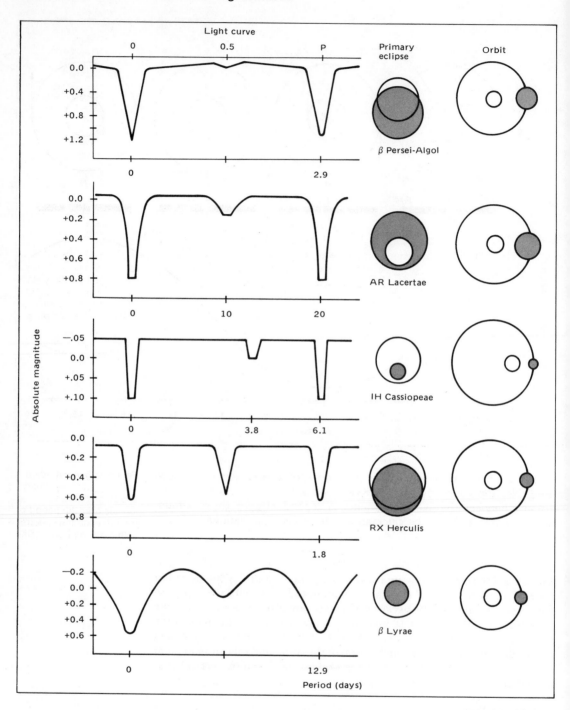

Figure 15.7 The brightness of eclipsing binaries varies with time. Shown here are some representative eclipsing binaries and their respective light curves.

spectroscopic techniques, confirmed what was suspected, that Algol was an eclipsing binary. Algol has a magnitude of 2.2, which decreases to about 3.4 in a few hours in a regular period of slightly under 69 hours (2d 20h 49m). If we plot the magnitude as a function of time, we obtain a *light curve* (Figure 15.7) which yields much valuable information about the nature of binary systems.

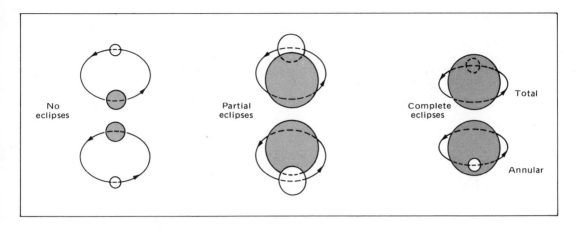

We need to recognize that the two components of a binary system are apt to be dissimilar in size and brightness, and this will result in a variety of light curves. If the stars are not in eclipse, the brightness is essentially constant and at a maximum. When the hotter star passes behind a cooler star, a *primary minimum* (in brightness) occurs. With the reverse there is a *secondary minimum* (Beta Persei) (Figure 15.7). An eclipse may be *total*, with one star completely hidden by the other, or there may be an *annular* eclipse, where the larger star is behind the smaller. A *partial* eclipse will result when one or the other of the components is partially hidden (Figure 15.8).

In the event of a total eclipse of a binary system with an orbit directly edge-on to the Earth, the eclipse is termed *central*. If the orbit is inclined slightly the eclipse may still be total but *noncentral*. This kind of information is useful in determining diameters of respective components and the orbital inclination. To illustrate, let us consider the light curve resulting from a central eclipse. In Figure 15.9 the limb of the smaller star is just tangent to the larger star in position 1 at the beginning of the eclipse. This is known as *first contact*. The brightness of the system decreases as the smaller star passes behind the larger star until completely hidden at *second contact*, when primary minimum begins. When the small star reappears at *third contact* the system begins to brighten, reaching maximum brightness on the light curve at *fourth contact*. The sequence is repeated as the smaller star passes in front of the larger and the light curve reveals a secondary minimum. Orbits do not generally occur edge-on, so central eclipses are infrequent. In a noncentral total eclipse the smaller star passes obliquely behind or in front of the larger companion, and therefore requires a longer period from first to second contact and third to fourth contact. At the same time the small star traverses a shorter chord rather than the diameter of the star. This results in a shorter period of totality and a shorter minimum on the light curve than if the eclipse were central.

The diameter of the components of an eclipsing binary may be determined if the eclipse is central, provided the velocity of the stars is known. The period from first contact to second contact and the velocity of the smaller star would yield its diameter. The period from first contact to third contact and the velocity of the smaller star would provide data for determining the diameter of the larger companion. If the system's orbit is inclined and has noncentral eclipses, the problem is somewhat more complex but can be solved nevertheless.

Figure 15.8 The type of stellar eclipse that may occur is dependent upon the inclination of the orbital plane to an edge-on position with respect to Earth.

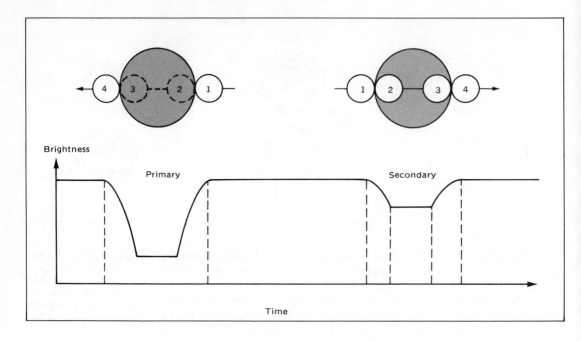

Figure 15.9 A central eclipse is useful in determining the diameters of the components. A circular orbit is assumed.

In the foregoing we have assumed an eclipsing binary system with a circular orbit. What would the light curve be like if the orbit were elliptical? We know that the motion of an object in an elliptical orbit is not uniform, and therefore the spacing of the light minimum in the light curve would also not be uniform (Figure 15.10). This would be the case if the eclipse occurred between *periastron* (point of smallest distance between components) and *apastron* (point of greatest distance between components). If the eclipses occur at or near periastron and apastron, the intervals of the light minimum would vary, because the stars would move more rapidly when close together at periastron than at apastron.

Binary and Multiple Systems

Many systems are composed of more than two components, for example, Alpha Centauri, the nearest star to the solar system. This star is resolved by the telescope as a double star, with a third, very faint component orbiting the others approximately every million years. Another interesting binary is Castor in the constellation of Gemini. Castor, seen as a single star with the unaided eye, has two components, A and B. These are separated by 86 AU and complete one revolution every 380 years. A third companion, C, lies approximately 1000 AU from the other two and is darker and cooler than its companions. Each of the three components making up Castor has been found to be spectroscopic binaries. Component C has a very short period of 0.81 days and is also an eclipsing binary. Thus Castor, seen initially as a single object, when closely examined by various techniques is found to be composed of six stars.

Many stars are multiple systems of one sort or another. In fact, it is likely that the majority of the identifiable stars are multiples, and with improved techniques more will be discovered no doubt.

A generation ago G. P. Kuiper expressed the idea that binary systems existed ranging from those composed of two components of equal mass to systems with a residual mass of less than 1 percent orbiting the main star as planets. According to Kuiper, if the angular momentum of the original dust and gas cloud was high, there was a good chance the cloud would split into equal parts, forming a binary of two components of approximately equal mass. A smaller angular momentum could result in one large and one small star, while a still smaller angular momentum could result in a star and a series of planets. It is possible, then, that many single stars may have planets associated with them, but like the planets of the solar system they may be too low in mass to be detected.

15.2 LARGER STELLAR GROUPS

We have so far described multiple-star systems that have up to six members in the system. However, this is not the upper limit of the number of stars that can occur in a group. Larger groups or clusters have been found that have a common bond other than an orbit around an identifiable barycenter. A lengthy study of the motion of large numbers of stars has revealed that stars occur in identifiable clusters by virtue of the fact they all have the same space velocity. This means they all move at approximately the same speed in the same direction. Several types of clusters are recognized; these include the open or galactic cluster, the globular cluster, and associations.

Open Clusters

Open clusters—so called because they appear as a loose assemblage of stars—are also called *galactic clusters* due to their location in the disk, particularly the spiral arms of our galaxy (see chapter 16). They lie almost without exception within a few degrees of the Milky Way. Open clusters contain from a few to 1000 stars and occupy a volume of space less than 10 pc in diameter.

Figure 15.10 The light curve at different points in the orbit of a binary pair with an elliptical orbit.

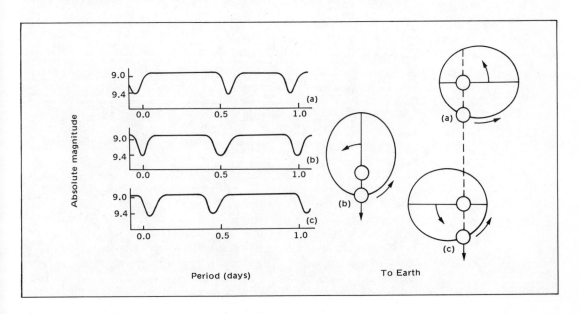

The *Pleiades* (Seven Sisters) is the best known of the open clusters, since it is easily visible to the naked eye in the winter sky. According to mythology, the Pleiades were the seven daughters of Pleione and Atlas. Jupiter changed them into a flock of doves to avoid their capture by Orion, and they flew heavenward and took up their position in the constellation Taurus. Only six can be seen without optical aids, but about 250 stars are visible in this cluster with a telescope. Some of these appear hazy because of the nebula in which the stars are imbedded. This, along with the fact that the largest star in the group is a spectral-type B5 and the absence of red giants in the cluster's H-R diagram, indicates that the cluster is reasonably young, perhaps only a few million years old (Figure 15.11).

Another famous open cluster making up the face of the constellation Taurus is the *Hyades,* about 40 pc from Earth. Only a few stars in the cluster are visible to the naked eye, but up to 200 stars comprise the group. All stars in the cluster are moving eastward and slightly away from us, traveling about $1°$ in 3×10^4 years. Aldebaran (Alpha Tauri), the brightest star in Taurus, is among the group but is not a member of the cluster since it is moving in a different direction. The Hyades appear to be converging toward a point a few degrees east of Betelgeuse but will not arrive there for some tens of millions of years. The stars in the Hyades are not imbedded in a nebula, which along with the fact that several red giants are found in the H-R diagram of the cluster, indicates that the Hyades is a much older (perhaps 10^8 years old) cluster than the Pleiades.

Figure 15.11 NGC 2682, an open type star cluster in Cancer taken with 200-inch telescope. (Courtesy of Hale Observatories)

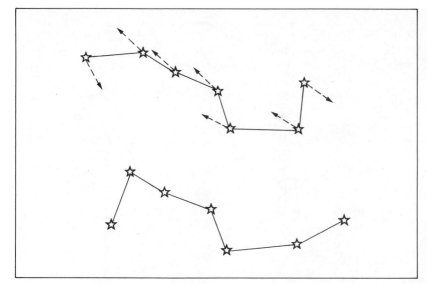

Figure 15.12 The Big Dipper as it now appears (top) and how it is expected to appear 10^5 years from now (bottom). The arrow indicates the direction of motion of the individual stars.

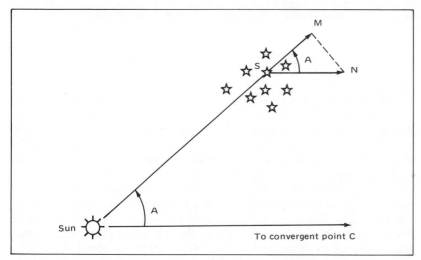

Figure 15.13 Stars in a cluster appear to converge to a point. This space motion is useful in determining distance to stars in the cluster.

The sun is not a member of an open cluster but is among stars that are. This group is known as the *Ursa Major* cluster and includes the five central stars of the Big Dipper (Figure 15.12), Alpha Coronae Borealis, Beta Aurigae, Sirius, and about 40 additional stars.

The space motion of stars in a relatively nearby cluster such as the Hyades can be measured, and this information is useful in measuring its distance. Although the stars are moving along parallel paths, they appear to be converging at a point. As a result, a line from the sun to the convergent point is parallel to the space motion *SN* of the stars in the cluster (Figure 15.13). A star in the approximate center of the cluster may be chosen and its radial velocity determined spectroscopically. Angle *A* formed by a line from the sun to the stars (*S*) and from the sun to the point of convergence at *C* may be measured, and is equal to angle *NSM*. Since angle *M* is a right angle, the tangential velocity *MN* may now be calculated

Figure 15.14 The Double Cluster in Perseus. (Lick Observatory photograph)

trigonometrically by $MN = SM \tan A$. Then, by use of the formula for determining tangential motion of a star (equation 13.10), it is possible to calculate the distance P in parsecs.

This technique has been found very useful, as it allows distance to be determined for stars too remote to permit the use of parallax with any degree of accuracy. However, its usefulness is limited by the distance to which the space motion of the individual star in a cluster may be measured. To overcome this limitation the *moving clusters* method, for which the point of convergence can be determined, is employed. The technique has been applied to the Hyades, the Ursa Major cluster, and the Coma cluster in the constellation Coma Berenices. Once the distances to the stars in a cluster have been found the technique may be used in conjunction with the H-R diagram of that cluster to provide a distance scale upon which distances of greater than 100 pc are based. By relating apparent magnitude of a distant cluster to the absolute magnitude scale on the H-R diagram of a cluster like Hyades, it is possible to determine how far the more distant cluster is from us. By this method distances to such clusters as Praesepe (Beehive) in the constellation of Cancer have been found to be 158 pc and the Double Cluster in Perseus, 2290 pc (Figure 15.14).

Globular Clusters

Far larger and more distant than the open clusters are the *globular clusters*— so named because of their shape. These clusters have highly elliptical orbits which take them varying distances from the galactic center that lies at one of the focal points of the ellipses. The globular clusters do not remain on the galactic disk but appear to populate a region 5° to 20° from the plane of the Milky Way. They have orbits that carry them well out from the galaxy into the *galactic halo*—a region above and below the plane of the galaxy (see chapter 16). Approximately 120 globular clusters have thus far been identified with the Milky Way galaxy, but many more may exist obscured by galactic dust and gas.

Most globular clusters are too far to be seen with the naked eye, but are beautiful to behold through the telescope. The two nearest, found in the southern sky, are 47 Tucanae and Omega Centauri at 4600 pc and 4800 pc, respectively. In the northern latitudes a globular cluster just barely visible to the naked eye is the Hercules cluster, M 13, seen in the constellation Hercules during the summer (Figure 15.15) at about 8000 pc. Most globular clusters are more than 10,000 pc from us, a few as much as 60,000 pc. Distance to globular clusters is generally determined from the apparent magnitudes of RR Lyrae stars (see section 14.4).

Once the distance to a globular cluster is established, it is possible by measuring its angular size to determine the actual diameter of the cluster. Most have been found to be between 10 and 30 pc in diameter, with the maximum about 100 pc. The exact outer limit is hard to determine because the cluster stars blend with those not a part of the cluster. Also, it is difficult to count the number of stars in a cluster as they cannot all be resolved. Allan Sandage counted about 44,000 stars in cluster M 3, but it was impossible to account for all in the closely packed central region. Estimates based on the size of the cluster range from 10^4 to 10^6 stars. Star density at the center of the cluster may be as high as 1000 stars per cubic parsec. Despite this apparent density the stars all moving in the same general direction are separated by thousands of astronomical units, and there is little danger of a collision.

Globular clusters are considered to be more stable as units than are open clusters for several reasons. The concentration of mass in the nucleus of the globular cluster prevents stars in the outer reaches of the cluster from escaping. In addition, globular clusters have orbits which carry them outside of the plane of the galaxy where they are less subject to the gravitational influences of the galaxy.

Figure 15.15 NGC 6205 (Messier 13) globular cluster in Hercules taken with 200-inch telescope. (Courtesy of Hale Observatories)

Associations

An *association* is a loose grouping of young stars with a presumed common origin and individual motions that cause them to expand outward from a common center. The stars making up the association are mostly young O- and B-type stars found primarily in the spiral arms of the galaxy. Unlike an open or globular cluster, the mutual gravitational force of the group is not strong enough to cause them to remain as a group for very long, and the association will gradually dissipate until it is no longer a recognizable entity. About 80 associations have been observed, with a population of 10 to 100 stars. These are distributed in a region of space up to 200 pc in diameter—a volume comparable to that occupied by large globular clusters. Because of the large volume of space inhabited by so few stars in an organized group, associations are difficult to define.

15.3 SUMMARY

Because of the fact that stars are formed in close association in interstellar dust and gas clouds, they will often occur in groups that may be identified by specific characteristics.

One recognizable type is the binary or double star made up of two components. Stars in a binary system orbit a common center of mass or barycenter and appear in several different forms. Visual binaries are those in which the components are visible through the telescope. The system where an unseen component may be detected through its influence on the visible component is known as an astrometric binary, and may be discovered only after detailed analysis of the motion of an individual star. The presence of the dark companion is detected from the wavy path followed by the visible component. In some instances the components are so close they cannot be visually distinguished but may be detected by the shift in their spectral lines. When the orbits of binary stars are edge-on to us, one component eclipsing the other will cause the light curve of the system to vary in specific ways. The light curve is a clue to the nature of the components in these eclipsing binaries.

Larger groupings include several types of clusters and an association. Open clusters contain from 50 to 1000 stars in a loosely related group occupying a volume of space less than 10 pc in diameter. Nearby clusters such as the Hyades have been found useful in determining distances to stars beyond the range where parallax is practical.

Far larger and quite remote from us are the globular clusters, which may contain up to 10^6 stars, and as a group follow a highly elliptical orbit around the galactic center. Globular clusters range from 4000 pc to 60,000 pc from Earth and cannot readily be seen without a telescope. Star density is quite high at the centers of globular clusters, but because they are moving in the same direction collision is unlikely to occur.

Another grouping of stars is the association. An association is a group of stars formed in the same nebula and is composed of from 10 to 100 stars. Because they are widely dispersed—the volume of space occupied by an association may be up to 200 pc in diameter—they have slightly different motions and will gradually move away independently from their point of origin.

QUESTIONS

1. Make a diagram showing the difference between optical and visual binaries.
2. Describe the relationship between two stars that exist as the components of a binary system.
3. Why is it not possible to always see the true relative orbit of a binary star?
4. Why would the astrometric binaries be of interest to the exobiologists?
5. What is a spectroscopic binary?
6. Describe the elements involved in detecting the motion of a spectroscopic binary.
7. Describe the light curve you would obtain from an eclipsing binary when the eclipse is total; when the eclipse is partial.
8. Describe the multiple nature of the star Castor which we see as a single star with the unaided eye.
9. How is it possible to identify stars as part of a cluster?
10. How do the characteristics of an open cluster and a globular cluster differ?
11. Where are the globular clusters most generally found?
12. Define what is meant by an association.

FOR FURTHER READING

AITKEN, R. G., *The Binary Stars.* New York: Dover Publications, 1964.

BINNENDIJK, L., *Properties of Double Stars.* Philadelphia: University of Pennsylvania Press, 1960.

KOPAL, Z., *Close Binary Systems.* New York: John Wiley & Sons, 1959.

PAGE, T., and L. W. PAGE, *Stars and Clouds of the Milky Way.* New York: The MacMillan Company, 1968.

TAYLOR, R. J., *The Stars.* New York: Springer-Verlag, 1970.

Sagittarius

CHAPTER 16   THE MILKY WAY— A GALAXY

I saw eternity the other night,
Like a great ring, of pure and endless light
All calm, as it was bright

—Henry Vaughan

If we stand outdoors on a particularly dark clear night away from the distracting glow of metropolitan lights, we will be treated to the unforgettable sight of the Milky Way. The Milky Way is a faintly luminous, irregularly shaped band of light that completely encircles the sky. People have viewed this phenomenon for centuries, but only in the past fifty years have we begun to fully comprehend the true size and structure of the Milky Way.

16.1 THE GALAXY DISCOVERED

Gala is a Greek word meaning milk, and the early Greeks thought the Milky Way represented milk flowing from the breast of Hera, wife of Zeus, the principal Greek god. Another ancient belief depicted the Milky Way as the stairway of the gods between Heaven and Earth. Ovid, the Roman poet (ca. 10 B.C.), wrote of it in his *Metamorphoses*:

> There is a way on high, conspicuous in the clear heavens, called the
> Milky Way, brilliant with its own brightness. By it the gods go to the
> dwelling of the great Thunderer and his royal abode.

Some astronomers guessed that the Milky Way was a multitude of distant stars, but Galileo was the first to demonstrate this truth (Figure 16.1). Through his newly built telescope he was able to see numerous stars at great distances from the Earth. In his *Siderus Nuncius* (Messenger from the Stars) published in 1610, Galileo stated:

> In addition to the stars of the sixth magnitude, a host of other stars are
> perceived through the instrument which escape the naked eye; these
> are so numerous as almost to surpass belief. One may in fact, see more
> of them than all the stars included among the first six magnitudes.

In the eighteenth century, astronomers began to visualize the true nature of the Milky Way. In 1750, Thomas Wright, a British astronomer and former sailor, questioned the arrangement of the stars and seeing what others had seen before him guessed that the Milky Way represented

Figure 16.1 Star cloud in
the constellation Sagit-
tarius photographed in
red light through 48-inch
telescope. (Courtesy of
Hale Observatories)

a region of the sky occupied by a dense population of stars. Wright
reflected upon whether the stellar density was really greater in the Milky
Way or similar to other regions of the sky. He felt that he was looking at a
cross–sectional view of a lens-shaped stellar structure which gave the
illusion of greater stellar density. He was able to see that the stars were not
a mere swarm scattered without order or design, but rather a system with a
definite structure and arrangement.

Immanuel Kant examined Thomas Wright's theory in 1753 and made
several unusual observations, which turned out to be prophetic. Kant
suggested that the galaxy as described by Wright was similar in some
respects to the solar system. Both were disklike and circular in structure.
Since by this time stars had been shown to move, Kant speculated upon
the possibility that the stars in the galaxy revolved around a galactic center
in much the same way that planets revolve around the sun. He viewed the
galaxy as a huge structure composed of stars turning like a gigantic wheel
in space. He also wondered whether the universe was limited to only one
such galaxy. Since a number of faint nebulosities had been observed
through telescopes, Kant suggested that these may represent distant
galaxies similar to the Milky Way. Reasoning of this kind greatly
expanded man's view of the universe. It may have prompted Johann
Heinrich Lambert, in 1761, to speculate that the scope of the universe was
unlimited, but the argument of whether the universe is finite or infinite
still has not been resolved.

Sir William Herschel hoped to solve the problem of the galaxy's shape,
which he also believed to be lens-shaped. Herschel systematically counted
stars in selected portions of the sky. His intent was to determine star
density and see if the star density varied with distance. Herschel used a 12-
inch reflecting telescope with which he sampled about 3400 points in the
sky. From his work he concluded that the galaxy was irregularly disk-
shaped (Figure 16.2). Although Herschel sought to visualize the shape of
the galaxy, he never attempted to guess at its size.

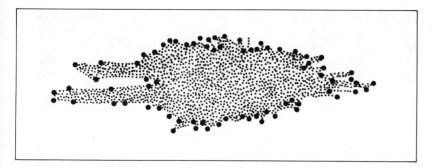

Figure 16.2 Sir William Herschel's view of the Milky Way.

16.2 SHAPE, SIZE AND STELLAR MASS

Shape

One of the first serious modern attempts to determine the shape and size of the galaxy was made by Jacobus C. Kapteyn in 1906. In his studies he observed that photographs bore out what had been noted in the seventeenth century, that "the *via lactea* (Milky Way) runs through the middle of the heaven's roundness." This meant that the solar system was located near the central plane in the center of a huge star system that is shaped like a flat disk. One of the problems that disturbed Kapteyn was the distribution of the globular clusters. Assuming a random distribution in nature, he expected the globular clusters to be evenly scattered in space. Every region of the sky seen from the center of the galaxy should therefore contain approximately the same number of clusters. Kapteyn found that the globular clusters appeared to be concentrated in the region of the sky near the constellations Sagittarius and Scorpius. It was Harlow Shapley of the Harvard Observatory who resolved the problem around 1920. He

Figure 16.3 A schematic view of the Milky Way. The plane view (left) as seen from the north shows Earth's revolution around the galactic center as clockwise.

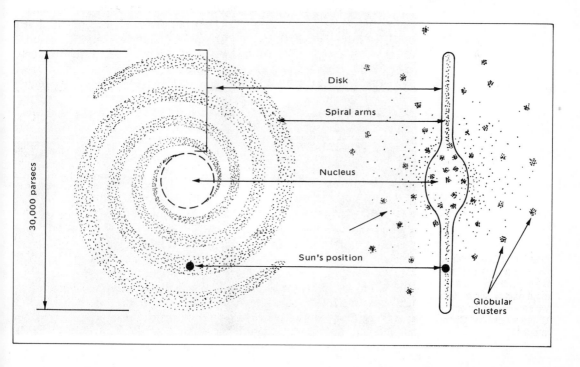

suggested that the solar system was not at the galactic center but rather was offset to one side. From this vantage point the globular clusters, which were reasonably uniformly distributed around the galaxy, would appear to be concentrated in the direction of the center of the galaxy. Shapley was able to derive their distribution in space by using the variable stars to determine distance to the clusters. From this it was possible to deduce that the Milky Way galaxy was a flat disk-shaped structure bulging somewhat at the center (Figure 16.3).

By 1952, details of the shape of the galaxy had been determined from the 21-cm radiation received from hydrogen clouds, which make up a large proportion of the mass of the galaxy. Observations made in the Netherlands and in Australia showed the radiation of the interstellar gas to be Doppler-shifted in accordance with the velocities of the points studied. In addition to information on the rotation of the galaxy, the data also yielded the distance to various points in the galaxy. In this manner the spiral structure of the galaxy was revealed, giving it the appearance of a giant pinwheel in space. This type of structure is not unusual in the universe, for as we shall see later many external galaxies also have this characteristic spiral appearance (Figure 16.4).

Size

In measuring the size of the galaxy, Kapteyn made a stellar density survey similar to that made by Herschel. Using photographic techniques Kapteyn counted the distant stars, ignoring the larger stars that were obviously nearby. He assumed most of the distant stars to be the same size, and tried to determine how far they must be to appear so dim. Kapteyn approached the problem on the basis that all stars of magnitude 1 or brighter were a given distance from the sun. He then extended the radius of the survey by counting all stars of magnitude 2, which are less bright than stars of the

Figure 16.4 NGC 224 (Messier 31), Great Galaxy in Andromeda. A spiral galaxy shown with two satellite galaxies. (Courtesy of Hale Observatories)

first magnitude by a factor of 2.5. This means the radius of the survey was increased by about 1.6 (approximate square root of 2.5), and therefore the volume of the space increased by a factor of about three. If stars are uniformly distributed in space the number of stars would triple each time stars of one magnitude greater were included. If the stars thinned out it was possible to conclude that the limits of the galaxy had been reached. Using data from this type of study, Kapteyn concluded that the galaxy was about 8000 pc in diameter and 6000 pc thick at the center.

In 1920, Harlow Shapley was able to derive a more accurate view of the position of the sun in the galaxy through his studies of distances to the globular clusters. He was able to ascertain that the sun was offset some distance from the center of the galaxy toward the edge. Shapley overstated the size of the galaxy, but at a later time this was corrected and the galaxy was determined to be a gigantic stellar structure with the disk approximately 3×10^4 pc in diameter and 3000 pc thick through the nucleus.

Surrounding the main disk of the galaxy is a spherical *halo,* consisting of globular clusters, a few isolated stars, and a tenuous cloud of hydrogen gas. The objects in the halo do not participate in the galactic motion but have independent movements. The orbits of these objects around the galactic center are thought to be elliptical in most cases, with one focal point coinciding with the galactic center. The sun is approximately 10^4 pc from the galactic center.

Stellar Mass

How many stars are there in the Milky Way galaxý? This was a question asked by Herschel as he sought to make a representative count in his attempt to determine the galactic structure. Gas and dust of which he was unaware prevented him from making an accurate estimate. Kapteyn repeated the experiment, but again the very distant stars were not visible. Now the mass of the galaxy may be approximately determined by applying Kepler's third law as modified by Newton (see equation 13.9). The distance from the galactic center to the sun (10^4 pc) is about 2×10^9 AU and the period of the sun's orbit is about 2.2×10^8 years. Since the mass of the sun compared to the galactic mass is relatively insignificant we may ignore it and we have

$$M_{galaxy} = \frac{a^3}{p^2} = \frac{(2 \times 10^9)^3}{(2.2 \times 10^8)^2} = \frac{8 \times 10^{27}}{4.84 \times 10^{16}} = 1.7 \times 10^{11} \text{ solar mass}$$

If we assume the sun to be an average star and its mass to be average for stars in the galaxy, we can accept in principle that there are up to 2×10^{11} stars in the Milky Way galaxy. The calculation shown above is not entirely correct, because the exact distance from the sun to the galactic center is not accurately known. In addition, the amount of mass beyond the sun in the outer reaches of the galaxy is not known for certain. However, despite these uncertainties the value above is accepted as being within reasonable limits of accuracy.

16.3 GALACTIC SPIRAL STRUCTURE

In the previous section mention was made of the spiral-like structure of the Milky Way galaxy. Within the past thirty-five years more details of this design have been emerging, which give astronomers a clearer picture

Figure 16.5 Early view of spiral arm structure in vicinity of sun. Numbers on the diagram indicate distance from the sun in parsecs.

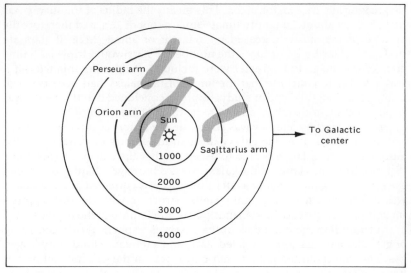

Figure 16.6 Diagram showing radio pattern of spiral arm structure of the Milky Way. Spiral arms are detected by hydrogen emission in 21-centimetre line.

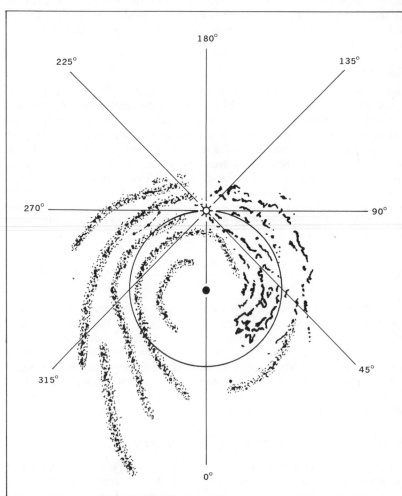

of the position of the sun in the galaxy. In 1943, Walter Baade of the Hale Observatory developed a technique that helped provide information on the spiral structure. He was studying the Andromeda galaxy, a neighboring spiral galaxy, and observed that the very large and bright type O and B stars served as markers for the spiral arms. He assumed that the same was true for the Milky Way galaxy, and that the spiral arms could be traced out by measuring the distance to O and B stars and plotting their distribution. This was accomplished, and by 1951 several sections of spiral arms in the vicinity of the sun (Figure 16.5) had been mapped. The sun appeared to be near the inner edge of what was called the *Orion arm.* In toward the center of the galaxy, approximately 2000 pc from the sun, is a segment of a spiral called the *Sagittarius arm.* Beyond the Orion arm from the galactic center, the *Perseus arm* could be traced. Unfortunately, the light from the type O stars cannot penetrate more than 6000 pc through the dust found in the galactic spaces, and optical study of the spiral arms is therefore limited.

More success has been achieved in the study of the galactic structure by use of radio telescopes. The hydrogen radio emission in the 21-cm line is able to penetrate the interstellar dust clouds, and astronomers have been able to detect more of the spiral design because hydrogen appears to be concentrated ten times more densely in the arms than in the interarm regions (Figure 16.6). However, results from optical and radio observations have not exactly coincided, and this led to a modification of the model of the galaxy by Bart Bok in 1959. In the new model the sun is located in the *Carina-Cygnus arm,* and the Orion arm is seen as a small spur along with the Vela spur (Figure 16.7).

Another feature of galactic structure is the type of stars found in the nucleus compared with those found in the spiral arms. During World War II, Walter Baade observed that the huge stars in the spiral arms of M 31 (Andromeda) were generally bluish, while those in the nucleus of the galaxy were red. This led him to suggest the two-population concept of stars. Population I stars are, for the most part, found in the upper part of the main sequence, include stars of type O, B, and A, and are generally a −4 absolute magnitude or brighter. These stars occur in regions of

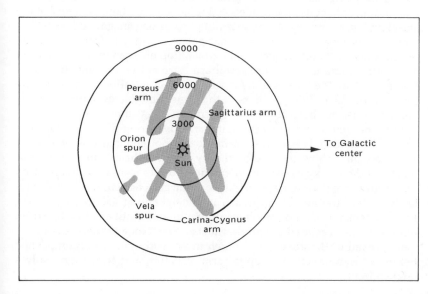

Figure 16.7 Recent model of nearby spiral arm structure shows sun in Carina-Cygnus arm with the Orion arm as a small spur.

abundant dust and gas, which appears to be the condition in the spiral arms. Population II stars are red giants and are found mainly in the galactic nucleus, an area mostly devoid of dust and gas, and in the globular clusters.

Baade's two-population concept was later found to be too simple, and at a conference in the Vatican Observatory in 1957 astronomers agreed on a system in which five rather than two populations were recognized. Briefly stated the five populations are summarized as follows:

1. Extreme Population I. Stars a few tens of millions of years old occurring in spiral arm regions rich in dust and gas are included. These stars are generally type O, B, and A types, of which Rigel in the constellation Orion is an example.

2. Intermediate Population I. These stars are a hundred million to a few billion years old and are found mainly on the galactic plane. Sirius is an example of this group.

3. Disk Population. These stars are located in the spiral arms and in the disk and are 3 to 5 billion years old. Most, like the sun which is in this group, include heavier elements in their chemical makeup and are probably second or third generation stars.

4. Intermediate Population II. Stars in this group are observed in the halo and in the nucleus and are older than 5 billion years.

5. Extreme Population II. These stars are up to 10 billion years old and are located in the older globular clusters and in the halo.

The system provides a convenient grouping for identifying types of stars, although some overlapping from one group to another does occur. For simplicity, the terms Population I and Population II are still frequently used to differentiate between stars in the spiral arms and disk and those in the galactic nucleus and globular clusters.

There are still a number of unanswered questions about spiral arms. Are spiral arms that now exist of recent origin or did they come into existence at the time the galaxy was first formed, perhaps 10 billion years ago? If they originated with the galaxy, why have they not wound down and disappeared eons ago? Some astronomers thought the existence of the spiral arm structure could be attributed to strong magnetic forces, but recent observations have shown that galactic magnetic forces are quite weak. A newer theory suggests that gravitation plays an important part in spiral-arm formation and that spiral arms are not permanent structures. Spiral arms are the locales of hydrogen gas concentrations, leading to the formation of O and B-type stars in the arms. These huge stars soon fade and new ones are born as the concentration of gas moves on.

Another question is how the spiral arms were originally created, since it may be assumed that their formation is not spontaneous or else all galaxies would show them. One theory suggests that a concentration of matter in the galaxy as it formed extended out to form into a spiral arm. Other ideas are that matter ejected from the galactic center formed spiral arms, or that the nearby approach of a neighboring galaxy such as the Large Magellanic Cloud may have drawn huge quantities of matter from our galaxy that resulted in spiral-arm structure. None of these ideas have been proved or disproved, and much more work must be accomplished before a clearer picture of spiral-arm structure and formation evolves (Figure 16.8).

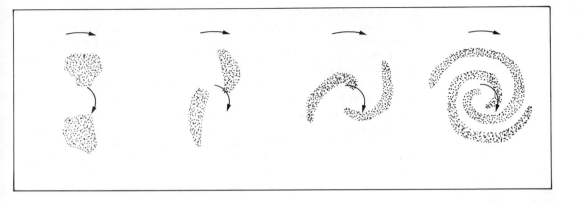

16.4 MOTION IN OUR GALAXY

In the late 1920s, Jan Oort of the Netherlands and others discovered from the motions of stars that the Milky Way was rotating in a manner suggested earlier by Immanuel Kant. Oort was able to show that the stars closer to the center of the galaxy moved faster than those further out, much like the motion of the planets around the sun. Had the stars in the galaxy described a rigid rotation as in a wheel, any motion would have been difficult to detect. Instead, *differential galactic rotation* occurs. This is the type of motion characteristic of circular motion around a central attracting mass. The axis of the galactic rotation is at right angles to the disk structure and lies in the direction of the constellations of Sagittarius and Scorpius. These constellations are acknowledged as marking the direction toward the center of the galaxy.

Figure 16.8 A suggested means whereby spiral arms are formed from irregular clouds of dust and gas.

The evidence for differential galactic rotation was gathered as a result of studying the motion of the stars that, along with the sun, are on the galactic disk. The radial velocities of these stars provide data about galactic rotation when the velocities are observed at different galactic longitudes. Longitude in the galaxy is measured eastward on the galactic plane with the direction toward the center of the galaxy marking 0° galactic longitude (Figure 16.9).

Stars at 0° and 180° will exhibit no radial motion with respect to the sun, as they are moving in essentially the same direction as the sun. However, proper motion will be evident with stars at 0° moving faster than the sun in orbit around the galactic center and stars at 180° moving slower and therefore appearing to move in the opposite direction. Stars at 90° and 270° will be in approximately the same orbit around the galactic center as the sun and will show no radial motion and very little proper motion with respect to the sun. Because of the presumably circular orbit of stars around the galactic center, stars some distance ahead of the sun will display a very slight proper motion toward the center of the galaxy, and those stars behind the sun will exhibit proper motion away from the galactic center.

Stars located at galactic longitude 45° and 225° will show a positive radial velocity, indicating recession. Stars toward the galactic center from the sun at 45° are moving faster than the sun, and stars at 225° galactic longitude are moving slower than the sun, although all are moving in the same general direction around the galactic center. Proper motion is also exhibited by these stars, as shown in Figure 16.9. Stars at 135° and 315°

galactic longitude have a negative radial velocity, indicating an apparent movement toward the sun and a proper motion, as shown in Figure 16.9. Interstellar clouds of dust and gas move in the same fashion as the stars do, a fact that may be determined from the Doppler shift of the spectrum of the interstellar material.

Motion of the stars in orbit around the galactic center has thus far been described with respect to the sun. To learn their velocities with respect to the galactic center, we must first determine at what speed the sun revolves around the galactic center. To accomplish this it is necessary to measure the sun's motion relative to objects which do not revolve around the galaxy's nucleus or at the very most do so very slowly. Stars of globular clusters in the halo are such objects, and the study of their motion with respect to the sun has provided data indicating that the sun's velocity in galactic orbit is about 240 km/sec.

When we plot radial velocities of stars at a given distance from the sun against their galactic longitude (Figure 16.10), we obtain a double sine curve that is helpful in ascertaining distances of remote stars from their radial velocities when other methods have failed. For example, if a star displays a radial velocity of –15 km/sec at 135° galactic longitude, its

Figure 16.9 (a) The motion of stars on the galactic plane as they appear to astronomers on Earth. (b) Because of galactic rotation, stars nearer than the sun to the galactic center move faster in orbit than stars further away.

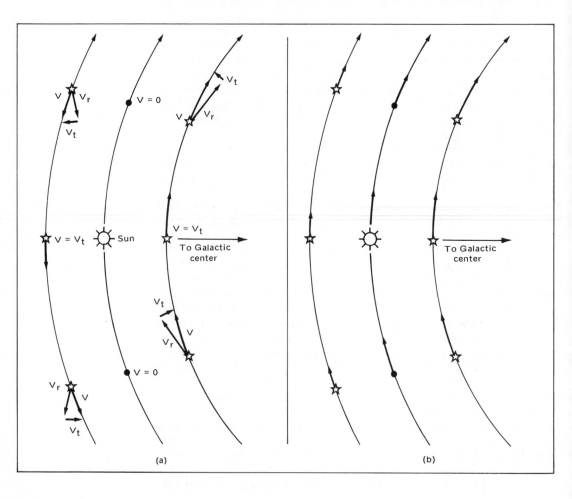

(a)　　　　　　　　　　　　　　(b)

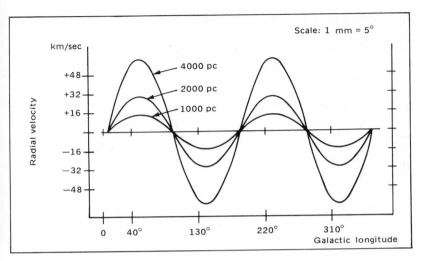

Scale: 1 mm = 5°

Figure 16.10 Distances to stars on the galactic plane may be determined from their radial velocity and galactic longitude.

distance from the chart is 1000 pc from the sun. Jan Oort, the Dutch astronomer, derived the following equation for the relationship:

$$V = rA\sin 2L$$

where V is the radial velocity of the object, r is its distance from the sun, A is a constant (Oort's constant) and L is the galactic longitude of the star. The constant A has been shown to have a value of 18.7 kilometres per second per 1000 pc. The equation is useful for determining distances to stars in the galactic plane where radial velocities may be measured. Distances cannot be measured for stars at 0° or 180° galactic longitude, because radial velocities are not available in these directions.

Studies made of the motions of stars and of dust and gas clouds have provided a fairly comprehensive view of the rotation of the galaxy. The central nucleus, with a radius of about 3000 pc, rotates essentially as a solid body, which results in those stars close to the center having lower orbital speeds than those near the outer edge of the galactic nucleus. Beyond the central nucleus rotation approximates Keplerian movement, in that the stars nearest the central nucleus have higher speeds than stars further out. This is similar to planetary motion within the solar system. The sun, at a distance of approximately 10,000 pc from the galactic center, has a velocity of 240 km/sec, and requires about 220 million years to make one revolution. This has become known as the *galactic* or *cosmic year*, and since its birth the sun has completed about 20 to 25 revolutions. Objects in the galactic halo, as already stated, do not share in this motion but rather have independent motions of their own. These objects generally follow elliptical orbits, with one of the foci of the ellipse located at the galactic center.

16.5 THE GALACTIC NUCLEUS

The galactic nucleus is the central portion of the galaxy that rotates as a solid and has a radius of about 3000 pc. This part of the galaxy is optically invisible because of interference by dust and gas found in the interstellar spaces. Only in the past several decades, with the aid of radio telescopes, has it been possible to study the central portion. Radio waves from the galactic nucleus are emitted by several mechanisms. One of the more important of these is the 21-centimetre wavelength radiation of neutral hydrogen. Neutral hydrogen comprises about 1 to 2 percent of the galactic mass, most of which is concentrated in a layer 50 pc above and below the galactic plane.

The Doppler shift in 21-cm wavelength radiation has revealed the presence of an expanding spiral arm at approximately $0°$ galactic longitude. The arm appears to be originating from the nucleus about 3000 pc from the center. The arm is expanding toward the sun with a velocity of approach of 50 km/sec. Another section of the arm can be detected beyond the center receding at 135 km/sec. It is not certain whether this feature is a true spiral arm or a ring originating in and expanding from the nucleus. It has been calculated that the expulsion occurred sometime between 10 and 100 million years ago, which is a relatively short period on a cosmological scale. In addition, there is evidence of massive amounts of neutral hydrogen having been ejected above and below the galactic plane, since neutral hydrogen clouds can be detected at high galactic latitudes. The source of energy responsible for ejecting hydrogen in this manner is unknown but may be the result of some explosions of gigantic proportions.

A second source of radiation from the galactic center is radiation from clouds of molecules in interstellar spaces. These have mostly been discovered in the past decade and include such compounds as water vapor (H_2O), ammonia (NH_3), carbon monoxide (CO), and formaldehyde (H_2CO) (see section 14.1). By 1974, a number of molecules had been discovered (see Table 19.1) in interstellar space, although not all in the galactic nucleus. The molecules appear to be grouped into great clouds but are less widely distributed than neutral hydrogen. N. Z. Scoville of the University of Minnesota has detected a ring of molecules with a radius of 300 pc around the center of the galaxy. The ring is expanding outward at a velocity of approximately 100 km/sec, a further indication of explosive forces at the galactic center.

A third source of radiation from the galactic nucleus is emission over a broad range of the spectrum by electrons accelerated in magnetic fields and moving at speeds near that of light. This is known as synchrotron radiation (see section 14.5). A strong source of this radiation has been identified as Sagittarius A detected in the direction of the galactic nucleus. There is evidence that Sagittarius A is in the nucleus, although the evidence is somewhat indirect since the exact source cannot be determined by the usual methods for measuring distance. The source does lie in a direct line between the sun and the galactic nucleus. Study of other spiral galaxies (see section 17.4, Seyfert galaxies) has revealed that such radiation does occur primarily in the nucleus of these external galaxies. Thus the conclusion that Sagittarius A is in the nucleus is not unreasonable. The spectral lines of Sagittarius A are not well-defined, so it is not possible to ascertain whether the source of the radiation is moving toward us, away from us, or is stationary. The source may also be related to explosive events at the galactic center such as supernovae.

A fourth form of radiation detected in the galactic nucleus is emission from H II regions. H II regions occur around hot stars where protons and electrons resulting from the ionization of hydrogen combine to reform hydrogen atoms. The existence of H II regions in the galactic nucleus indicates that new stars are being born. Several H II regions have been located within a few hundred parsecs of the center and appear also to be associated with large molecular clouds.

Infrared radiation from the galactic nucleus provides additional information on the central region of our galaxy. Recent measurements in the 2.2 micron region from the center of the galaxy have been found to compare favorably with light distribution from the core of the Andromeda galaxy. The central core of the Andromeda galaxy has a high stellar density, and on this basis it is possible to conclude that the nucleus of our galaxy has a stellar population of about one million stars per cubic parsec. Comparing this with approximately one star per cubic parsec in the vicinity of the sun leads to some interesting speculation as to the nature of the night sky in the galactic nucleus. The night sky of a planet orbiting one of these stars would be over 5 magnitudes brighter than the full moon, and would limit optical astronomy to the observation of nearby stars. However, the prospects of such observations being made are remote, because such high stellar densities would lead to unstable orbits for any planets that may exist. For life to develop on a planet, it is necessary that a stable orbit be maintained for very long periods within the life zone of the star (see chapter 19).

Additional data on the galactic nucleus has been provided by infrared radiation at wavelengths between 4 and 20 microns. At these wavelengths radiation appears to come from several distinct points close to the galactic center. The source of the radiation is thought to be dust particles surrounding huge hot stars near the nucleus, which may have a mass equivalent to 50 solar masses or more. There has also been a great outpouring of energy detected at approximately 100 microns. This radiation was found to extend about two degrees across the galactic center and is thought to originate from dust particles surrounding a dense assemblage of normal stars in that region.

The galactic nucleus is also found to be a source of X radiation. This was discovered by X-ray detectors aboard the artificial satellite Uhuru. The source of the X ray appears to coincide with the 100-micron radiation, although the manner in which the X ray is emitted is not known. It may be related in some way to infrared emission from hot gases in the galactic nucleus.

All the factors thus far mentioned appear to indicate an immense amount of activity in the galactic nucleus, although it may be some time before a total and coherent view of it is available. The outpouring of energy and gases provides evidence that an explosion or series of explosions have occurred. Such explosions, it is thought, would only influence the distribution of the gases and not the stars. A computer model suggests the possibility that the 3000 pc arm actually oscillates in and out like a ring over a period of several hundred million years. This motion is the result of the opposing forces of gravity, causing the ring to move inward, and centrifugal force generated by the inward momentum of the cloud, setting up a rapid rotation and forcing the ring outward again. Eventually, the ring would reach an equilibrium state and the galaxy would experience a period of quiescence until another explosion occurred. Whether this model is correct or not is a matter of speculation.

Where does the energy come from that is responsible for the violent activities at the galactic center? One suggested source of energy is a supermassive star, at least one million times as massive as the sun, located at the galactic center. Such a star would have collapsed as a result of the gravitational force generated by the huge mass and in the process would have generated sufficient heat to initiate a thermonuclear explosion. If the explosion did not occur, the alternative would have been the formation of a tremendous black hole. At present there does not appear to be sufficient mass at the galactic center to form such a massive object, either by the wholesale collision of large numbers of normal stars or by the accretion of gases by gravitational attraction. It may be, as has been proposed, that a past explosion has depleted the galactic nucleus of mass and that a gravitational inflow of gases and dust would be necessary to resupply the core for a future explosion. Where would the gas come from? Possibly from the galactic plane or even from outside the galaxy. It has even been suggested that mass appears spontaneously in the galactic nucleus by a mechanism as yet unknown in physics. Thus far astronomers have few answers to questions about conditions and events in the center of the galaxy.

16.6 THE CHANGING MILKY WAY

Has our galaxy changed since it was first formed? We know that stars within the galaxy are in motion, but does this change the general organization of the galaxy? Generally, little change is noted in stellar positions during the lifetime of one person, and even during a period of several thousand years changes do not appreciably alter galactic structure. If, however, these changes continue for periods of many galactic years, substantial alterations in galactic structure can result.

In the initial stages of formation around 10 to 15 billion years ago (40 to 60 galactic years), the galaxy is considered to have been a huge shapeless cloud of gas. Condensation of portions of the gas cloud into stars and star clusters occurred throughout the cloud, and as time progressed the cloud flattened as a result of its rotation into a relatively thin galactic disk. The stars and clusters that originally formed in the galaxy now occupy the halo, a region where it is believed little star formation is currently taking place. Because most of that dust and gas occupies a region a few hundred parsecs above and below the galactic plane, most star formation is now confined to this region.

The galactic structure is still evolving, although what its ultimate form will be is not known. Small changes in form still occur, and when continued for eons of time, become substantial. For example, we may consider changes that occur in such open clusters as the Hyades and the Pleiades. The stability of these clusters is maintained principally by the mutual gravitational force of the constituent stars, but all are influenced in turn by the gravitational force of the galactic nucleus. Those stars in the cluster closest to the galactic center will be subject to greater gravitational strain than those stars farther out from the galactic center. At the same time stars nearer the galactic center will have a slightly higher orbital velocity around the center of the galaxy, resulting in a tendency for the cluster to very gradually shear apart.

Whether or not this shearing takes place appears to be dependent upon the density of the stars within the cluster. A cluster with a star density equivalent to 10 solar masses in a volume of space three parsecs on a side at the distance of the sun from the galactic center should be stable if not interfered with by the passing of random field stars. *Field stars* are individual stars that are not identified with any group. It has been calculated by I. R. King that stars in the interior of such a cluster will be little disturbed by external forces. However, there is a boundary, the *tidal limit*, where stars in the cluster will not be held together against the external gravitational forces. These stars will escape from the cluster and after a few galactic years will no longer be recognized as members.

The influence of external stars upon members of a cluster must also be noted. A star need not approach close to a cluster for its influence to be felt. Even at one parsec the gravitational influence of a passing field star will alter slightly the path of a star in the cluster. This slight change may be sufficient to eventually cause the star to leave the cluster, thereby weakening the cluster's ability to retain the remaining members by lowering the total mass of the cluster. The Hyades cluster, which is not a compact cluster, may be stable for 4 or 5 galactic years before the combined gravitational influence of the galactic nucleus and the disruption of orbits by passing stars cause the stars in the cluster to scatter. By 10 galactic years the Hyades cluster will no longer exist as a unified group.

The Pleiades is a much more compact cluster, with a stellar density up to 10 times greater than the Hyades, and therefore is far more stable in structure. The Pleiades may be able to withstand the disruptive forces of gravity from the galactic nucleus for up to 50 galactic years, but on the other hand another process may cause the Pleiades cluster to collapse. In 1937, the Russian astronomer V. A. Armburtsumian thought that the loss of individual stars from the periphery of the cluster eventually would result in the remaining stars grouping closer together in the center of the cluster, ultimately causing its collapse. For a cluster like the Pleiades this could take up to 20 galactic years. In addition to the loss of mass by the removal of stars from the cluster, individual stars also lose mass by the escape of gases, especially during the later stages of their life cycle. The less massive stars are thus more susceptible to orbital disruptions by field stars and as a result eventually escape from the cluster. Dust and gas clouds passing near or through a cluster are thought to have a severe effect on them. Dust and gas clouds have masses equivalent to thousands of solar masses and therefore disrupt a cluster much more drastically than a single star.

The galactic population of single stars is constantly changing. We have already discussed the fact that spectral type O and B stars have a relatively short life cycle, enduring for only a few tens of millions of years (see chapter 14). These huge stars are continually being formed and dying out. Not only are new stars born only to subsequently die in the galaxy, but they are in constant motion and are being rearranged with respect to each other. As small a difference in velocity as 0.5 km/sec will in 10 million years separate two stars by 5 parsecs. In one galactic year whole communities of stars will have become completely disorganized.

Close approaches of one star to another are extremely uncommon even in the closely packed globular clusters. The chances of a star in the vicinity of the sun colliding with another star is remote and may be expected to occur only once in every billion galactic years. Even in the dense galactic nucleus or within a closely packed globular cluster, collisions between stars occur on the average of only once in a thousand years. A near encounter of the sun and another star within range of the outer planets is unusual and expected once in 10,000 galactic years. This would result in a change in the direction taken by the sun, but such changes are likely to occur on a very small scale over long periods of time by stars approaching within an average of one parsec. These kinds of changes may not result in an overall drastic alteration of galactic structure over the cosmic short term. However, the constant reshuffling of individual stars and the disruption of organized groups of stars such as the clusters must have contributed to some changes during the approximately 50 galactic years of the galaxy's existence.

16.7 SUMMARY

The Milky Way was seen by ancient peoples as an irregular band of light stretching across the sky. It was first observed to be a multitude of distant stars by Galileo in the early seventeenth century. Later the Milky Way was described as a lens-shaped arrangement of many stars; most recently it has been recognized as a huge, flattened disk in a spiral structure perhaps 30,000 pc in diameter, containing 100 billion stars.

The spiral structure of the galaxy was determined by means of the radio telescope only in the past 25 years, but the structure was previously guessed at from observation of external galaxies. Surrounding the disk of the galaxy is a halo composed of globular clusters, a few isolated stars, and a tenuous cloud of hydrogen gas.

Radio telescope study has detected concentrations of hydrogen gas making up the spiral arms of the galaxy and located the approximate position of the sun within the structure. The sun is approximately 10,000 pc from the galactic center in what is now recognized as the Carina Cygnus arm. The general population of stars in the spiral arms differs somewhat from those in the central nucleus. The term Population I is used for stars in the spiral arms and Population II for stars in the nucleus. A more detailed subdivision of stellar populations was developed in 1957, but the two-population concept is still generally used. The cause of the spiral structure and its evolution is being studied but as yet no answers are forthcoming.

Studies of motion within the galaxy indicate that the galaxy is rotating and that beyond the galactic nucleus the Keplerian laws apply. The sun requires about 220 million years to make one revolution; this is known as the cosmic or galactic year.

The galactic nucleus appears to rotate as a solid wheel and has a radius of about 3000 pc. This part of the galaxy is optically invisible, so all information is based on radiation detected by radio telescopes. A great deal of activity in the central nucleus indicates the existence of enormous sources of energy. Whether these are intermittent bursts of energy or whether the activity is continuous, and what is the source of all this energy, is not known.

Another question for which there is yet no answer is, What was the galaxy like in the past and what will it be like in the future? Does it evolve, and if so what form does this evolution take? During the 50 or so galactic years of its existence it is thought some changes must have occurred.

QUESTIONS

1. Briefly outline the development of understanding the nature of the Milky Way.
2. What is the shape, size, and mass of the Milky Way as a galaxy? Where is the sun located in this scheme?
3. How is it possible to determine the mass of the Milky Way?
4. How do Population I and Population II stars differ? In which category is the sun?
5. What are some of the questions about spiral arms that are as yet unanswered?
6. What type of motion does the galaxy describe? How is the motion determined?
7. How is a galactic year measured? How long is the galactic year?
8. What are the types of radiation coming from the galactic nucleus?
9. What are the suggested sources of energy thought to be responsible for the violent activity at the galactic center?
10. In what ways may changes in the configuration of the galaxy be brought about?

FOR FURTHER READING

Bok, B. J., and P. F. Bok, *The Milky Way.* Cambridge, Mass.: Harvard University Press, 1974.

Gingerich, O., *New Frontiers in Astronomy.* San Francisco: W. H. Freeman and Company, 1975.

Hodge, P. W., *Galaxies and Cosomology.* New York: McGraw-Hill, 1966.

Mihalas, D., and P. M. Routly, *Galactic Astronomy.* San Francisco: W. H. Freeman and Company, 1968.

Shapley, H., *Galaxies*, 3rd ed. Cambridge, Mass.: Harvard University Press, 1972.

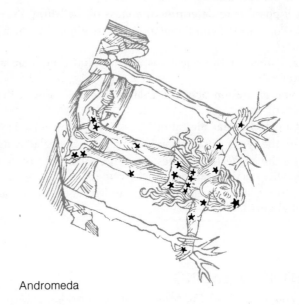

Andromeda

CHAPTER 17　　THE EXTERNAL GALAXIES

One gets such wholesale returns of conjecture out of such a trifling
investment of fact.

—Mark Twain

After discussing the Milky Way galaxy, the natural·question that follows
is: What lies beyond? This question has been answered with a limited
degree of certainty only within the past fifty years, although there had
been considerable speculation for centuries. In the nineteenth century,
some astronomers felt that the Milky Way galaxy was a unique object in
the universe, an extension of the ancient idea of man as the center of the
universe. This contrasts sharply with the current view that other galaxies
exist and that man is in all probability himself not unique in the universe.

17.1 BEYOND THE MILKY WAY

Immanuel Kant was right in his musings on the nonuniqueness of the
Milky Way. He had correctly guessed that the numerous nebulosities seen
through telescopes in his day were indeed other systems of stars. In the
eighteenth century, William Herschel was the first to see these nebulae as
"island universes" with his giant 48-inch reflecting telescope. He found
many of the nebulae to be located well above and below the galactic disk,
but very few were actually seen in the disk. From this Herschel concluded
that the objects were external to the Milky Way galaxy and very distant
from it. However, the confusion engendered by his attempt to equate
these distant nebulae with such nebulosities as the Orion nebula caused
Herschel to abandon the island universe theory (Figure 17.1).

The concept of the existence of external galaxies was never completely
discarded, and by the twentieth century a distinction was recognized
between the symetrically shaped elliptical and spiral nebulae found well
away from the disk of the Milky Way and those with no well-defined
structure found within the galactic disk. At approximately the same time
(1914), Sir Arthur Eddington suggested that the distribution of the nebulae
away from the galactic disk was only apparent, and that possibly dust
lying in the galactic disk obscured the view of nebulae in this region of the
sky. The existence of the dust was established in 1930, but in the

Figure 17.1 NGC 4594 (Messier 104), also known as Sombrero galaxy, is a spiral galaxy in Virgo seen edge-on. Photo taken with 200-inch telescope. (Courtesy of Hale Observatories)

meantime astronomers explored the question of whether these nebulae were external to the Milky Way and therefore equal to our galaxy in physical characteristics. The problem was debated by H. D. Curtis and Harlow Shapley in 1920.

Harlow Shapley was interested in Cepheid variables and their use for determining distances to globular clusters. As a result of his studies he was led to assume that the globular clusters defined the galactic center and from this concluded that the Milky Way had a diameter of 3×10^5 light years. Other astronomers were reluctant to accept this size for the galaxy, because it precluded the possibility that the nebulae were external galaxies. Shapley argued against the island-universe concept, suggesting instead that the nebulae were a part of our galaxy. Shapley also made use of data obtained on a nova in M31, the Andromeda nebula. In 1885, a nova had occurred that became as bright as the entire Andromeda nebula, thereby supporting his contention that the external spiral nebulae were much smaller than the Milky Way.

H. D. Curtis believed that the size of the Milky Way galaxy was grossly overstated. He insisted that the external nebulae were comparable to the Milky Way galaxy and that the novae observed in the external nebulae may actually be brighter than those seen in our galaxy.

The controversy was essentially settled by Edwin P. Hubble during 1924–26, when he discovered the presence of Cepheid variables in M31 (Andromeda) and M33. Using the same techniques employed earlier by Henrietta Leavitt in her study of the Magellanic clouds, Hubble was able to determine the distances of M31 and M33 by means of the period-luminosity relationship of the Cepheid variables. He found their distances to be 240,000 pc, which definitely established these nebulae as external to the Milky Way. Later this distance was found to be even greater. Other data from such objects as bright stars and novae, used as criteria for measuring distance, were consistent with the results obtained with the

Cepheid variables and provided confirming evidence. In this way the nebulae described by Kant one hundred and fifty years earlier as island universes were found by Hubble to be external galaxies. Once again man was displaced from a central position in the universe when it was discovered that the Milky Way was merely another galaxy among many (Figure 17.2).

17.2 GALACTIC DISTANCES

Measuring the distance to external galaxies is complicated by the factor of interstellar dust and gas within the Milky Way. There is a *zone of avoidance* that coincides with the visible portion of the Milky Way where almost no external galaxies are to be seen due to dust and gas. In part it was this dust and gas that caused Hubble to err in his distance determination to external galaxies, but his techniques were of great value for they provided a means of understanding more of the structure of the universe. In his work Hubble made an assumption which he referred to as his *assumption of uniformity.* He believed, as do other physical scientists, that the laws of nature which operate on Earth apply equally to all parts of the universe, and it was this assumption that caused him to consider the Cepheid variables to be a reliable tool for measuring distances. The luminosity of the Cepheid variables is related to the period of variability from which the absolute magnitude may be determined and the distances in turn may be calculated. This technique is useful only to a distance of about 6×10^6 pc, which is the limit to which Cepheid variables can be observed. Only about 30 galaxies of the approximately 150 within this radius have observable Cepheid variables.

The bright blue supergiant stars found in the upper lefthand quadrant of the H-R diagram are 10 to 15 times brighter than the Cepheid variables and allow the astronomer to somewhat extend the range to which he can measure distances to galaxies. The brightness of the occasional novae and of clusters associated with external galaxies has also been useful in

Figure 17.2 NGC 598 (Messier 33), spiral galaxy in Triangulum. Photographed in red light with 48-inch telescope. (Lick Observatory photograph)

measuring distances out to about 25 million pc. Supernovae offer some promise as a means of measuring distances to external galaxies since they have been observed as far as 300 million pc, but as yet there is insufficient data on their characteristics to make them reliable. The intrinsic brightness of galaxies may also be found useful for distance determination. However, it is difficult to know if a dim galaxy is a large, distant galaxy or a nearby small one. Only when in association with neighboring galaxies can the true value of its brightness be obtained. At present this type of information is only of statistical value, and much work is needed before more precise data can be obtained.

After Hubble made his announcement of the distances to M31 and M33, certain problems concerning extragalactic distances arose to plague astronomers. For example, the Milky Way appeared to be several times larger than any other galaxy observed. In addition, globular clusters associated with the Milky Way appeared brighter than those associated with neighboring galaxies, and novae in the Milky Way seemed to be of a different type than those observed in other galaxies. With the discovery in 1930 of the existence of dust and gas in the interstellar spaces some corrections in distances were made, but this did not entirely solve the problem.

In 1952, Walter Baade, an American astronomer, discovered that there were actually two types of Cepheid variables, each with its own period-luminosity relationship. He found that distances to globular clusters associated with the Milky Way had been determined from what became known as type II Cepheids and were essentially correct. Cepheids observed by Baade in external galaxies were the same as the classical or type I Cepheids found in the galactic disk. These were seen to be one to two magnitudes brighter than the type II Cepheids. However, this had not been recognized prior to Baade's discovery, and therefore the brightness of the type I Cepheids in external galaxies had been valued too low. As a consequence, the distances to external galaxies had been underestimated.

The brightness of type I Cepheids had been underestimated by a factor of almost 5, and as a result the corrected distances were about three times greater than originally determined by Hubble. Not only were the external galaxies further away, but it was realized that they also must be much larger in order to have the observed brightness. At the same time, the Milky Way was reduced from its once exalted position of the largest galaxy in the universe to a size comparable to other observed spiral galaxies. Thus M31 was determined to be approximately 690,000 pc from Earth and twice as large as formerly thought, making it slightly larger than the Milky Way.

We mentioned that the determination of distance to galaxies by the use of intrinsic brightness of various types of stars was limited to relatively nearby galaxies. However, another technique for indirectly measuring such distances began to be developed in 1888 when H. C. Vogel, a German astronomer, demonstrated that radial motion of stars could be detected by means of the Doppler shift in starlight (see section 5.5). In 1912, V. M. Slipher used this technique to measure radial velocities of approximately 40 nebulae, then thought to be part of the Milky Way. By the 1920s he had measured the radial velocities of many more nebulae and found most of them to be *receding* from us at speeds up to 1800 km/sec. When Hubble made the discovery that the nebulae were actually external galaxies, he compared distances of the galaxies with Slipher's velocities for these galaxies and came up with the remarkable fact that the velocities increased in direct proportion to their increased distance. This relationship became known as *Hubble's law* or the *Red Shift law*.

Edwin Hubble and his colleague Milton L. Humason continued to work on the velocity-distance relationship of galaxies into the mid-1930s. Humason laboriously collected data from galactic spectra on radial velocities, and Hubble measured the apparent brightness of the galaxies to determine their relative distances. Assuming an average luminosity for all galaxies, he was able to determine distances by comparing the brightness of a galaxy of known distance with the brightness of a more distant galaxy. By 1936, Hubble and Humason had reached the limit of the 100-inch telescope's ability to penetrate space but had firmly established the validity of Hubble's law (Figure 17.3).

An empirical relationship between distance and velocity as provided by the red shift is given in the following equation:

$$r = \frac{v}{H}$$

where *r* is the distance to a remote galaxy in megaparsecs, *v* is the velocity of recession in kilometres per second and *H* is a value known as *Hubble's constant*. Early values for the Hubble constant were as high as 500 km/sec/

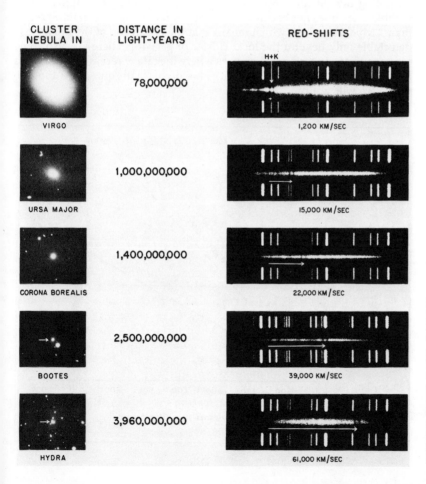

CLUSTER NEBULA IN	DISTANCE IN LIGHT-YEARS	RED-SHIFTS
VIRGO	78,000,000	1,200 KM/SEC
URSA MAJOR	1,000,000,000	15,000 KM/SEC
CORONA BOREALIS	1,400,000,000	22,000 KM/SEC
BOOTES	2,500,000,000	39,000 KM/SEC
HYDRA	3,960,000,000	61,000 KM/SEC

Figure 17.3 Relationship between red shift velocity in kilometres per second versus distances in light years. Each of the galaxies on the left is part of a cluster of galaxies in the constellation indicated. Distances in light years are based on the expansion rate of 50 kilometres per second per million parsecs. (Courtesy of Hale Observatories)

Mpc, but more recent work has established a range from less than 50 km/ sec/Mpc to 140 km/sec/Mpc, which indicates a degree of uncertainty by a factor of 3. The value arrived at by A. R. Sandage in a current analysis is 55 km/sec/Mpc with a possible error of $\pm$ 7 km/sec/Mpc; for convenience the value may be rounded off at 50 km/sec/Mpc.

By using 50 and 100 km/sec/Mpc we can examine some distance relationships as derived from Hubble's law and shown in Figure 17.4. The velocity of several galaxies are used (Figure 17.3) that are representative of the clusters of galaxies in which they occur. Thus the Virgo cluster, receding at 1200 km/sec, is 12 million pc or 24 million pc, depending upon which Hubble constant is used. When combined with distances similarly determined for the other representative galaxies, a straight line relationship is revealed in which it appears that distance is directly proportional to the velocity of recession. Use of the two Hubble constants does result in lines with different slope.

Regardless of which Hubble constant is used the interpretation is the same: as far as can be determined the universe is *expanding*, with the rate of expansion increasing with distance from our point of observation. Actually, the Hubble law cannot be applied universally, since nearby galaxies do not appear to participate in the expansion. These galaxies, part of a cluster of galaxies known as the Local Group (see section 17.5), appear to be moving randomly, with most exhibiting a negative radial velocity indicating movement toward rather than away from us. It appears that Hubble's law does not become an important factor until a distance greater than 1 Mpc is reached, which means that the expansion of the universe is detectable only beyond the local cluster of galaxies. There are also some discrepancies appearing i distances where the rate of recession becomes a large fraction of the speed of light. This occurs with quasars, a topic to which we will return.

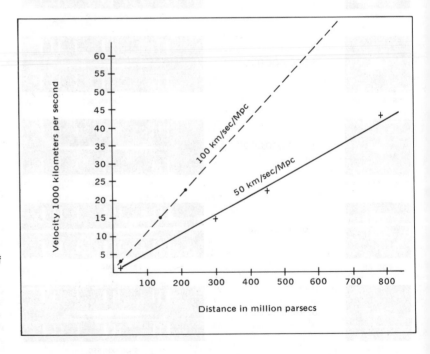

Figure 17.4 Expansion of galaxies results in a straight-line relationship when red shift velocity versus distance is plotted. The slope of the plots will differ.

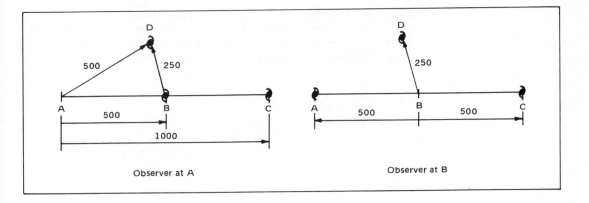

Observer at A

Observer at B

Figure 17.5 An observer will see all objects receding from him regardless of the position he occupies in the universe.

One conclusion that may be incorrectly drawn from the expanding universe concept is that the Milky Way occupies the center of the universe. An examination of Figure 17.5 reveals that an observer in any galaxy will see all other galaxies receding from him, resulting in no detectable center to the universe. The observer in a galaxy at A may see a galaxy at B receding at a rate of 500 km/sec. At the same time, a galaxy at C may be receding at 1000 km/sec from the observer, but from the vantage point of the observer at B, both A and C would be receding at 500 km/sec. There is no way the observer can determine if these apparent motions are the result of motion on the part of galaxy A, B, or C, or the result of their combined motion. The action has been described as being similar to what happens when one bakes a raisin cake: as it is being baked the cake expands and the raisins all recede from one another. Unfortunately, this action can only be depicted in two dimensions, but by imagining the motion in three dimensions one can easily visualize an expanding universe.

An interesting aspect of galactic distance is its relationship to time. Light, the basis for measuring distance, has a finite velocity. Therefore, time is implied in any expression of distance, such as a light year or parsec. What we are seeing when viewing a galaxy 2 billion ly (610 Mpc) from us is the galaxy as it existed 2 billion years ago, not the galaxy as it appears now. As a result, the farther out into space an astronomer peers, the more remote past he is probing.

We may also speculate upon the length of time required for the galaxies or clusters of galaxies to reach their present position, assuming a constant outward speed. In other words, if we consider the Virgo cluster to be receding at 1200 km/sec or Bootes at 39,000 km/sec, we may ask how long it took for them to reach their respective distances. Since the relationship of distance and red-shift velocity form a straight line (Figure 17.4), the calculated time using 100 km/sec/Mpc is approximately 10 billion years in all cases. If we use the lowest value for the Hubble constant of 50 km/sec/Mpc, the calculated duration of the expansion period would be about 20 billion years. Thus it is fair to say that the expansion process was initiated at some point 10 to 20 billion years ago, which is in fair agreement with estimates of the age of other objects in the Milky Way. However, it must be remembered that this is only an estimate, because there is no assurance that the expansion has been uniform throughout the ages. It may be that expansion began slowly and has accelerated or that the initial rate of expansion has continuously slowed down. It appears possible that the latter of these two choices may be correct, and that the expansion of the universe was initiated by a tremendous explosion of energy and matter 10 to 20 billion years ago.

17.3 TYPES OF GALAXIES

Numerous galaxies have been observed, and with characteristic thoroughness astronomers have classified them into a few convenient categories. Hubble initially classified galaxies into three basic types: spiral, elliptical, and irregular. The majority of the galaxies fall into the first two categories, and only a few are classed as irregular.

Spiral Galaxies

Spiral galaxies are subdivided into two types: normal spirals and barred spirals. *Normal spirals,* of which the Milky Way and the Andromeda galaxies are examples, have a lens-shaped central nucleus populated by older stars. Attached to the nucleus are spiral arms that coil around the nucleus on the same plane, giving the structure a pinwheel appearance. The arms contain young stars and considerable dust and gas. Because of the variation in the manner in which the arms are coiled, the normal spirals have been subdivided into several classes (Figure 17.6). The class Sa spiral has a large, bright central nucleus and very tightly wound arms. Class Sb spirals have more loosely coiled arms, and the brightness is more evenly distributed between the arms and the nucleus. The Milky Way is in this class. The Sc type spiral has large, loosely wound arms and a relatively small nucleus. Class S0 spirals are considered by some astronomers to represent a transition from spiral to elliptical galaxies. At first view they may appear to resemble ellipticals, but an edge-on view approximates a Sa spiral. There is no indication of arms or of the dust and gas usually found in the spiral arms region. Surrounding the nucleus is a thin disk of stars in the region generally occupied by the arms. Some astronomers feel that the S0 galaxies are those in which the dust and gas has been swept clean by the passage of one galaxy through another, leaving these galaxies mainly with Population II stars (Figure 17.7).

Somewhat less common than the normal spirals are the *barred spirals,* which are characterized by two spiral arms passing as bars through the nucleus. These also are subclassified according to the degree of openness of the spiral arms (Figure 17.8). The bars of a class SBa are rather tightly coiled, with those of SBb having an intermediate and SBc an open

Figure 17.6 Normal spiral galaxies are classified according to the degree of openness of the spiral arms.

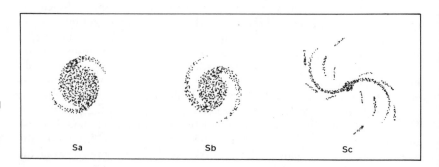

Sa   Sb   Sc

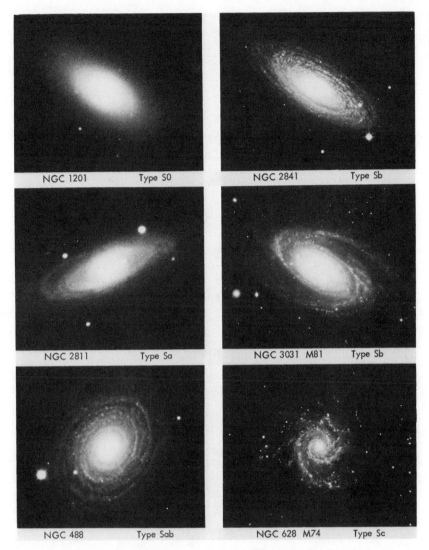

Figure 17.7 Representative examples of normal spiral galaxies. (Courtesy of Hale Observatories)

NGC 1201   Type S0

NGC 2841   Type Sb

NGC 2811   Type Sa

NGC 3031 M81   Type Sb

NGC 488   Type Sab

NGC 628 M74   Type Sc

structure. The straight portion of the bar out to a point where it begins to coil over appears to rotate, as the spoke of a solid wheel. This enables the bar to maintain its characteristic shape rather than winding up. How this comes about is not completely understood (Figure 17.9).

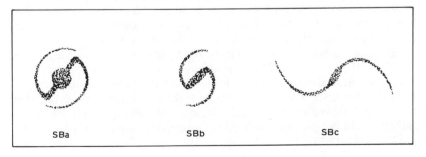

SBa   SBb   SBc

Figure 17.8 Barred spiral galaxies are classified according to the degree of openness of the spiral arms.

Figure 17.9 Representative examples of barred spiral galaxies. (Courtesy of Hale Observatories)

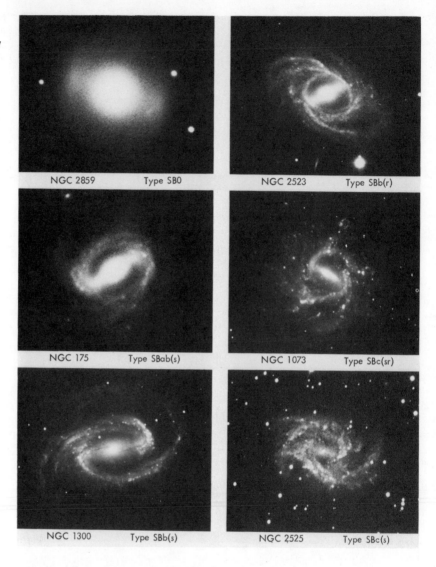

NGC 2859   Type SB0

NGC 2523   Type SBb(r)

NGC 175   Type SBab(s)

NGC 1073   Type SBc(sr)

NGC 1300   Type SBb(s)

NGC 2525   Type SBc(s)

Elliptical Galaxies

Elliptical galaxies are the most common galaxies in the universe. They vary in shape from spheroidal to elliptical and are generally smaller than spirals. The small size of the elliptical galaxies, and therefore relatively low brightness, makes them difficult to detect at great distances. On the other hand, some of the largest galaxies that have been observed are elliptical, extending to more than twice the size of the Milky Way.

Elliptical galaxies (E) are classified according to their degree of ellipticity (Figure 17.10). The relative measure of the semiminor axis is subtracted from the semimajor axis and divided by the semimajor axis to yield the ellipticity. The resulting value is then multiplied by 10, and that number is assigned as a numerical subclass of the elliptical galaxies. An E0 galaxy would be spherical with others designated successively E1, E2, up to E7 (Figure 17.11).

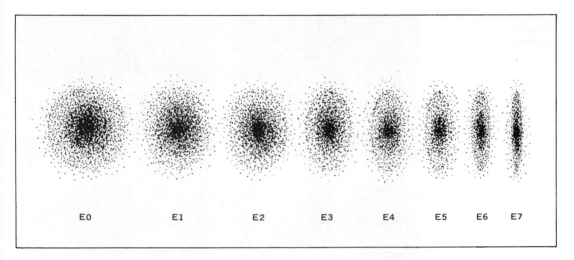

E0   E1   E2   E3   E4   E5   E6   E7

Elliptical galaxies look like the nuclei of spiral galaxies without spiral arms. They appear to contain only Population II stars, with few if any young stars, and seem to be devoid of dust and gas.

Figure 17.10 Elliptical galaxies are classified according to their degree of ellipticity.

Irregular Galaxies

Irregular galaxies make up about 3 percent of the total galactic population and are, as the name implies, lacking in symmetry. The best known of this group are the Large and Small Magellanic Clouds visible only in the

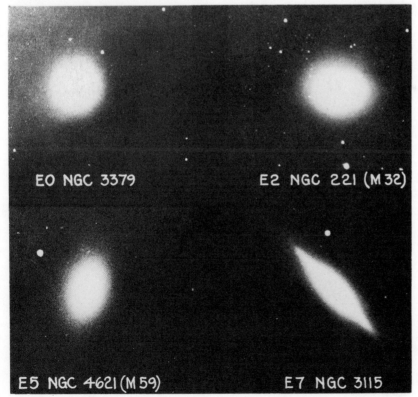

E0 NGC 3379

E2 NGC 221 (M 32)

E5 NGC 4621 (M 59)

E7 NGC 3115

Figure 17.11 Representative examples of elliptical galaxies and their classification. (Yerkes Observatory photograph)

Figure 17.12 The Large Magellanic Cloud is typical of an Irregular I galaxy . (Lick Observatory photograph)

southern sky near the south celestial pole (Figure 17.12). These galaxies, designated *Irr I* galaxies, contain Population I objects—type O and B stars and interstellar dust and gas. Some globular clusters and RR Lyrae are also present, so they are not composed of young stars only. A second type of

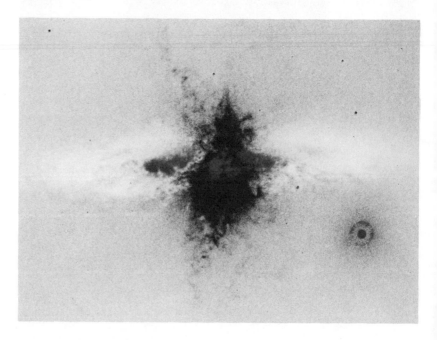

Figure 17.13 NGC 3034, an example of an Irregular II galaxy. Messier 82 in Ursa Major. (Courtesy of Hale Observatories)

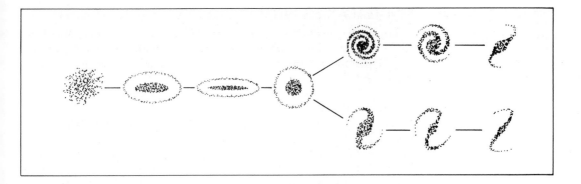

irregular galaxy, *Irr II*, is also asymmetrical in shape, but no stars have been resolved. In NGC 3034 the spectrum resembles that of A5 stars, and the Doppler shift indicates a rotation and the fact that matter is being ejected at high speed from the center of the galaxy (Figure 17.13).

Figure 17.14 Hubble's view of the evolutionary sequence of galaxies.

After Hubble developed his classification of galaxies based on their variation in structure, some astronomers were led to think that galaxies may go through a sequence of evolutionary stages as the galaxies age. They suggested that the sequence started with elliptical E0 types, which gradually flattened into elliptical disks and then formed into the S0 type spirals. Subsequently, arms took shape and the galaxies were transformed into either normal or barred spirals, each type progressing as shown in Figure 17.14. Other astronomers suggested a sequence beginning with irregular galaxies that became either normal or barred spirals. With time the spiral arms coiled tighter, so that the sequence of spiral stages was in reverse of the earlier model (Figure 17.15). After reaching the S0 type the galaxy becomes elliptical and progresses from type E0 to E7.

Not all astronomers subscribe to this evolutionary view of galaxies. Some objections to an evolutionary process are, for example, that no mechanism is known that would permit spiral galaxies to unwind, nor is there a means of accounting for the increase in mass from irregular galaxies to spiral galaxies. Spiral galaxies are perhaps ten times more massive than irregulars, and the giant elliptical galaxies are 10 to 100 times the mass of spirals. Some astronomers are of the opinion that the length of time required to proceed through such a sequence is far too great for the calculated age of the universe, and that galaxies just do not evolve from one form to another. The chances are good that a galaxy forms into one type or another based on its mass and angular momentum.

Figure 17.15 Shapley's view of the evolutionary sequence of galaxies.

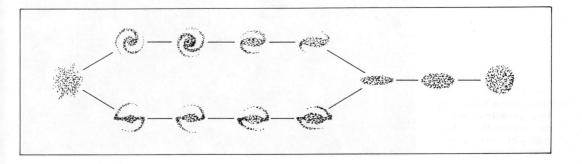

17.4 GALAXIES WITH PECULIAR CHARACTERISTICS

Although all galaxies thus far observed fall into one of the categories listed above, there are some with interesting characteristics worthy of particular note. These include the radio galaxies, Seyfert galaxies, exploding galaxies, and the quasars.

Radio Galaxies

These objects were first identified in 1951 as sources of radiation in the radio frequencies. A strong radio source, Cygnus "A," detected in 1948, was identified by Walter Baade as originating from a pair of galaxies in a cluster in the constellation of Cygnus. These galaxies are about 215 million pc from us and appear to be in contact or colliding (Figure 17.16). The radio emissions seem to emanate from two sources, each approximately 30,000 pc from the center. Since that time many radio sources have been detected, and about 100 have been identified with external galaxies.

Galaxies may be divided into *normal radio galaxies* and *peculiar radio galaxies* with respect to their radio characteristics. Many galaxies are numbered among the normal radio galaxies, including the Milky Way. They all emit some radio energy, generally about one-millionth as much as is emitted in the visible range of the spectrum. Radiation in the radio portion of the spectrum is detected coming from the nucleus of many galaxies and from the halo and arms of most. The source of the radio energy may be from the collision of interstellar matter with cosmic rays. In the process electrons are displaced which emit energy as they move along magnetic lines of force.

Figure 17.16 Cygnus A, a radio source detected in 1948, appears to be a pair of galaxies in collision. (Courtesy of Hale Observatories)

Figure 17.17 NGC 4151 is a Seyfert-type galaxy photographed with the 200-inch telescope. (Courtesy of Hale Observatories)

Peculiar radio galaxies emit 100 times as much or more radio energy than energy in the visible portion of the spectrum. Much speculation on the source of such radiation has not yielded any satisfactory answers to date. Stellar collision, a series of supernovae, or the gravitational collapse of large amounts of galactic matter have been suggested as sources. The apparent collision of galaxies as in Cygnus "A" as a possible radio source was considered but has now been rejected.

Seyfert Galaxies

Seyfert galaxies were first discovered in 1943 by Carl K. Seyfert. He found that a few galaxies with a superficial resemblance to spiral galaxies had small, very bright nuclei that were almost starlike in appearance. Extremely strong energy in all wavelengths was being emitted from the nucleus in amounts equivalent to the total energy output of the Milky Way. Typical of the Seyfert galaxies is NGC 4151 (Figure 17.17), which in short exposures appears like a fuzzy star, but in long exposures the spiral structure is seen fully developed. This galaxy is also recognized as an X-ray source.

The exact amount of energy emitted is not known, but Seyfert galaxies have some characteristics in common with planetary nebulae and supernovae. Some are sources of infrared radiation and exhibit emissions from synchrotron radiation. Some of the activities detected in the central nucleus of the Milky Way resemble those of Seyfert galaxies, and it is suspected that all galaxies may react like Seyfert galaxies a small percentage of the time.

Another group of galaxies quite similar to Seyfert galaxies are the *N galaxies*. These objects also appear as small, very bright nuclei which seem to be related to a nebulous background. The exact nature of the objects is not known.

Exploding Galaxies

Exploding galaxies are peculiar galaxies that look as though they have been disrupted by a tremendous explosion. Photographs of M82 made in red light (6300 Å) reveal filaments of hydrogen gas being ejected outward at up to 1000 km/sec to distances of 4300 pc from the central nucleus (Figure 17.13). The amount of energy emitted in the infrared portion of the spectrum of this galaxy exceeds the total energy output of the Milky Way in all wavelengths. This activity is not restricted to irregular galaxies only. The giant elliptical galaxy NGC4486 (M87) displays a jetlike appendage which, along with the nucleus, emits strong synchrotron radiation. Conventional sources of energy are not adequate to explain the phenomenon, and some astronomers think that a new energy principle would have to be formulated.

Quasars

Quasars, a contraction of quasi-stellar-radio-sources, are objects that appear to be more energetic than any of the galaxies thus far described. They have received considerable attention in the past two decades. Beginning in 1960, four radio sources of great energy were found to coincide with starlike objects in photographs made with the 200-inch telescope. One of the objects, 3C 48, was found to be similar in color to white dwarfs or old novae. Its general appearance led to the conclusion that 3C 48 was a star within the Milky Way, perhaps 100 pc from Earth. Photoelectric data also indicated that the brightness of 3C 48 varied with a period of about one day. Such a short period meant that the object was quite small, lending additional support to the view that it was a relatively nearby star rather than a distant galaxy.

In 1962, the position of another 3C source (3C for Third Cambridge Catalogue of Radio Sources), 3C 273, was determined with great reliability by Cyril Hazard and his associates of Parkes Observatory in Australia. From Parkes the moon could be seen passing in front of 3C 273, blocking its signal. Since the position of the lunar limbs is known precisely, the position of 3C 273 could be determined within 1 second of arc. From these observations it was determined that the quasar was composed of two components, A and B, about 20 seconds apart, and that 3C 273B, the smaller of the two, coincided with the position of a blue star (Figure 17.18). The implication was that, like 3C 48, the object was a star within the Milky Way.

In 1960, A. Sandage of the Hale Observatories had obtained a spectrum of 3C 48, but it consisted of emission lines that were unfamiliar and could not be identified. In 1963, M. Schmidt obtained a spectrum of 3C 273B, also with emission lines that could not be identified. He made an assumption that the object had a substantial red shift, despite the fact that it was a star in the Milky Way. As a result of this assumption he was able to successfully identify the lines to be those of hydrogen with a red shift of 0.158. This meant that the object was receding from us at 16 percent of the

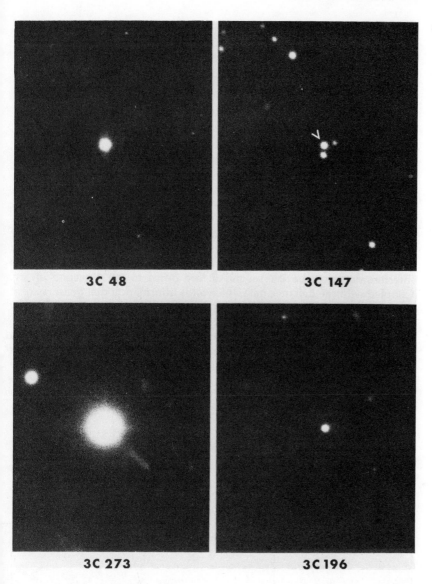

3C 48   3C 147

3C 273   3C 196

Figure 17.18 Quasi-stellar radio sources taken with the 200-inch telescope. (Courtesy of Hale Observatories)

speed of light or 48,000 km/sec, which, with the application of Hubble's law, indicated a distance well beyond the realm of our galaxy. About the same time J. L. Greenstein, working on the spectrum of 3C 48, found it to have a red shift of 0.367, which indicated the object was receding at 110,000 km/sec, placing this object at over 2 billion pc from the Earth.

By 1965, a number of quasars had been located, and Schmidt had determined red shifts for some of them. He found in 3C 9 the basic hydrogen line Lyman α with a wavelength of 1216 Å shifted into the visible range at 3666 Å, thereby yielding a red shift of 2.012. Such a shift is so large it requires a relativistic interpretation, and the velocity of recession works out to be about 80 percent of the speed of light (Table 17.1). A graphic relationship of the data in Table 17.1 is shown in Figure 17.19.

TABLE 17.1   CONVERSION OF RED SHIFT TO VELOCITY OF RECESSION

Relativistic Red Shift	Percent of Speed of Light	Velocity of Recession (km/sec)
0.105	.1	30,000
0.225	0.2	60,000
0.363	0.3	90,000
0.528	0.4	120,000
0.732	0.5	150,000
1.000	0.6	180,000
1.381	0.7	210,000
2.000	0.8	240,000
3.359	0.9	270,000

From the above we can enumerate some characteristic properties of quasars. They appear to be essentially starlike in appearance with relatively small diameters. The diameters have been measured with radio interferometers to be 0.1 second of arc down to 0.001 second of arc. Coupled with the distance data, this yields information that quasars are at most 100 pc in diameter. The quasars vary in radio and light energy emission, and the fact that they vary in brightness has been used to conclude that the dimensions of quasars may be as small as a fraction of a parsec. Huge red shifts indicative of great distance have been measured for over 300 quasars. None show a blue shift. In order for quasars to be visible

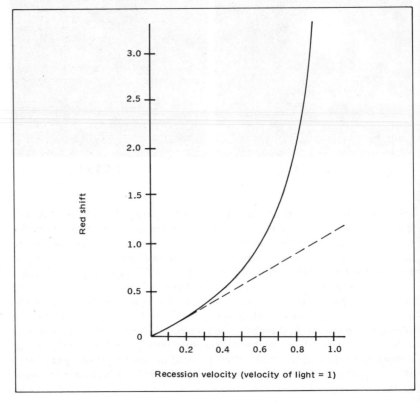

Figure 17.19 Relationship between red shift and velocity according to special relativity. For small velocities with a red shift of up to 0.35 (35%) the relationship is linear. As velocity of the object approaches the speed of light, the red shift grows without limit.

over these distances, they must be the brightest known objects in the universe. However, this cannot be because of the small sizes measured for quasars, and therein lies the paradox that has resulted in considerable controversy as to their true nature.

As stated above, the red shifts for quasars are large, with many in excess of even the most distant galaxies. Accordingly, in order to be seen the distant quasars would need to be extremely luminous, perhaps as much as a hundred times brighter than the largest galaxy. Applying Hubble's technique of using an average brightness for galaxies to determine their distance does not work for quasars. The dimmest quasar does not have the largest red shift thereby indicating the greatest distance. At the same time data indicate that quasars are quite small and, considering their size, should be incapable of emitting such prodigious amounts of energy, at least by means now understood by astronomers and physicists.

One proposed solution to the problem borders on astronomical heresy. It suggests the possibility that quasar red shifts are not the result of the Doppler effect and therefore do not indicate great distance. In this way there is no need to provide for a new energy source, nor do the relatively small objects need to be at cosmic distances as determined by the large red shift. If this reasoning is accepted, then it is necessary to seek a means other than the Doppler effect to explain the quasar red shifts.

There is some evidence that the red shift-distance relationship is correct and applies to quasars as reliably as it does to galaxies. Four quasars have been located in close association with small clusters of galaxies. The distances to the clusters have been determined by the Red Shift law, and because of the association these distances are assumed to be the same for the quasars. In each case the red shift of the quasars turns out to be similar to that of the adjacent galaxies, lending support to the idea that the Red Shift law is a valid technique for determining distances to quasars as well as galaxies.

Evidence also exists that the red shift-distance relationship does not apply to quasars. Halton Arp of the Hale Observatories has photographed nearby galaxies with relatively low red shift that seem to be associated with quasars exhibiting a high red shift. He has observed at least five such associations, and in one there appears to be luminous material extending between a galaxy and a quasar. It may be argued that the galaxies and quasars are not really associated but merely appear in the same line of sight. This reasoning could apply equally well to the relationship described in the previous paragraph. There have also been dissimilar red shifts discovered between two galaxies that appear to be physically associated. The spiral galaxy NGC 7603 has a red shift corresponding to a speed of recession of 8700 km/sec. At the same time a companion galaxy seemingly attached by a spiral arm has a speed of recession of 16,700 km/sec, which appears in clear violation of the Red Shift law.

Some suggestions have been made to account for the origin of such unusual red shifts. The belief of many astronomers is that the quasars conform to the Red Shift law and are expanding with the rest of the objects in the universe. Another view suggests that the high red shift is the result of the quasar having been blown out of some relatively local galaxy by a violent explosion. The red shift would be the result of the velocity achieved by the quasar as it moves away from the source of the explosion. If this view were correct then we should expect as many quasars moving toward us with a blue shift as away from us, but no quasars with a blue shift have been detected. It has also been proposed that gravitation is responsible for the huge red shift—in effect, that a massive object with a great gravitational potential is influencing the light emitted by the quasar. If this were the case the possibility exists that the quasar would not be able to support itself against the gravitational force and would collapse in a short time into a black hole. A fourth alternative is that a law of physics exists that is as yet unknown.

Thus we are faced with a paradox. If, as some astronomers believe, the Hubble law is correct and the quasars are very distant, then it is necessary to find a mechanism that would provide the tremendous energy required to cause the small objects that are the quasars to shine so brightly at such great distances. If, on the other hand, the quasars are associated with nearby galaxies, then it is necessary to find a new fundamental mechanism to explain the huge red shifts of the quasars. If the Red Shift law is not valid, it would upset our current view of the universe. Jesse E. Greenstein, astronomer at California Institute of Technology, summed it up in the following poem:

> Horrid quasar
> Near or far,
> This truth to you I must confess;
> My heart for you is full of hate
> O super star,
> Imploded gas,
> Exploded trash,
> You glowing speck upon a plate,
> Of Einstein's world you've made a mess!*

17.5 DISTRIBUTION OF GALAXIES

We have referred earlier to clusters of galaxies, a recognizable grouping of galaxies which appears to occur when the distribution of galaxies is carefully studied. Hubble, in plotting position and distance for galaxies, found them to be more or less uniformly distributed in space except for a region—the zone of avoidance—where viewing is blocked by dust and gas in the disk of the Milky Way. Hubble found that when he doubled the distance of his observations by increasing the exposures through the telescope, he increased the number of galaxies seen in space by a factor of 8. This coincides with the fact that doubling the radius of a sphere increases its volume by 8 times. It was assumed that a similar distribution of galaxies was to be found in the zone of avoidance. This has now been proved correct by observation with radio telescopes through the less dense portions of the Milky Way.

Later studies revealed that the galaxies were not independent groups of stars, but rather that they were organized into clusters ranging in number from a few galaxies to thousands. The Milky Way is a member of such a group known as the *Local Group* (Figure 17.20). This group now includes 21 known galaxies of which the Milky Way and Andromeda (M31) are the largest. In Figure 17.20 the galaxies are plotted on a two-dimensional surface with the approximate center of mass at the center. The entire cluster lies within an area having a diameter of somewhat more than

---

*Reprinted by permission of Jesse E. Greenstein.

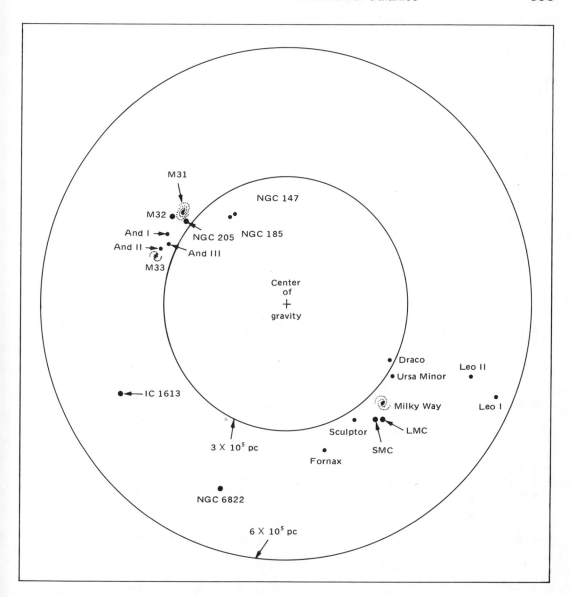

600,000 pc. Additional galaxies may be part of the Local Group but hidden by galactic dust and gas. In addition, some galaxies thought to be at great distances because they appear small and dim may actually be dwarf galaxies in the Local Group. Table 17.2 provides data on the physical characteristics of members of the Local Group.

A number of galactic clusters have been observed beyond the Local Group, some quite rich in galaxies. The Coma cluster of galaxies includes over 1000 visible galaxies, and there may be many more too small to be detected. The cluster is about 130 million pc from the Milky Way and occupies an area with a diameter of approximately 3 million pc. Over 2700 such clusters have been catalogued, with the most distant at over 1.2 billion pc. The general impression when viewed on a large scale is that the

Figure 17.20 The Local Group with + at approximate center of gravity of the Group. Galaxies appear to be loosely concentrated in two groups: one around M 31 (Andromeda) and the other around the Milky Way.

TABLE 17.2   Galaxies Generally included in the Local Group

Name	Type	Distance (kpc)	Apparent Magnitude (mv)	Right Ascension (1950)	Declination (1950)	Galactic Coordinates Long.	Galactic Coordinates Lat.	Radial Velocity (km/sec)	Mass Solar Mass
Milky Way	Sb					0°	0°	–	2×10^{11}
Andromeda M31, NGC 224	Sb	680	3.5	00h 40m	+41°	121°	–20°	–266	3×10^{11}
M32, NGC 221	E3	680	8.2	00 40	+41°	121°	–22°	–214	2×10^{9}
M33, NGC 598	Sc	720	5.6	01 31	+30°	135°	–31°	–189	8×10^{9}
Large Magellanic Cloud	Irr I	48	0.9	05 24	–70°	280°	–33°	+276	2.5×10^{10}
Small Magellanic Cloud	Irr I	56	2.5	00 50	–73°	303°	–45°	+165	1.5×10^{9}
Sculptor System	dE3	83	8.0	00 57	–34°	286°	–83°	–	3×10^{6}
Draco System	dE2	100	–	17 19	+57°	86°	+35°	–	10^{5}
Fornax System	dE3	250	8.4	02 38	–35°	237°	–66°	+39	2×10^{7}
Leo I	dE4	280	12.0	10 06	+12°	226°	+49°	–	3×10^{6}
Leo II	dE0	230	–	11 11	+22°	219°	+67°	–	10^{6}
Ursa Minor System	dE5	70	–	15 08	+67°	103°	+45°	–	10^{5}
NGC 6822	Irr I	460	8.9	19 42	–15°	26°	–20°	–32	1.4×10^{9}
NGC 147	E5	570	9.7	00 30	+48°	120°	–14°	–	–
NGC 185	E2	570	9.4	00 36	+48°	121°	–14°	–305	–
NGC 205	E5	680	8.2	00 38	+41°	121°	–21°	–239	–
IC 1613	Irr I	680	9.6	01 02	+ 2°	129°	–60°	–238	4×10^{8}
And I	dE0	680	(14)	00 43	+37°	(124°)	–	–	–
And II	dE0	680	(14)	01 13	+33°	(131°)	–	–	–
And III	dE3	680	(14)	00 33	+36°	(128°)	–	–	–
No. 21		(30)							

clusters are fairly uniformly distributed in space, but careful examination reveals that even clusters seem to be organized into groupings of clusters. For example, the several thousand galaxies, including those in the Local Group, found within a radius of 21 million pc are all organized into clusters of galaxies of which the Virgo cluster is the largest. G. deVancouleurs discovered that a scarcity of galaxies exists for some distance beyond this grouping of clusters, which constitute what may be considered a supercluster or supergalaxy approximately 30 million pc in diameter. The supercluster appears to be shaped like a flattened disk that is rotating with the Milky Way about 12 million pc from the center, where the Virgo cluster is located. Another supercluster has been detected in the direction of the constellation Hercules, where a number of overlapping galaxies can be seen.

Distant clusters of galaxies provide valuable information on luminosities, sizes, and types of galaxies. It may be assumed that members of a distant cluster are all approximately the same distance and therefore true relationships may be observed. The densely populated clusters like the Coma cluster are of interest because they appear to contain few spiral galaxies. However, it is thought that many of the galaxies seen are armless spirals, with the lack of arms attributed to collisions between galaxies. An event such as the collision of two galaxies is not necessarily cataclysmic because stars occupy a relatively small fraction of the total galactic volume, so that collisions between individual stars are extremely unlikely occurrences. Dust and gas within the galaxies is swept out, and since the arms of the spiral galaxies are rich in dust and gas, the arms are generally dissipated. Also, a collision of galaxies is not an instantaneous event. Estimates reveal that as long as 10^5 years may elapse from the time of initial contact until the galaxies separate again.

17.6 SUMMARY

The idea that many more galaxies existed beyond the Milky Way arose in the 1700s, but it wasn't until the twentieth century that Edwin Hubble proved it correct. External galaxies may be seen in all directions except through the disk of the Milky Way, but it is assumed that galaxies also occur in this region.

Distances to external galaxies have been determined by means of the distance-luminosity relationship of Cepheid variables and type O and B stars. Novae and the intrinsic brightness of galaxies have also been used and are considered to be statistically reliable for this purpose. Since Hubble did not recognize the existence of two types of Cepheid variables, he erred in his distance calculations, an error later corrected by Walter Baade in 1952.

A relationship between distance and velocity of recession was established that is useful in estimating distances to remote galaxies. The relationship, expressed as Hubble's law, shows most galaxies to be receding at rates that are in direct proportion to their distance from us. Almost all galaxies are receding from us, with the rate of recession increasing with distance. This led to the concept of the expanding universe, a process in which an observer in any part of the universe would see similar expansion.

Several types of galaxies are recognized. The spiral galaxies are of two types, including the normal spiral with spiral arms emanating from a central nucleus and coiled like a pinwheel and the barred spiral with a single bar through the nucleus coiled over at the ends. Elliptical galaxies range in shape from spherical to a fairly flattened ellipse. Irregular galaxies are those lacking in symmetry.

There are some galaxies with unusual characteristics that vary from the normal. Some are strong radio sources with radio emissions thought to be generated by a variety of mechanisms. Seyfert galaxies have a superficial resemblance to spiral galaxies, but the Seyfert galaxies have extremely bright nuclei, each with an energy output equivalent to that of the entire Milky Way. Exploding galaxies that are being, or appear to have been, disrupted by tremendous explosions have been observed. The explanation of quasars is an apparent paradox: either they are small, very distant objects emitting tremendous amounts of energy by an unknown means, or they are nearby objects and the great red shift from which their distance is determined is not due to the Doppler effect.

Galaxies seem to be organized into groups of few to thousands of galaxies in what is called a cluster. The Milky Way is part of such a grouping of about 20 galaxies called the Local Group. On an enlarged scale the clusters appear to be grouped in superclusters or supergalaxies. Because galaxies within such distant groupings are assumed to be at approximately the same distance from us, true relationships between them may be observed.

QUESTIONS

1. What are the opposing points of view of H. D. Curtis and Harlow Shapley concerning whether or not the nebulae were external to the Milky Way?
2. How was the debate finally resolved?
3. What was Hubble's assumption of uniformity?
4. What methods were used to determine distance to far-off galaxies?
5. What were some of the problems that arose after making some of the early distance determinations to external galaxies?
6. How were the problems in question 5 eventually solved?
7. Describe how distance may be measured making use of the Hubble law.
8. What is the Local Group?
9. What is the implication of the time-distance relationship over very large distances using light years or parsecs as a unit of measure?
10. Briefly describe the three recognized types of galaxies. Include a brief account of the subdivisions within each type.
11. Briefly describe: radio galaxies, Seyfert galaxies, exploding galaxies.
12. What is known about quasars? What are the problems related to quasars?
13. How do galaxies appear to be organized beyond the Local Group?

FOR FURTHER READING

BAADE, W., *Evolution of Stars and Galaxies.* Cambridge, Mass.: Harvard University Press, 1963.

BOK, B. J., and P. F. BOK, *The Milky Way.* Cambridge, Mass.: Harvard University Press, 1974.

GINGERICH, O., *New Frontiers In Astronomy.* San Francisco: W. H. Freeman and Company, 1975.

HODGE, P. W., *Galaxies and Cosmology.* New York: McGraw-Hill, 1966.

MIHALAS, D., and P. M. ROUTLY, *Galactic Astronomy.* San Francisco: W. H. Freeman and Company, 1968.

SHAPLEY, H., *Galaxies,* 3rd ed. Cambridge, Mass.: Harvard University Press, 1972.

Virgo

CHAPTER 18   **COSMOLOGY**

Refrain from illusions, insist on work and not on words. Patiently search divine and scientific truth.

—Mendeleyev's mother

Our efforts to comprehend the nature and the scope of the celestial environment have led to the development of techniques and instrumentation from simple devices for measuring angles to the complex systems currently used to measure electromagnetic radiation and to enhance images of celestial objects. The improved capability for making precise measurements has permitted us to change our view of the universe from a sphere containing stars and planets whose motions were centered on the Earth to an expanding universe whose limits have not yet been ascertained. Again we ask such questions as: From where did all this start? Where will it end? To answer these questions we turn to *cosmology,* the study of peoples' view of the evolution and organization of the universe. Cosmology is not an exact science: it is difficult to collect exact data, and therefore the interpretation of the available data will vary. This has led to the development of several models of the universe which appear to fit the physical properties as they are now understood. Work is continuing to establish a model which will explain all known facts about the expanding universe.

18.1 EARLY COSMOLOGY

We tend to think of cosmology as a modern addition to the study of astronomy, but the properties of the universe have been contemplated for several millennia. The earliest Chinese astronomers limited the size of the universe but the Taoists several thousand years ago subscribed to the *empty infinite space* or Hsuan Yeh concept. According to the Chinese Taoists, space was a formless void and the Earth and other celestial objects were propelled through this infinite space by hard winds.

    The early Greeks, who developed an earth-centered concept of the universe, believed that all objects orbited around a stationary, centrally located Earth. This cosmology evolved as a result of their inability to detect the Earth's motion. The sun, moon, and other planets could be observed apparently moving against the background of more distant stars.

Aristarchus, in replacing the Earth with the sun at the center, explained that the inability of astronomers to detect stellar parallax was due to the great distance of the stars from Earth. He was of the opinion that the stars were an infinite distance from Earth.

The earth-centered concept of the universe dominated thinking through the Middle Ages: the universe, it was believed, was composed of concentric spheres bounded by heaven on the outer sphere. During the Renaissance new modes of speculative thought about the universe were developed, and the possibility of its infinity was entertained. Nicolaus Cusanus, Bishop of Brixen in the early part of the fifteenth century, expressed some ideas on the nature of the universe that are not unlike modern-day thinking. He believed the universe was infinite and the heavens were populated by creatures similar to those on Earth. He also thought that regardless of what location one occupied in the universe, that position would appear to be the center and all objects would be in motion around it.

In the late sixteenth century, Giordano Bruno enthusiastically supported the Copernican concept of a sun-centered system. He was neither scientist nor astronomer, but rather a philosopher who endorsed the Copernican idea with such vigor that he was heartily supported or bitterly opposed, depending upon his audience's viewpoint. Bruno did not believe the sun to be the center of the universe but considered it one star among many in an infinite universe. He believed that the planets orbited these stars and that some of the planets were inhabited. Bruno had no data to support his contention but arrived at his notions entirely through metaphysical speculation. In 1600, he was burned at the stake for his beliefs.

Thomas Digges, a contemporary of Bruno's, was the recognized leader of the English supporters of the Copernican theory. Not only did Digges translate sections of Book I of *De Revolutionibus*, in which Copernicus set forth the essential features of his system, but Digges expanded the idea of the Copernican system to include an infinite universe. Digges was mainly responsible for influencing the scientific thinking of sixteenth-century England in support of the Copernican theory.

In the mid-eighteenth century, Thomas Wright suggested a structural form for the Milky Way. Immanuel Kant speculated on the existence of similar structures, and Herschel endorsed the "island universe" concept. These ideas led to the concept of an infinite universe, but the concept was not seriously pursued until the twentieth century. Hubble was able to identify the nebulae as distant external galaxies and developed an empirical relationship that was useful in determining distances. This led to the discovery of the expanding universe, and as a result several cosmological theories were formulated to explain the new view of the universe.

18.2 THE EVOLUTIONARY UNIVERSE

All cosmological models currently considered important are expanding-universe models. They are based on two assumptions: that the observed red shift is in reality a Doppler effect; and that the universe is everywhere the same—that is, we are not in a unique region of the universe different from all other regions. The latter is known as the cosmological principle. It is recognized that small-scale differences may exist, but the large-scale view of the universe would be uniform regardless of where it is observed.

After Einstein developed his General Theory in 1916, he attempted to apply it to cosmology. He assumed a static universe and included a cosmological constant in his equations to counter the contracting force of gravity. In the late 1920s, Georges LeMaitre, a Belgium astronomer, and others found that without the cosmological constant general relativity would predict an expanding universe, which was consistent with what was being observed.

In 1931, LeMaitre published a cosmological hypothesis in which he proposed to establish the radioactive origin of cosmic rays and all existing matter, as well as to account for the present structure of the universe. He theorized that the entire universe originated as a *primeval atom* containing all the matter in the universe in a volume of space having a radius no larger than one astronomical unit. The primeval atom expanded rapidly in all directions, according to LeMaitre, thinning out and cooling. Density of the gas was not uniform, so regions of greater density formed nucleii of condensation, developing into nebulae or galaxies. The galaxies were assumed by LeMaitre to have all been formed at one time, and so were considered by him to be the same age—about 2×10^9 years.

The calculated age of LeMaitre's universe was one of the first problems to be faced by the new theory. This figure was contradicted by the discovery in the late 1930s that some rocks on Earth were in excess of 3×10^9 years old. The problem was later resolved with Baade's discovery in 1952 that the galaxies were actually more than twice the distance apart they had been thought to be, and therefore the universe was twice as old.

When LeMaitre presented his theory he stated that not too much importance should be attached to the description of the primeval atom. However, the theory is historically important because it was the first cosmological theory based on observations of the behavior of the universe. Prior to LeMaitre's presentation cosmological theories were mathematical or philosophical in nature.

In 1946, George Gamow offered a modification of the evolutionary universe model. Gamow assumed that the universe was initially in a very dense state, and that in the early stages of expansion radiant energy predominated over matter. However, in an expanding system the density of the radiant energy is inversely proportional to the fourth power of the distance expanded. Thus a doubling of the radius of expansion results in a decrease of the density of radiant energy to one-sixteenth. Density of matter decreases as the third power; therefore doubling the radius results in an eightfold decrease in density of the matter. On this basis the density of matter would eventually become greater than the mass density of radiant energy.

Gamow assumed the initial mass of *ylem* (a term applied by Aristotle to primordial matter) to be radiant energy at $15 \times 10^{9\circ}$K. By dividing this temperature by the square root of the age of the universe in seconds, he was able to obtain a chronological picture of conditions in the universe at any age. Five minutes after expansion began the temperature had dropped to $10^{9\circ}$K, which he considered cool enough for hydrogen, helium, and the other elements to begin forming. In thirty minutes or, as Gamow expressed it, "in less time than it takes to roast duck and potatoes," the temperature had dropped to approximately $3.7 \times 10^{8\circ}$ K, too cool for these reactions to continue. At that time, according to Gamow, the universe was composed of roughly equal parts of hydrogen and helium and about 1 percent of all the other elements. The expansion continued until at an age of 2.5×10^8 years temperature had reached 170°K and the density of matter equaled that of radiant energy. At this time matter became gravitationally more important than radiant energy, and the gas became concentrated into huge clouds. These were the *protogalaxies*, initially cold and dark but in which the gas soon began to condense into stars.

The *Big Bang*, as the evolutionary theory has become known, remains essentially unchanged from Gamow's original statement. The galaxies once formed continue to recede and the universe continues to expand, but some of the details have been altered. Hydrogen and helium still comprise 99 percent of matter in the universe, but hydrogen is now known to be by far the dominant element of the two. The heavier elements are now thought to be formed during the evolution of a star (chapter 14). The emergence of hydrogen as the principal ingredient in the universe is now thought to have occurred at a temperature of $3 \times 10^{3\,\circ}$K when the universe was 10^6 years old. At this stage radiant energy and matter decoupled and evolved independently. Matter, now independent of radiant energy, began to clump to form discrete clouds of gas that became the protogalaxies. The exact details of galaxy formation are not known, but it is thought that galaxies formed within the first billion years, and that the minimum size required for a gas cloud to form is about 10^5 solar masses. This corresponds to the mass of a typical globular cluster.

There are now three cosmological models of the *evolutionary* or *expanding universe* under investigation, based on Einstein's General Theory but without the cosmological constant. In the development of these models certain assumptions were made: (1) that the cosmological principle as previously stated is valid; (2) that the universe is expanding according to Hubble's law, and (3) that matter in the universe is uniformly distributed. In all the models the initial creation from a singularity or point of infinite density is accepted. The Hubble constant provides a means whereby the time back to the Big Bang may be determined. If the Hubble constant of 50 km/sec/Mpc is used, then the *maximum age* of the universe is calculated to be 20×10^9 years. Actually, the age will be less than this because the gravitational influence of matter will slow down the rate of expansion.

The rate of expansion is controlled by the gravitational influence of matter in the universe. It has been calculated that if the mean density of all matter in the universe is 5×10^{-30} g/cm^3, we would have a *flat* model of the universe, which would follow the principles of Euclidian geometry (Figure 18.1). The flat model of the universe would be analogous to a rocket sent from Earth at escape velocity. The rocket would leave the Earth in a parabolic curve to infinity, where at an infinite time it would come to rest. The age of the universe for this model is calculated to be about 13×10^9 years.

Should the matter in the universe prove to be a density of less than 5×10^{-30} g/cm^3, we would have an *open* model or ever-expanding universe. The geometry of the universe would be hyperbolic, much as the path of a rocket launched with a speed greater than escape velocity. The curvature of space would be positive and can be visualized by a sphere causing two beams of light traveling parallel paths to eventually cross. In this event the universe would be finite in volume and the age of the universe greater than 13×10^9 years but less than 20×10^9 years.

A third cosmological model of the evolutionary universe is the *closed* universe. In this instance the density of matter would be greater than 5×10^{-30} g/cm^3 and the universe would reach a point of maximum expansion and then contract, presumably returning to the primeval atom again. Based on this model there is the possibility that the universe alternately expands and contracts, which would give us a pulsating or oscillating universe. If we compared this model to a rocket launched from Earth, it

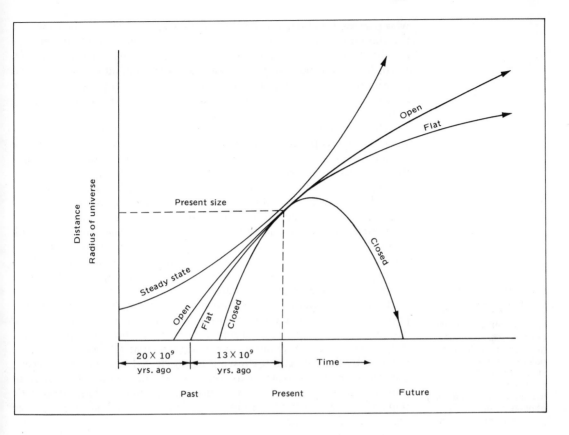

Figure 18.1 A plot of distance versus time for the various models of the universe. The curves are based upon solutions for the various models of the universe.

would be as though the rocket were fired at less than escape velocity and fell back to Earth. Space would be negatively curved, and could be represented by a saddle-shaped surface with parallel lines ultimately diverging. The age of the universe in this model would be less than 13×10^9 years.

The critical factor in developing a cosmological model of the universe appears to be the density of the universe. The density is extremely difficult to measure. It is possible to determine the number of galaxies in a given volume of space, but the exact mass of the galaxies is not known. Then, too, the amount of matter (presumably hydrogen) if any in the intergalactic spaces cannot yet be measured. Estimates that have been made indicate a density less than 5×10^{-30} g/cm^3, supporting the open or everexpanding universe theory. However, the data are quite tentative and not sufficiently reliable to permit definite conclusions to be drawn.

18.3 THE STEADY STATE UNIVERSE

Some cosmologists opposed to the evolutionary theory have suggested an alternate view of the universe, which is attractive because it is relatively simple and, to a certain extent, philosophically appealing. The steady state concept, initially proposed by Thomas Gold and Hermann Bondi of England in 1948, is based upon the perfect cosmological principle that the large-scale features of the universe are the same throughout space and time. This concept was supported by Fred Hoyle of Cambridge University who provided mathematical rigor for the steady state hypothesis within the framework of general relativity. The three viewed the universe as essentially unchanging over large distances and extensive periods of time. Accordingly, the population of galaxies is constant, although individual galaxies may in time disappear. From whatever vantage point the universe is observed, the features remain constant. From this we may infer a *steady state*—a state of being without change (Figure 18.2).

The steady state cosmologists recognize the expansion of the universe—they accept the validity of the Red Shift law and the Hubble constant. How is it possible, then, to maintain a steady state condition if the galaxies are receding and the universe expanding? Here a basic assumption is made that hydrogen has been and is being spontaneously created at a constant rate throughout time. The hydrogen forms into new galaxies which replace those that have receded, thus maintaining a constant density of the galactic population. Individual galaxies are formed, a stellar population evolves, and as the galaxy ages and galactic hydrogen is spent the stars gradually disappear from view. If this does happen, then we might expect to see young, middle-aged, and old galaxies randomly scattered in a given region of space.

Hoyle has described the steady state universe as analogous to the population of a small town. Viewed from a distance it is possible to distinguish the different age groups that make up the town's population. Assuming a static population, we recognize that the old die and are replaced by the young. If the town is again viewed in one hundred years, the same population would exist but it would be composed of different individuals. Hoyle suggests that the universe is similar in terms of galactic population, has existed for an infinitely long time, and is infinitely large with no beginning and no end.

Figure 18.2 In the evolutionary universe (a) continued expansion results in greater distances between the individual galaxies or clusters of galaxies. In the steady state model (b) galaxies continue to form maintaining a constant density of galaxies despite expansion of the universe.

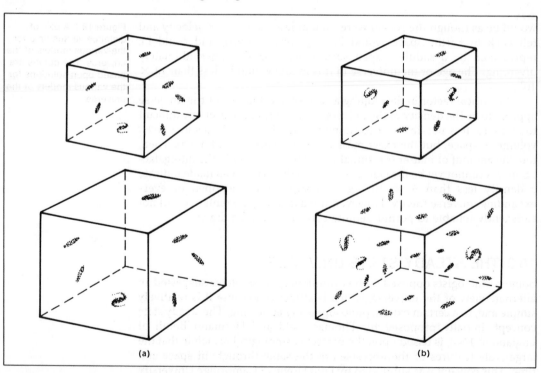

(a)   (b)

The spontaneous creation of hydrogen is a concept that is difficult to accept until, as Hoyle points out, it is realized that the creation of matter is also a basic tenet of the evolutionary universe. The steady state proponents hold that hydrogen is formed spontaneously and continuously throughout the entire universe. The evolutionists assume that all matter was formed as a consequence of the initial explosion. It appears that the act of creation is not in dispute, only its timing.

One of the objections to the steady state theory is that the spontaneous creation of matter violates the law of conservation of energy. However, Hoyle points out the difference between an open infinite system that is expanding and a closed system. In an open expanding universe, local concentrations of energy are related to the energy of expansion of the whole universe. Hoyle believes that the energy of expansion can in some way lead to the formation of hydrogen on a local scale. Within a closed system, as for example within the immediate Earth environment, creation of hydrogen would not be noticeable. Creation of hydrogen within a small volume of space would be extremely small; Hoyle suggests that in an average-size room it may be on the order of one hydrogen atom in 1000 years. This rate of creation would not be recognizable even over an extended period of time, so for practical purposes we may consider the law of conservation of energy inviolate, at least within our ability to detect any change. In section 18.2 we noted that the density of matter in the universe was a critical factor in determining the type of evolutionary universe we live in. Hoyle has suggested that for a steady state condition to exist, it is necessary that the average density be about 10 times greater than is necessary for the evolutionary universe.

18.4 TESTING THE THEORIES

Changes in the universe occur over extremely long periods of time, so the usual method of testing a theory by noting how predictions made from the theory coincide with actual changes cannot be used. However, by observing the universe at great distances it is possible to see changes that have occurred in the past and in this manner check the validity of the cosmological models. In addition, it may be possible to observe phenomena that would be related to one type of universe but not the other.

Cosmic Background Radiation

Gamow was the first to describe the physical consequences of his evolutionary universe theory, and several of his students in 1949 calculated that the residual cosmic radiation after expansion had reached the current point would be approximately $5°K$. In the early 1960s, an attempt was made by personnel of the Bell Telephone Laboratory to evaluate all sources of noise in the equipment used in relation to the Echo satellite project. A small amount of residual noise remained, and this was investigated by A. A. Penzias and R. W. Wilson. The noise showed no sidereal, solar, or directional variation, thus causing them to suspect the radio receiver as the source. After careful examination and reassembly of the equipment, they found that a $3°K$ background radiation at 7.3-cm wavelength persisted. A conversation with B. F. Burke of M.I.T. and R. H. Dicke of Princeton revealed that work was being conducted at Princeton on the same problem. In 1965, a joint paper was published proposing that this background radiation was in fact an echo of the initial Big Bang. Six months later the radiometer at Princeton was completed, and it verified the existence of the 3°K background radiation that had previously been predicted as originating with the evolutionary universe. Measurements of microwave radiation at other wavelengths confirmed the existence of the isotropic (radiation of equal intensity from all directions) thermal background radiation, which at present is more accurately assessed to be 2.7°K. These findings have done much to increase support for evolutionary theory as the most probable cosmological model of the universe.

However, Fred Hoyle has provided an alternate explanation for the background radiation that fits into the steady state hypothesis. He has suggested that the universe is composed of a number of regions of alternating mass polarity. He is of the opinion that the mass of a body is not constant but varies, depending upon the distribution of the rest of the matter in the universe. Thus some aggregates of mass have positive values and some negative, and these can exist separated by distinct time and space boundaries. The universe is a checkerboard of such regions of alternating mass polarity. The limits of astronomical observation have so far all been in one region, and this defines the limits of the universe as described by the evolutionary universe theory. According to Hoyle, light crossing the boundary between two polarity regions is detected as the 3°K thermal background radiation. At present there is no evidence to support this concept.

Quasars

With the discovery of quasars it was thought that additional support for the evolutionary theory was at hand. Quasars appeared to exist only at great distances according to their red shift. It was suggested that if the steady state universe were the true model, quasars should appear uniformly distributed throughout the observable universe. As described earlier (section 17.4), quasars are strange objects appearing to be great distances from us and emitting tremendous amounts of energy by an unknown mechanism. More recently, discrepancies have been found in this definition, and some astronomers think quasars may be nearby objects but still displaying enormous red shifts. If this is true, then quasars may no longer serve as a means of negating the steady state hypothesis.

Density of Matter

Both the evolutionary and steady state theories recognize the expansion of the universe. The evolutionary theory reverses the expansion process to arrive at time zero of the initial Big Bang when matter is flung at high speed in all directions. With time the density of matter decreases, until at some point in the future the density of the universe will approach zero, if we accept the open universe as the existing condition. Those who support the steady state theory also admit to the expansion of the universe but not to a decrease in density. The creation of hydrogen and the formation of new galaxies results in maintaining density at a uniform level.

The question that may be asked is, Can a change in density be detected? Not during the short period of time that astronomers have been observing the universe. However, it is possible to observe objects that are 5 to 10 billion light years from us, which means that we are seeing objects as they were 5 to 10 billion years ago. In this case density of matter at these great distances could be compared with densities in the vicinity of the

Milky Way. If densities are less in the vicinity of the Milky Way, it would help prove the evolutionary theory correct. If densities are similar, support for the steady state theory would be indicated. Determining densities in this manner is not easily accomplished, and it may be some time before such a test is possible.

Age of Matter

Steady state theorists expect galaxies of various ages to exist in any given region of space. Thus in viewing galaxies nearby we would expect to see the same age population as would be seen at great distances. In the evolutionary model of the universe galaxies are presumed to have formed at approximately the same time and therefore are about the same age. Recall that when observing great distances in space we are also looking backward in time. As a consequence, the more distant galaxies would appear younger. The means for determining ages of galaxies have not yet proved too satisfactory and until better methods are developed to determine age this data cannot serve to prove the validity of one cosmological model over the other.

Red Shift vs Distance

The rate of expansion of the universe is influenced by gravitational forces between galaxies, and as a result the ratio of the radial velocity of very distant galaxies to their distance is not exactly a straight line, as is the case with nearby galaxies. The amount of deviation from a straight line is related to the type of universe we occupy, so this information would be of value in solving the problem of open vs closed universe or evolutionary vs steady state universe (Figure 18.3). Unfortunately, determining distances of remote galaxies is very difficult, since the exact luminosities of these objects are not known. Therefore it is only possible to approximate the distances at between 2 and 5×10^9 pc away. Some very tentative data appear to favor the closed evolutionary universe model but are as yet too uncertain to reach definite conclusions.

Observations made to test the various cosmological models have not yet proved sufficiently exact to enable astronomers to draw definite conclusions. At the present time the evolutionary theory is favored over

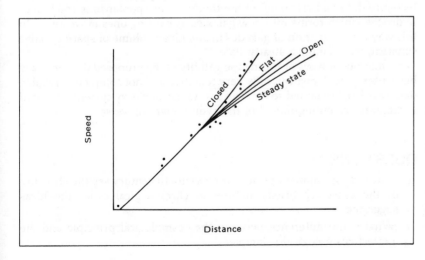

Figure 18.3 A plotting of distance versus speed of recession of very distant galaxies may not produce a straight line as proposed by Hubble. Such a plot is superimposed on ideal curves of distance versus speed of recession for the various models of the universe. Tentative data seems to favor the closed evolutionary model.

the steady state, and of the choices within the evolutionary theory the open universe is considered the most likely model. Many technical problems remain to be solved. The radio telescope has been a major advance, and the establishment of orbiting satellites which eliminate atmospheric interference are of significant value. Perhaps a future observatory on the moon where atmospheric disturbances and distortions are practically nonexistent will provide the ultimate in observation and bring the problem of the origin of the universe closer to a final solution.

18.5 SUMMARY

Our view of the universe has gradually evolved from a limited universe in which the Earth was the central object to a vast universe where the Earth revolves around one of billions of stars located near the outer regions of one of billions of galaxies in the universe. Thousands of years ago, people sought explanations for the nature of the world and devised many theories to explain what they saw. Even in ancient times the universe was described in terms of infinite space. From the beginning theories included physical characteristics of the universe as they are now seen. The universe appears to be expanding, with galaxies receding at increasing speed with increasing distance. The current models of the universe are based on the cosmological principle that the universe is the same regardless of the vantage point from which it is viewed.

Currently, the most favored cosmological model is the evolutionary universe in which all energy and matter was originally compressed into an infinitely dense mass that exploded and expanded into the universe that we see today. Hydrogen formed and gradually gathered by gravitational force into density irregularities which formed the protogalaxies. Matter within the protogalaxies condensed to form stars, resulting in the galaxies we see today. Will the expansion of the universe continue indefinitely or will the universe at some point contract and revert back to the primitive atom? This can be answered if the density of matter in the universe can be accurately measured.

An alternate theory, the steady state theory, suggests that the universe is infinitely old and infinitely large. On a broad scale it has always been and will forever remain as we see it now. The expansion of the universe is recognized, but a basic tenet of the theory is the spontaneous creation of hydrogen which forms into new galaxies as existing ones move away. In this way the population of galaxies in any given volume of space remains constant, maintaining a steady state.

A number of ways of testing the validity of the proposed theories have been attempted. However, only very tentative and not completely reliable data have been obtained so far. The cosmological theory currently favored is the ever-expanding model of the evolutionary universe.

QUESTIONS

1. How did astronomers prior to the twentieth century see the structure of the universe? Briefly outline in chronological order the ideas suggested.
2. What is the difference between the cosmological principle and the perfect cosmological principle?

3. How did LeMaitre's version and Gamow's version of the evolutionary universe differ?

4. What are some of the changes that have occurred in Gamow's version of the evolutionary universe theory since he first suggested it?

5. What are the three basic models of the evolutionary universe and what is the basis for each?

6. Why is it difficult to determine which model is correct?

7. What is the basis for the steady state concept of the universe?

8. How is the steady state condition maintained over long periods of time?

9. Discuss very briefly each of the ways in which the theories concerning the universe may be tested.

10. In outline form list what the results of each test discussed in question 9 must be to favor each theory.

FOR FURTHER READING

BONDI, H., *Cosmology*. New York: Cambridge University Press, 1960.

GINGERICH, O., *New Frontiers in Astronomy*. San Francisco: W. H. Freeman and Company, 1975.

HODGE, P. W., *Galaxies and Cosmology*. New York: McGraw-Hill, 1966.

KAUFMANN, W. J., *Relativity and Cosmology*. New York: Harper & Row, 1973.

SCIAMA, D. W., *Modern Cosmology*. New York: Cambridge University Press, 1972.

WOLTJER, L., ed., *Galaxies and the Universe*. New York: Columbia University Press, 1968.

Taurus

CHAPTER 19   LIFE IN THE UNIVERSE

A Recipe for Making Mice

If a dirty undergarment is squeezed into the mouth of a jar containing wheat within a few days, say twenty-one, a ferment drained from the garments and transformed by the smell of the grain, encrusts the wheat itself with its skin and turns it into mice. And what is more remarkable, the mice from corn and the undergarments are neither weanlings nor sucklings nor premature, but they jump out fully formed.

—J. B. van Helmont (1577–1644)

There is a tendency to deal with the origin of life and its evolution as a unique biological event unrelated to any physical phenomena. Can we separate the evolution of life from the physical activities in the universe, or is life simply one stage in the total evolutionary process of the universe? To answer this question let us see if there is a possible route that takes us from the formation of the universe to the evolution of intelligent life.

19.1 THE BEGINNING

In the last chapter we discussed several theories of the origin of matter in the universe. According to the evolutionary universe theory, matter was formed from an infinitely dense mass that exploded and expanded in all directions. As a result, hydrogen, the simplest element, formed and was gradually gathered into huge clouds by gravitational forces to form protogalaxies. Hydrogen within the protogalaxies condensed into stars and the galaxies as we see them today gradually took shape.

An alternate theory, the steady state hypothesis, recognizes the spontaneous creation of hydrogen, the formation of new galaxies, and the subsequent condensation of hydrogen into stars. It posits a different means of beginning hydrogen formation, but that is not at issue here. What we are concerned with is the fact that initially hydrogen was the predominant element forming the stars, and this factor is common to both theories.

Since hydrogen was the predominant element during the early stages of the universe, it is assumed that the first stars formed in the universe were composed almost entirely of hydrogen. This appears to be the case in the Milky Way. Older stars formed of hydrogen are found in the halo and in the globular clusters where it is thought stars first appeared in the galaxy. Once the star is formed hydrogen is converted to helium through the proton-proton reaction as the star attains a position on the main sequence. As the star ages hydrogen is expended and helium becomes the fuel at the stellar core. By means of the triple-alpha process helium is converted to carbon; in the later stages of stellar development carbon and helium are believed to combine to form the heavier elements. Heavier elements are built up within aging stars and through the supernova mechanism are distributed in space. These elements combine with hydrogen in the interstellar spaces and subsequently form into second generation stars such as the sun. Stars formed from this material contain a very small percentage of the heavier elements. Table 19.1 lists elements that compose the sun and solar system. It indicates that only a small percent of heavier elements are necessary for planets to form and life to evolve.

Do all the heavier elements exist as atoms in the interstellar spaces until they become part of the gas and dust cloud that forms a new star? The answer to this appears to be negative, since molecules have also been discovered in the interstellar spaces. These molecules are primarily organic in nature—that is, they are composed principally of carbon, hydrogen, oxygen, and nitrogen. The existence of the organic molecules in the interstellar spaces provides part of the evidence that permits us to answer the question, Is it possible for life to exist in areas of the universe other than the Earth?, in the affirmative.

The first of these molecules to be discovered was cyanogen (CN) in 1937. Shortly thereafter two unnamed molecules were detected, one composed of carbon and hydrogen (CH) and the other of carbon and ionized hydrogen (CH^+). In 1963, after a period of more than twenty years, the hydroxyl ion (OH^-), a combination of hydrogen and oxygen, was found. Five years later, in 1968, water and ammonia (NH_3) were discovered in the interstellar spaces, and this prompted a more vigorous effort in the search for additional molecules. Through 1975, forty compounds had been discovered, in addition to several as yet unidentified molecules (Table 19.2).

The discovery of these compounds has lent considerable support to the belief that life may be fairly common in the universe. In fact the question has been raised: Could life on Earth have evolved more expeditiously as a result of the existence of these organic molecules than from the

TABLE 19.1   SOLAR COMPOSITION

Element	Percent of Total Composition
Hydrogen	87.0
Helium	12.9
Oxygen	0.025
Nitrogen	0.02
Carbon	0.01
Magnesium	0.003
Silicon	0.002
Iron	0.001
Sulfur	0.001
Others	0.038

MOLECULES DISCOVERED IN INTERSTELLAR SPACE   TABLE 19.2

Name of Molecule	Symbol	Year Discovered
Cyanogen	CN	1937
(Unnamed)	CH	1940
(Unnamed)	CH$^+$	1941
Hydroxyl	OH	1963
Water	H$_2$O	1968
Ammonia	NH$_3$	1968
Formaldehyde*	H$_2$CO	1969
Carbon monoxide*	CO	1970
Hydrogen	H$_2$	1970
Hydrogen cyanide	HCN	1970
X-ogen	?	1970
Cyanoacetylene*	HC$_3$N	1970
Methyl alcohol*	CH$_3$OH	1970
Formic acid*	HCOOH	1970
Carbon monosulfide	CS	1971
Formamide*	NH$_2$COH	1971
Silicon monoxide	SiO	1971
Acetonitrile*	CH$_3$CN	1971
Carbonyl sulfide	OCS	1971
Isocyanic acid*	HNCO	1971
Hydrogen isocyanide	HNC	1971
Methylacetylene	CH$_3$CCH	1971
Acetaldehyde*	CH$_3$CHO	1971
Thioformaldehyde	H$_2$CS	1972
Hydrogen sulfide	H$_2$S	1972
Methanimine	CH$_2$NH	1972
Sulfur monoxide	SO	1973
(Unnamed)	HD	1973
(Unnamed)	DCN	1973
Dimethyl ether	(CH$_3$)$_2$O	1974
Methylamine	CH$_3$NH$_2$	1974
Vinyl cyanide	CH$_2$CHCN	1974
Acetylene radical	CCH	1974
(Unnamed)	N$_2$H$^+$	1974
Silicon sulfide	SiS	1974
Ethyl alcohol	CH$_3$CH$_2$OH	1974
Heavy water	HDO	1974
Sulfur dioxide	SO$_2$	1975
Cyanamide	NH$_2$CN	1975
Formyl radical	HCO	1975

*Building blocks of life produced in laboratory.

"primordial soup"? Some astronomers believe that the organic molecules could not survive the heat generated by a newly forming star. Others believe that the density of the dust and gas in the disk from which the planets were formed may be sufficiently high to protect the organic molecules from excessive heat and prevent their breakdown from occurring.

How do the molecules form in space? The exact mechanism is not known, but the elements making up the organic molecules are formed in the dying stars (see chapter 14). The elements are scattered in the interstellar spaces by supernovae and drift about aimlessly. Very occasionally, it is thought, the atoms collide to form molecules or the atoms accumulate on particles of dust. The properties of the dust particles are not known but enough of the atoms may collect to allow the complex molecule to form. It is also possible that cosmic rays strip electrons from the atoms, forming ions that ultimately combine into organic molecules. As yet these problems have not been solved, but they are currently receiving a great deal of attention.

Planets formed in association with second generation stars may have relatively the same composition as planets in our solar system, since we can assume that the same laws of physics and chemistry we recognize on Earth apply throughout the universe. We might then expect life to occur upon any planet where temperatures and energy sources are such that these molecules can react to form the complex materials that evolve into living systems. The form of life that we might expect would be based on the carbon atom as it is here on Earth.

There have been some suggestions that life may possibly have evolved on some distant planets with silicon as the basic element instead of carbon. This element, it has been pointed out, is quite versatile and could possibly develop the complex molecules which seem to be necessary for life. However, there is no evidence to support such an idea. Carbon is unique in that it has the capability of bonding to itself into very long chains, a characteristic that appears to be a necessity in the formation of living substances. In addition, carbon is far more abundant in the universe than silicon and therefore is more likely to be the basis of life.

19.2 LIMITATIONS

For life to evolve on a planet a certain combination of conditions must exist. First, a planetary system, a variation of a binary system (chapter 15), must be associated with a star. For life to occur on any of the planets it is necessary that the planet be of sufficient mass to permit the gases that form an atmosphere to be retained. The atmosphere must be composed of gases from which life may evolve. Temperatures need to be within an appropriate range to permit the evolution of large molecules, which means that the planet be not too near or too far from the parent star. Time is an additional factor to be considered, as it required up to one billion years for life to develop on Earth and an additional 3.5 billion years for intelligent life to evolve. As a result of the time factor, it is presumed that life capable of communication may only evolve on planets associated with long-lived stars.

We previously discussed the possible existence of planets associated with Barnard's star, and in our search for intelligent life it will probably be necessary to study many such systems. These planets cannot be directly seen but only detected by their influence on the parent star. Of the fifty-nine stars within 5 pc of the sun there are good data available indicating that at least nine stars have dark companions, and an additional number are suspected of having a planetary system (Table 19.3). It is necessary that one be reminded of the difficulty in detecting the presence of dark companions associated with stars. However, based on data of nearby stars it is reasonable to assume that planets are fairly commonplace throughout the galaxy and the universe.

The part played by gases and their distribution during the formation of the solar system was discussed in chapter 6. The gas giants have approximately the same composition as the sun, with additional gases

TABLE 19.3

Nearby Stars with Dark Companions

Star	Distance (Parsec)	Spectral Type	Mass of Star (Sun = 1)	Mass of Dark Companion (Jupiter = 1)
Proxima Centauri	1.31	M5	0.1	1.8
Barnard's Star	1.83	M5	0.15	1.1 and 0.8
Lalande 21185	2.51	M2	0.35	10
Epsilon Eridani	3.3	K2	0.6	6
61 Cygni A	3.4	K5	0.58	8
Kruger 60 A	4.0	M3	0.27	9
70 Ophiuchi	5.2	K0	0.89	10
Cin 2354	5.5	M4?	0.2?	?
Cin 2347	8.3	M1	0.33	20

such as ammonia and methane. The terrestrial planets are composed almost entirely of the heavier elements, with relatively little hydrogen and helium. Much of the hydrogen on Earth is incorporated in water.

The variation in atmospheric composition of planets in the solar system is due to such factors as planetary mass and distance from the sun. The ability of a planet to retain atmospheric gases depends upon the energy required for a gas molecule to escape the planet's gravitational field. Lighter elements such as hydrogen and helium require less energy and can therefore escape more easily than the heavier gases such as oxygen and nitrogen. None of the terrestrial planets have sufficient mass to permit them to retain hydrogen and helium, although these were the most abundant gases present during the formation of the solar system. Table 19.4 lists the length of time some gases may be retained by representative planets. Note that the smaller planets such as Mercury and Mars are not capable of retaining gases as readily as Earth and Venus.

Proximity to the sun will influence the nature of the atmosphere as well as the amount of radiation received by a planet. We have a vivid example of this in Venus and Earth. Both planets are approximately the same size and mass. It may be assumed that they are composed of the same material, although at some later date exploration of Venus may show this to be untrue. Presumably both planets evolved in the same manner and had the same opportunity for similar atmospheres. However, the present atmosphere of Venus is equivalent to 90 Earth atmospheres and is composed principally of carbon dioxide with very little water vapor (chapter 9). The high carbon dioxide composition has produced an intense greenhouse effect, resulting in extremely high temperatures on the surface. On the other hand, Earth's atmosphere, composed principally of nitrogen and oxygen, permits moderate temperatures at the surface. Venus, approximately 0.7 AU from the sun, receives twice as much solar radiation as the

TABLE 19.4

Length of Time in Years Planets Are Able to Retain Representative Gases in Their Respective Atmospheres

Gas	Moon	Venus	Earth	Mars	Jupiter
Hydrogen	10^{-2}	10^{3+}	10^{3+}	10^{3}	10^{200}
Helium	10^{-1}	10^{8}	10^{8}	10^{6}	$>10^{200}$
Oxygen	10^{2}	10^{32}	10^{32}	10^{20}	$>10^{200}$
Argon	10^{10}	10^{70}	10^{70}	10^{49}	$>10^{200}$

Earth. Proximity to the sun may have influenced the formation of oceans on one and not the other of these two planets and resulted in their atmospheres being so dissimilar.

Conditions for the development and continuance of life are dependent upon a moderate temperature on the planet. The temperatures should range from -5°C to 60°C, because chemical activity is very sluggish at the lower end of the scale and certain molecules important in the development of life break down at temperatures much above 60°C. Most organisms can tolerate temperatures outside these limits but only for short periods of time. Humans have been able to survive greater extremes because of their ability to modify the environment to suit their needs.

On the basis of temperature requirements it is possible to establish an *ecosphere* or life zone within which, it has been suggested, conditions are suitable for the development of life. The limits of a planet being too close to a star and therefore too hot or too far away and therefore too cold are not known exactly. Within the solar system several possible limits for the ecosphere have been proposed. S. H. Dole chose 0.86 AU and 1.25 AU as the inner and outer limits. Solar radiation at these distances is 135 percent and 65 percent, respectively, of that received on Earth. The greater the ratio of solar radiation between the inner and outer limits, the greater the probability of finding one or more planets of a planetary system within the ecosphere. Based on the planetary spacings within the solar system, Dole computed a 66 percent probability of one favorably situated planet and a 5 percent chance of two planets within the ecosphere. Another slightly different boundary for an ecosphere was suggested by S. I. Rassool, who thinks the inner limit is about 0.9 AU with a solar flux of 123 percent of that received by Earth. The outer limit would be 1.5 AU with 43 percent as much radiation. This compares quite well with the limits given by S. H. Dole.

The suggested limits apply to the solar system only, since, to date, this is the only planetary system available for study. It must be recognized that as we consider larger stars with greater luminosity the ecosphere is enlarged. A. G. W. Cameron has suggested that the spacing of the planets increases in proportion to the size of the star. This means that as the ecosphere increases the spacing between the planets increases, thus giving a constant number of planets per ecosphere.

Is the development of life entirely dependent upon a planet's position within the ecosphere? Some astronomers think not. Conditions on Jupiter, for example, may be conducive to the development of organic compounds. Water and the gases thought to make up the early atmosphere of the Earth have been detected on Jupiter and temperatures are within the limits that would permit the interaction of the basic precursors of life. However, as yet there is no evidence to support the theory that life may be forming in this manner on the outer planets.

Although a planet may exist within the ecosphere of a star, this is not assurance that intelligent life has evolved. Life forms, it is presumed, progress up the evolutionary scale from the lower forms to the higher forms over an extended period of time. On Earth this may have required as long as 3.5 billion years. If the current theories of star evolution are correct, then some stars do not have a sufficiently long life cycle to permit life to develop (see section 13.8). Type O and B stars are large, bright stars that endure for only a few tens or hundreds of millions of years. This is too short a period to permit life to evolve to higher forms on planets associated with such stars. If it requires 4.5 billion years for intelligent life to appear from the time a planet is formed, then type A and early type F stars may also have too short a life span. Stars of late type K and type M stars, although enduring for very long periods, may have such a restricted ecosphere that only a planet with a perfect circular orbit would be within the life zone and capable of supporting life. Thus it realistically appears that only late type F, type G and early type K stars have sufficiently large ecospheres and life cycles long enough to permit intelligent life to evolve. The sun, it may be recalled, is a type G2 star.

If the presence of intelligent life capable of communicating from some distant planet is established, it would be natural to speculate on the form this intelligence would take. Are there limitations of structure that influence the development of intelligence? It is not possible to determine the direction evolution may have taken on another planet, but it is assumed that certain physical characteristics are essential to a being possessing the intelligence to communicate. A high degree of mobility is needed to enable the creature to collect raw materials necessary to build and a high degree of manipulative ability to convert these raw materials into useful products. This would imply structural features similar to hands and legs. Since communication equipment and all the auxiliary equipment necessary for its functioning are complex, the being would need to possess the ability to hear and see. Some biologists believe that these alien creatures would possess a degree of symmetry, if for no other purpose than to maintain balance, and would therefore not be too dissimilar to humans. However, other scientists believe that the vicissitudes of evolution have caused humans to be unique in the universe and that other intelligent beings could not be classified on the same basis.

19.3 HOW MANY?

Once we accept the idea that life exists in other parts of the universe, it is only a step to inquire how many civilizations there are that have the capability for extraplanetary communication. We cannot answer this question now, nor in the foreseeable future. In 1961 a group of scientists met in Green Bank, West Virginia, to discuss the prospects of communicating with other intelligence and to estimate the number of possible societies in the galaxy capable of communicating. One outcome of the meeting was an equation, attributed to Frank Drake, which is generally presented as:

$$N = R_* f_p n_e f_l f_i f_c L$$

The letter N represents the number of societies in the galaxy currently capable of communicating with other societies. This number is dependent upon the factors on the right-hand side of the equation, which when multiplied together provide an estimate of the number of communicative societies in the galaxy. An answer to this equation is based on very limited data (one solar system) and must be viewed critically, as it is subject to many interpretations. A small number would indicate only a few societies scattered throughout the galaxy and therefore separated by thousands of parsecs. A large answer would mean that nearby stars may have planets occupied by civilizations capable of communicating. Nevertheless such societies could still be tens of parsecs from us, and these distances are not conducive to idle interstellar chit-chat.

The factors that are the basis for N do not represent hard data but rather are the best estimates made by those considered to be expert in each factor area. The values used are very approximate and possibly optimistic. The factor R_* represents the rate of star formation, which throughout the life of the galaxy may have averaged about 20 stars per year. This is based upon the number of stars estimated to be in the galaxy, about 2×10^{11} stars, and the age of the galaxy, about 10^{10} years. It is assumed that star formation at present is much lower than in the past; at the time the sun was formed star formation may have been about 350 per year or approximately one per day. This would provide an indication of the number of stars with planets upon which life may be approaching maturity—that is, life capable of communicating, as is life on Earth.

The second factor in the equation, f_p, represents the percentage of stars with planetary systems. Planetary systems are now considered to be more common than was thought a few decades ago. Even within the solar system there are several planets with a number of satellites simulating the structure of a planetary system, indicating that the formation of smaller bodies in association with a larger object is a fairly common occurrence. It is felt that multiple-star systems may be excluded, since any planets associated with such a system would have complex, frequently changing orbits, taking the planets beyond the limits of the ecosphere. Although planets of other star systems cannot be seen from Earth, studies previously cited on the presence of dark companions associated with stars provide sufficient evidence to tentatively conclude that planets are a common occurrence in the universe. Since approximately half of all the stars studied appear to be multiple-star systems, we may assume 50 percent are stars with planets.

Factor three, n_e, represents the number of planets in a planetary system that have an environment capable of sustaining life. This refers to planets within an ecosphere, although some planets beyond this limit may have a greenhouse effect that warms them to a life-sustaining level. Some astronomers are of the opinion that the spacing of planets in the solar system is representative of spacing in other systems, which means that at least one planet may be found within the ecosphere (Figure 19.1). S. H. Dole was cited (section 19.2) as believing 66 percent of the planetary systems have at least one planet in the ecosphere. Other proposals suggest that as many as five planets may exist within the ecosphere of some stars. Thus we can see that there is no unanimity of opinion on the number of planets associated with a star that might support life, and we may therefore presume one planet to be a reasonable number for factor n_e.

The next factor, f_l, is the symbol used for the fraction of suitably placed planets on which life appears and evolves to more complex forms. The process or some modification of the process by which life evolved on Earth could result in life on other planets. Carl Sagan has suggested that the production of self-replicating systems is a "forced process." Given the appropriate mixture of gases and a source of energy as occurred on the primitive Earth, the basic molecules prerequisite to life would result and some form of life would evolve. It is felt that Darwinian evolution would be inevitable, although it could most certainly follow some alternate path than that which occurred on Earth. Because the prospect of life evolving is considered to be certain, a value of 1 or 100 percent was ascribed to factor f_l.

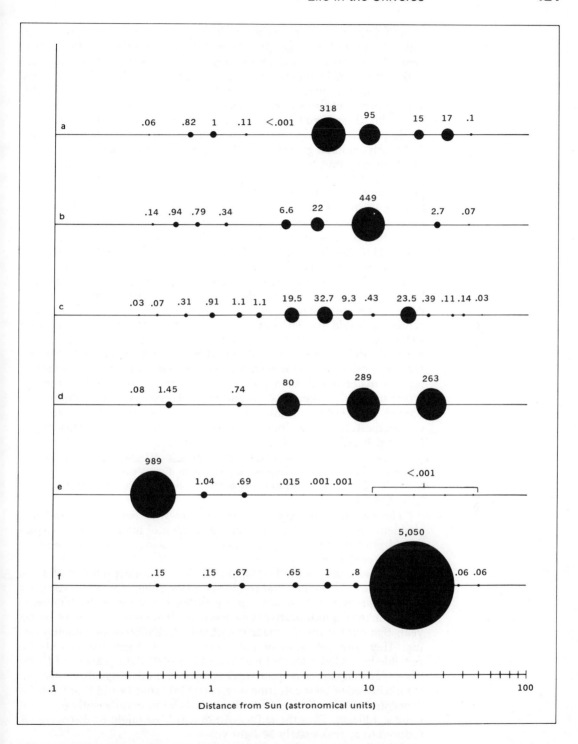

Figure 19.1 Several alternative planetary models suggested by Carl Sagan and Richard Isaacman of Cornell. Solar system is model (a). Value over each planet indicates mass in multiples of Earth's mass. (By permission of Carl Sagan and *Scientific American* from "The Solar System," by Carl Sagan. Copyright © 1975 by Scientific American, Inc. All rights reserved.)

The fifth factor, f_i, refers to the fraction of life-bearing planets upon which intelligence emerges. Intelligent life was considered by the participants of the Green Bank conference to be an inevitable outgrowth of the initial development of life on a planet. Life on Earth had filled almost every kind of environment and prospered and sooner or later, it was felt, intelligence would appear because of its marvelous survival value. John C. Lilly, who has studied the possibility of communication between man and the dolphin, is of the opinion that the dolphin is an intelligent creature and that much could be gained from the experience of learning to communicate with dolphins, for it would be analogous to learning to comnunicate with alien forms of intelligence. The possible presence of more than one intelligence on Earth prompted the participants at Green Bank to give factor f_i a value of 1, indicating that intelligence would surely occur once life appeared on a planet.

Factor six, f_c, is the fraction of intelligent societies that evolve to the point where they have both the ability and the desire to communicate with outside worlds. A developing society may have the ability to accomplish such technical feats, but after having conquered the problems of disease, hunger, and poverty, would the desire to look further afield for new experiences still linger? Or, after a comfortable existence has been achieved, would the temptation to vegetate be too great? It has been suggested that science has, in part, been advanced by the desire for material ease and once this is accomplished interest in science will wane. However, others feel that even if it were possible to reach a satisfactory level of material well-being, society will continue to have problems for scientists to solve, thereby constantly stimulating a continuing interest in science. The resulting technology will stir further interest in interstellar communication for as long as humans survive. Fridtjof Nansen, the Norwegian explorer, perhaps stated it best:

> The history of the human race is a continuous struggle from darkness
> toward light. It is therefore of no purpose to discuss the use of
> knowledge—man wants to know and when he ceases to do so he is no
> longer man.

The Green Bank conferees and others who have considered this problem vary in their estimates of the number of intelligent species that would attempt interstellar communication, ranging from one-tenth to one-half of the total.

Now we come to the last factor, which is the most difficult of all to assess. The other factors had some basis, albeit slim, for an estimate, but the seventh factor, L, which represents the longevity of the intelligent society in the communicative state, has none. It has been only in the last 75 years that our society has made use of radio frequencies for communicating. Therefore detection of radio signals from Earth would only be possible by societies located within a radius of 75 light years from Earth. Those early signals were quite weak compared to other signals in space, so the likelihood of their detection is remote. Only after World War II did the concentration of radio and television signals become sufficiently powerful to be significant. Thus the radius within which we might be discovered is reduced to approximately 30 light years.

What are the prospects of these signals being detected and our position in the universe discovered? There is no consensus on the number of societies capable of communicating, but one estimate places the number at about one million. If these are uniformly distributed throughout the galaxy, then the nearest society capable of receiving and interpreting radio signals would be about 300 light years away. In that event our presence would probably still be unknown, and the society would have to wait 250 to 300 years before signals from Earth reached them.

Does an intelligent society with highly advanced technological abilities survive that long, or does it destroy itself only a few generations after achieving such a technology? This is the question posed by the seventh factor. Is our civilization to be ended by nuclear holocaust, by mass starvation due to overpopulation, or by pollution? Or will we overcome these problems to enter a new era wherein people are raised to a higher level of existence and interstellar communication becomes an established fact?

Several scenarios have been suggested. Fred Hoyle envisions a series of cycles of alternating catastrophes and recoveries during a 5000 year period, with intelligence and sociability improving upon each cycle. Have we reached the peak of a cycle where man is capable of communicating and benefiting from alien contact? Or will we encounter another catastrophe, such as worldwide starvation due to overpopulation or a breakdown in distribution of resources, before the desired scientific breakthrough is achieved?

In attempting to solve the problem of longevity, Sebastian von Hoerner, like Hoyle, suggests the possibility of a succession of societies emerging on a planet. However, in his view the dominant species is destroyed at each stage in a catastrophe only to be replaced by a new society evolving from an unaffected lower form. Through this process, it is estimated, it would require hundreds of millions of years to develop a society capable of communicating, instead of the 5000 proposed by Hoyle. Von Hoerner suggests that if such a society endured 6500 years, and during the existence of a planet there were four successive technological societies, then only about one in 3×10^6 stars would currently have a planet with a society with communication capabilities, and these would be about 1000 light years apart. This is somewhat more pessimistic than other estimates, but von Hoerner feels that if a society endures a minimum of 4500 years there is a good possibility that communication can be established. If successful in making contact, the chance for greater longevity is improved because of the possible increase in wisdom and knowledge such contact engenders.

Many elements enter into the computation of factor L and many interpretations result. There is no precedent for longevity of a communicative society and no basis upon which to measure it. We can only hope that our society will endure sufficiently long to determine whether or not we are alone in the universe.

19.4 WHAT IF . . . ?

Once we have accepted the possibility of intelligent life existing on other worlds, and our communicating with them, the question arises: What if we are successful in making contact? No doubt extraterrestrial contact could be advantageous to the inhabitants of Earth, but we cannot automatically assume that such contact would be beneficial. We must also consider the potential risk to being exposed to an alien society. Since we have had the capability of communicating for only a very short period of time, it is reasonable to expect that our society would be the youngest and therefore the least sophisticated. Our Earth experience has been that contact between two cultures has inevitably resulted in the domination of the more backward society by the stronger, although this has never been accomplished at long range. The danger to us of establishing radio contact may be the cultural shock of realizing that a higher order of intelligence exists elsewhere and represents a potential threat to our existence. However, the realization may be gradual rather than sudden, since the information from such contact would probably occur over generations. We must recall that even to planets associated with nearby stars, messages may require twenty years for a round trip.

A greater danger than cultural shock may be the possibility of subversion by an alien society. Under the pretense of teaching us or providing us with aid in solving our problems, an alien society could gradually gain control. This could only occur through the collusion or gullibility of persons in our own society who were charged with the responsibility of carrying out communication with an outside group. Adequate safeguards would be required to protect our society against this threat.

Invasion of Earth by superior beings is considered unlikely because of the problems arising from the great distances involved. However, escape from the solar system by man has been a recurrent theme in science fiction and at some point in the future this may become science fact. There is the possibility that other intelligent beings more advanced than us have already accomplished this feat. If so, we need to look for an answer to Enrico Fermi's question: "Where are they?"

19.5 IS IT WORTH THE EFFORT?

Once having started on the search for extraterrestrial life it is likely that we will continue until contact is made or we become discouraged through lack of results. The search may periodically be discontinued until new techniques are discovered that may renew interest. For this reason an endeavor such as Project Cyclops (see chapter 1) is advantageous because it can be designed to operate indefinitely with a minimum of supervision. The problem may be one of cost. Society may look at efforts like Project Cyclops and the recent Apollo space program and question the huge expenditures of money necessary for their operation.

Many have viewed the lunar landings and related projects as nothing more than stunts and felt that the money and effort could have been put to better use in solving the many critical earthly problems. Arnold J. Toynbee, the British historian, compared the lunar landings with the building of the pyramids or the palace at Versailles at times in history when much of the population was underfed. Toynbee felt that it was a foolish mismanagement of human affairs that allowed such an enormous gap to develop between social problems and technology. Others, like Reinhold Niebuhr the theologian, disputed the priorities that had become established for space exploration and human problems. Niebuhr pointed out that the space program was being conducted by a nation afflicted by water and air pollution and decaying inner cities.

It is difficult to evaluate such efforts as the Apollo program or the search for intelligent life in the universe in terms of its influence on society unless these efforts are viewed from a historic prospective. Exploration of the universe by whatever method must be viewed as a continuing effort by humankind to broaden its horizons and understand itself. Father Gerald Smith of Marquette University expressed this idea best:

> Man is curious by nature. He has to explore whatever strikes his
> curiosity and he undoubtedly will never stop exploring.

Sir Bernard Lovell, director of Jodrell Bank Observatory, claims that a nation decays unless it continues to progress to the limit of its technological ability, and cites the Roman Empire as an example of a civilization that fell as a result of a loss of interest in further growth and progress. Margaret Mead, the anthropologist, suggests that unless we pursue changes humanity will wither and die. She believes that it is human nature to be venturesome and that opportunities for adventure on Earth decrease as the Earth becomes overpopulated. An alternative, then, becomes the exploration of space.

Can a project like the search for life in the universe or landing on the moon be validly compared to the solving of social problems? Some feel that such a comparison cannot be justified. Society has evolved through people's effort to control their baser nature and provide for their material well-being, and in the process an increasingly complex social order has developed. This in turn has produced more complex problems with which people are obliged to cope. The social problems are never clear-cut and are complicated by inborn prejudices and ancient customs generated by the very social structures through which people seek order.

An endeavor such as the search for extraterrestrial life or a technological problem of sending a man to the moon is relatively simple compared to solving a problem like poverty. The goal is clear-cut, the universal laws of science are understood, and the technology is available. These ingredients, when combined with individual creativity and courage and the necessary funds, are adequate to accomplish the task.

No one can determine in advance what the effect of such endeavors will be on the future of humankind, any more than observers at the beginning of the sixteenth century could foresee the outcome of Columbus' voyages to the New World. History tells us that many social problems were solved, but this is, of course, hindsight. Exploration of the universe is in its infancy and has as yet provided few direct benefits. However, the efforts involved have produced some new approaches and techniques to the problem-solving process. We can hope that the application of these techniques may bring about desirable results by providing new perspectives on old social problems.

19.6 SUMMARY

Alternate theories of cosmology describe the formation and evolution of the universe. Hydrogen and helium form into stars wherein nuclear processes are thought to be responsible for the development of the heavier elements. These elements, by some as-yet-unknown mechanism, combine to form molecules in the interstellar spaces. The presence of molecules causes many scientists to feel that life may form in many parts of the universe, since the molecules represent the precursors of living material.

There are some limitations that regulate the evolution of life on a planet. Temperatures appropriate for the development of organic compounds are necessary, as well as sufficiently long periods of time for life to evolve. The planet must be of sufficient mass to enable an atmosphere to accumulate. The gases of the atmosphere must be suitable to the development of life, and water in liquid form appears to be a prerequisite to life. The star with which the planet is associated must be massive enough to have an ecosphere sufficiently large to enable a planet to orbit within a zone of suitable temperature. The star must also have a life span long enough to permit life to evolve on the planet. Late type F, type G and early type K stars have large enough ecospheres and a sufficiently long life span.

Once the concept of life existing in other parts of the universe is accepted, the question that arises is how many civilizations evolve that are able and willing to communicate. A method of estimating this, although producing only tentative answers, was developed by participants at the conference in Green Bank, West Virginia, in 1961. Their interest was to determine the number of intelligent societies now in existence with the capability of communicating with other worlds. An equation was presented which included variables that the conferees felt were essential in making such an estimate.

Making contact with extraterrestrial intelligence entails a certain amount of risk as well as benefits, and the risks must be considered seriously. Since we have only recently acquired the ability to participate in interplanetary communication, it is thought that a responder would be somewhat more technologically advanced than we are. Therefore a danger exists that we could be subverted, exploited, or even invaded. While these contingencies are remote, they must be considered in any future contact.

Any search for extraterrestrial life will require great expenditures of time, money, and energy, and the question is raised: Will it be worth it? We cannot tell. Only the future can reveal the worth of such an endeavor.

QUESTIONS

1. Very briefly describe how the heavy elements are formed in the stars.
2. What have been the suggested means by which the molecules in space have been formed?
3. Why does the presence of these molecules mean that extraterrestrial life may exist?
4. Why does carbon appear to be so uniquely suited to the development of life?
5. What are the conditions necessary for life to evolve on a planet?
6. What are thought to be the structural requirements of an extraterrestrial being capable of communicating?
7. Briefly define the factors that make up the equation suggested by Frank Drake.
8. What are the factors that limit the chances of our being discovered by some extraterrestrial intelligence?
9. What are the risks involved in making contact with an extraterrestrial society?
10. After reading this chapter, what are your thoughts on the search for extraterrestrial life?

EXERCISES AND PROJECTS

1. Write a term paper on the pros and cons of the exploration of space or the search for extraterrestrial life. Support one side or the other. This topic may be developed as a classroom debate.

FOR FURTHER READING

DRAKE, F. D., *Intelligent Life in Space.* New York: The Macmillan Company, 1962.

FIRSOFF, V. A., *Life Beyond the Earth—A Study in Exobiology.* New York: Basic Books, 1963.

PONNAMPERUMA, C., and A. G. W. CAMERON, *Interstellar Communication: Scientific Prospectives.* Boston: Houghton Mifflin Company, 1974.

SAGAN, CARL, ed., *Communication With Extraterrestrial Intelligence.* Cambridge: M.I.T. Press, 1973.

SAGAN, CARL, *The Cosmic Connection.* Garden City, N. Y.: Doubleday and Company, 1973.

SULLIVAN, W., *We Are Not Alone.* New York: McGraw-Hill, 1966.

APPENDIX 1

UNITS OF MEASUREMENTS AND PHYSICAL CONSTANTS

Length

1 metre (m) = 100 centimetres (cm) = 1000 millimetres (mm) = 1.094 yards = 39.37 inches

1000 metres = 1 kilometre (km) = 0.6214 miles

1 mile = 1.6093 km

1 inch = 2.540 cm

1 micron (μ) = 10^{-6} m

1 angstrom unit (Å) = 10^{-8} cm = 10^{-10} m

1 astronomical unit (AU) = 1.49598×10^8 km $\cong 1.5 \times 10^8$ km $\cong 9.3 \times 10^7$ miles

1 light year (LY) = 6.3235×10^4 AU = 9.46×10^{12} km = 5.878×10^{12} miles

1 parsec (pc) = 206265 AU = 3.26 LY = 3.086×10^{13} km

Mass

1000 milligrams (mg) = 1 gram (g)

1000 grams = 1 kilogram (kg)

1000 kilograms = 1 metric ton = 2204.6 pounds

1 pound = 453.6 g

1 ounce = 28.3495 g

The following prefixes indicate the power of 10 to be used with the basic unit of measurement in the metric system:

Tera = 10^{12} deci = 10^{-1}
giga = 10^9 centi = 10^{-2}
mega = 10^6 milli = 10^{-3}
kilo = 10^3 micro = 10^{-6}
hecto = 10^2 nano = 10^{-9}
deca = 10^1 pico = 10^{-12}

Time

1 second (sec) = 3.161×10^{-8} of the tropical year 1900

Tropical year 1900 = 31,556,926 sec = 365d 5h 48m 46s

1 mean solar day = 86,400 seconds

1 sidereal day = 23h 56m 4.091s mean solar time

1 sidereal year = 365d 6h 9m 10s

Temperature

Kelvin (°K), Celsius (°C), Fahrenheit (°F)

Absolute zero = 0°K,   −273.18°C,   −459.72°F

Freezing point of water = 273°K,   0°C,   32°F

Boiling point of water = 373°K,   100°C,   212°F

Conversions between scales:

$°C = \frac{5}{9}(°F − 32)$, $°K = °C − 273$, $°F = \frac{9}{5}°C + 32$

Constants

Velocity of light = 2.99793×10^5 km/sec = 1.86×10^5 miles/sec

Gravitational constant (G) = 6.67×10^{-8} dynes-cm^2/g^2

Dyne = force necessary to impart acceleration of 1 cm/sec^2 to a mass of 1 g

Solar constant (S) = 1.37×10^6 ergs/cm^2-sec = 1.97 calories m^2/min

Erg = work done by a force of 1 dyne acting on a body through a distance of 1 cm. 1 erg = 1 dyne-cm

Pi (π) = $3.1416 = \frac{22}{7}$

Radian (R) = 206265 arcsec = 57°17′36″ (57.°3)

Acceleration due to gravity (g) = 980 cm/sec^2,   9.8 m/sec^2,   32 feet/sec^2

Mass of Earth M $\oplus$ = 5.977×10^{24} kg = 5.977×10^{27} grams

Mass of sun M $\odot$ = 1.991×10^{30} kg = 1.991×10^{33} grams

Mass of hydrogen atom = 1.673×10^{-24} grams

Mass of proton = 1.67×10^{-24} grams

Mass of electron = 9.11×10^{-28} grams

APPENDIX 2

ANNUAL METEOR SHOWERS

Name of Shower	Dates	Radiant °R.A.*		°Dec	Associated Comets	Hourly Rate
		h	m			
Quadrantids (Bootids)	Jan 3	15	20	+50		40
Lyrid	Apr 19–23	18	4	+33	1861 I	15
Aquarids	May 1–6	22	16	−2	Halley ?	20
Aquarids	Jul 26–31	22	36	−11		20
Perseids	Aug 10–14	3	8	+58	1862 III	50
Giacobinids	Oct 9	17	40	+55	1946 V	Variable
Orionids	Oct 18–23	6	8	+15	Halley	25
Leonids	Nov 14–18	10	0	+22	1886	Variable
Geminids	Dec 10–13	7	32	+32		50

*R.A., right ascension.

APPENDIX 3   SOLAR SYSTEM
SATELLITES

Planet	Satellite	Discovery	Diameter in Kilometres	Sidereal Period Days*	Distance from Planet (km)
Earth	Moon	—	3476	27.322	384,405
Mars	Phobus	A. Hall 1877	15†	0.319	9,380
	Deimos	A. Hall 1877	8†	1.262	23,500
Jupiter	V Amalthea	E. Barnard 1892	150†	0.498	180,500
	I Io	Galileo 1610	3659	1.769	421,600
	II Europa	Galileo 1610	3100	3.551	670,800
	III Ganymede	Galileo 1610	5265	7.155	1,070,000
	IV Callisto	Galileo 1610	4910	16.689	1,882,000
	VI Hestia	C. D. Perrine 1904	120†	250.57	11,500,000
	VII Hera	C. D. Perrine 1905	50†	259.65	11,800,000
	X Demeter	S. Nicholson 1938	24†	263.55	11,850,000
	XIII Leda	C. Kowal 1974	8†	—	12,000,000
	XII Adrastea	S. Nicholson 1951	24†	631.1 r	21,200,000
	XI Pan	S. Nicholson 1938	32†	692.5 r	22,600,000
	VIII Poseidon	P. Melotte 1908	40†	735 .0 r	23,500,000
	IX Hades	S. Nicholson 1914	30†	758.0 r	23,700,000
	XIV	C. Kowal 1975 (unconfirmed)	—	—	—
Saturn	Janus	A. Dollfus 1966	350†	0.749	168,700
	Mimas	W. Herschel 1789	520†	0.942	185,800
	Enceladus	W. Herschel 1789	600†	1.370	238,300
	Tethys	Cassini 1684	1200	1.888	294,900
	Dione	Cassini 1684	800†	2.737	377,900
	Rhea	Cassini 1672	1350	4.518	527,600
	Titan	C. Huygens 1655	5800	15.945	1,222,600
	Hyperion	Bond 1848	400†	21.277	1,484,100
	Iapetus	Cassini 1671	1300†	79.331	3,562,900
	Phoebe	W. Pickering 1898	300†	550.45 r	12,960,000
Uranus	Miranda	G. Kuiper 1948	550†	1.414	128,000
	Ariel	Lassell 1851	800†	2.520	191,000
	Umbriel	Lassell 1851	400†	4 .144	266,000
	Titania	W. Herschel 1787	1000†	8.706	436,000
	Oberon	W. Herschel 1787	800†	13.463	583,400
Neptune	Triton	Lassell 1846	4000	5.877 r	355 ,500
	Nereid	G. Kuiper 1949	300†	359.881	5,560,000

*r indicates retrograde motion with respect to planet's rotation.
†Diameter of the satellite is estimated from the amount of light it reflects.

Constellation Name	Description	Position in Sky	
		R.A.†	Dec.†
Andromeda	Princess of Ethiopia	1ʰ	+40°
Antilia	The Air Pump	10ʰ	−35°
Apus*	The Bird of Paradise	16ʰ	−75°
Aquarius	The Water Bearer	23ʰ	−15°
Aquila	The Eagle	20ʰ	+5°
Ara*	The Altar	17ʰ	−55°
Aries	The Ram	3ʰ	+20°
Auriga	The Charioteer	6ʰ	+40°
Boötes	The Bear Driver	15ʰ	+30°
Caelum	The Sculptor's Chisel	5ʰ	−40°
Camelopardus*	The Giraffe	6ʰ	−70°
Cancer	The Crab	9ʰ	+20°
Canes Venatici	The Hunting Dogs	13ʰ	+40°
Canis Major	The Greater Dog	7ʰ	−20°
Canis Minor	The Lesser Dog	8ʰ	+5°
Capricornus	The Sea Goat	21ʰ	−20°
Carina*	The Keel (of Argo Navis)	9ʰ	−60°
Cassiopeia	Queen of Ethiopia	1ʰ	+60°
Centaurus*	The Centaur	13ʰ	−50°
Cepheus	King of Ethiopia	22ʰ	+70°
Cetus	The Sea Monster	2ʰ	−10°
Chamaeleon*	The Chameleon	11ʰ	−80°
Circinus*	The Compasses	15ʰ	−60°
Columba	The Dove (of Noah)	6ʰ	−35°
Coma Berenices	Berenice's Hair	13ʰ	+20°
Corona Austrina	The Southern Crown	19ʰ	−40°
Corona Borealis	The Northern Crown	16ʰ	+30°
Corvus	The Crow (or Raven)	12ʰ	−20°
Crater	The Cup	11ʰ	−15°
Crux*	The Southern Cross	12ʰ	−60°
Cygnus	The Swan	21ʰ	+40°
Delphinus	The Dolphin	21ʰ	+10°
Dorado*	The Swordfish	5ʰ	−65°
Draco	The Dragon	17ʰ	+65°
Equuleus	The Foal	21ʰ	+10°
Eridanus	The River	3ʰ	−20°
Fornax	The Laboratory Furnace	3ʰ	−30°
Gemini	The Twins	7ʰ	+20°
Grus	The Crane	22ʰ	−45°
Hercules	Hercules	17ʰ	+30°
Horologium	The Clock	3ʰ	−60°
Hydra	The Water Serpent	10ʰ	−20°

*Declinations between −50‡ and −90‡ are difficult or impossible to see from the United States.
†R.A., right ascension; Dec., declination.

THE CONSTELLATIONS (Continued)

Constellation Name	Description	Position in Sky	
		R.A.†	Dec.†
Hydrus*	The Water Snake	2ʰ	−75°
Indus*	The American Indian	21ʰ	−55°
Lacerta	The Lizard	22ʰ	+45°
Leo	The Lion	11ʰ	+15°
Leo Minor	The Lion Cub	10ʰ	+35°
Lepus	The Hare	6ʰ	−20°
Libra	The Beam Balance	15ʰ	−15°
Lupus	The Wolf	15ʰ	−45°
Lynx	The Lynx	8ʰ	+45°
Lyra	The Lyre	19ʰ	+40°
Mensa*	The Table Mountain	5ʰ	−80°
Microscopium	The Microscope	21ʰ	−35°
Monoceros	The Unicorn	7ʰ	−5°
Musca*	The Fly	12ʰ	−70°
Norma*	The Carpenter's Square	16ʰ	−50°
Octans*	The Octant	22ʰ	−85°
Ophiuchus	The Serpent Holder	17ʰ	0°
Orion	The Great Hunter	5ʰ	+5°
Pavo*	The Peacock	20ʰ	−65°
Pegasus	The Winged Horse	22ʰ	+20°
Perseus	The Hero	3ʰ	+45°
Phoenix*	The Phoenix	1ʰ	−50°
Pictor*	The Painter's Easel	6ʰ	−55°
Pisces	The Fishes	1ʰ	+15°
Piscis Austrinus	The Southern Fish	22ʰ	−30°
Puppis	The Stern (of Argo Navis)	8ʰ	−40°
Pyxis	The Compass Box (of Argo)	9ʰ	−30°
Reticulum*	The Net	4ʰ	−60°
Sagitta	The Arrow	10ʰ	+10°
Sagittarius	The Archer	19ʰ	−25°
Scorpius	The Scorpion	17ʰ	−40°
Sculptor	The Sculptor's Workshop	0ʰ	−30°
Scutum (Sobieski)	The Shield	19ʰ	−10°
Serpens	The Serpent	17ʰ	0°
Sextans	The Sextant	10ʰ	0°
Taurus	The Bull	4ʰ	+15°
Telescopium*	The Telescope	19ʰ	−50°
Triangulum	The Triangle	2ʰ	+30°
Triangulum Australe*	The Southern Triangle	16ʰ	−65°
Tucana*	The Toucan	0ʰ	−65°
Ursa Major	The Greater Bear	11ʰ	+50°
Ursa Minor	The Lesser Bear	15ʰ	+70°
Vela*	The Sail (of Argo Navis)	9ʰ	−50°
Virgo	The Maiden	13ʰ	0°
Volans*	The Flying Fish	8ʰ	−70°
Vulpecula	The Fox	20ʰ	+25°

*Declinations between −50° and −90° are difficult or impossible to see from the United States.
†R.A., right ascension; Dec., declination.

M	NGC*	Right Ascension (1950)		Declination (1950)		Apparent Visual Magnitude	Description
		(h)	(m)	(°)	(')		
1	1952	5	31.5	+21	59	8.4	Crab Nebula in Taurus; remains of super
2	7089	21	30.9	−1	02	6.4	Globular cluster in Aquarius
3	5272	13	39.8	+28	38	6.3	Globular cluster in Canes Venatici
4	6121	16	20.6	−26	24	6.5	Globular cluster in Scorpius
5	5904	15	16.0	+2	16	6.1	Globular cluster in Serpens
6	6405	17	36.8	−32	10	5.3	Open cluster in Scorpius
7	6475	17	50.7	−34	48	4.1	Open cluster in Scorpius
8	6523	18	00.1	−24	23	6.0	Lagoon Nebula in Sagittarius
9	6333	17	16.3	−18	28	7.3	Globular cluster in Ophiuchus
10	6254	16	54.5	−4	02	6.7	Globular cluster in Ophiuchus
11	6705	18	48.4	−6	20	6.3	Open cluster in Scutum
12	6218	16	44.7	−1	52	6.6	Globular cluster in Ophiuchus
13	6205	16	39.9	+36	33	5.9	Globular cluster in Hercules
14	6402	17	35.0	−3	13	7.7	Globular cluster in Ophiuchus
15	7078	21	27.5	+11	57	6.4	Globular cluster in Pegasus
16	6611	18	16.1	−13	48	6.4	Open cluster with nebulosity in Serpens
17	6618	18	17.9	−16	12	7.0	Swan or Omega Nebula in Sagittarius
18	6613	18	17.0	−17	09	7.5	Open cluster in Sagittarius
19	6273	16	59.5	−26	11	6.6	Globular cluster in Ophiuchus
20	6514	17	59.4	−23	02	9.0	Trifid Nebula in Sagittarius
21	6531	18	01.6	−22	30	6.5	Open cluster in Sagittarius
22	6656	18	33.4	−23	57	5.6	Globular cluster in Sagittarius
23	6494	17	54.0	−19	00	6.9	Open cluster in Sagittarius
24	6603	18	15.5	−18	27	11.4	Open cluster in Sagittarius
25	(4725)†	18	28.7	−19	17	6.5	Open cluster in Sagittarius
26	6694	18	42.5	−9	27	9.3	Open cluster in Scutum
27	6853	19	57.5	+22	35	7.6	Dumbbell Planetary Nebula in Vulpecu
28	6626	18	21.4	−24	53	7.6	Globular cluster in Sagittarius
29	6913	20	22.2	+38	21	7.1	Open cluster in Cygnus
30	7099	21	37.5	−23	24	8.4	Globular cluster in Capricornus
31	224	0	40.0	+41	00	4.8	Andromeda galaxy
32	221	0	40.0	+40	36	8.7	Elliptical galaxy; companion to M31
33	598	1	31.0	+30	24	6.7	Spiral galaxy in Triangulum
34	1039	2	38.8	+42	35	5.5	Open cluster in Perseus
35	2168	6	05.7	+24	21	5.3	Open cluster in Gemini
36	1960	5	33.0	+34	04	6.3	Open cluster in Auriga
37	2099	5	49.1	+32	33	6.2	Open cluster in Auriga
38	1912	5	25.3	+35	47	7.4	Open cluster in Auriga
39	7092	21	30.4	+48	13	5.2	Open cluster in Cygnus
40	—	12	20	+58	20	—	Close double star in Ursa Major

*New General Catalogue.
†Index Catalogue (IC) number.

MESSIER CATALOGUE OF NEBULAE AND STAR CLUSTERS (Continued)

M	NGC*	Right Ascension (1950) (h)	(m)	Declination (1950) (°)	(')	Apparent Visual Magnitude	Description
41	2287	6	44.9	−20	41	4.6	Loose open cluster in Canis Major
42	1976	5	32.9	−5	25	4.0	Orion Nebula
43	1982	5	33.1	−5	19	9.0	Northeast portion of Orion Nebula
44	2632	8	37	+20	10	3.7	Praesepe; open cluster in Cancer
45	—	3	44.5	+23	57	1.6	The Pleiades; open cluster in Taurus
46	2437	7	39.5	−14	42	6.0	Open cluster in Puppis
47	2422	7	34.3	−14	22	5.2	Loose group of stars in Puppis
48	2458	8	11	−5	38	5.5	Open cluster in Hydra
49	4472	12	27.3	+8	16	8.5	Elliptical galaxy in Virgo
50	2323	7	00.6	−8	16	6.3	Loose open cluster in Monoceros
51	5194	13	27.8	+47	27	8.4	Whirlpool spiral galaxy in Canes Venatici
52	7654	23	22.0	+61	20	7.3	Loose open cluster in Cassiopeia
53	5024	13	10.5	+18	26	7.8	Globular cluster in Coma Berenices
54	6715	18	51.9	−30	32	7.3	Globular cluster in Sagittarius
55	6809	19	36.8	−31	03	7.6	Globular cluster in Sagittarius
56	6779	19	14.6	+30	05	8.2	Globular cluster in Lyra
57	6720	18	51.7	+32	58	9.0	Ring Nebula; planetary nebula in Lyra
58	4579	12	35.2	+12	05	8.2	Barred spiral galaxy in Virgo
59	4621	12	39.5	+11	56	9.3	Elliptical spiral galaxy in Virgo
60	4649	12	41.1	+11	50	9.0	Elliptical galaxy in Virgo
61	4303	12	19.3	+4	45	9.6	Spiral galaxy in Virgo
62	6266	16	58.0	−30	02	6.6	Globular cluster in Ophiuchus
63	5055	13	13.5	+42	17	10.1	Spiral galaxy in Canes Venatici
64	4826	12	54.2	+21	57	6.6	Spiral galaxy in Coma Berenices
65	3623	11	16.3	+13	22	9.4	Spiral galaxy in Leo
66	3627	11	17.6	+13	16	9.0	Spiral galaxy in Leo; companion to M65
67	2682	8	48.4	+12	00	6.1	Open cluster in Cancer
68	4590	12	36.8	−26	29	8.2	Globular cluster in Hydra
69	6637	18	28.1	−32	24	8.9	Globular cluster in Sagittarius
70	6681	18	40.0	−32	20	9.6	Globular cluster in Sagittarius
71	6838	19	51.5	+18	39	9.0	Globular cluster in Sagitta
72	6981	20	50.7	−12	45	9.8	Globular cluster in Aquarius
73	6994	20	56.2	−12	50	9.0	Open cluster in Aquarius
74	628	1	34.0	+15	32	10.2	Spiral galaxy in Pisces
75	6864	20	03.1	−22	04	8.0	Globular cluster in Sagittarius
76	650	1	38.8	+51	19	11.4	Planetary nebula in Perseus
77	1068	2	40.1	−0	12	8.9	Spiral galaxy in Cetus
78	2068	5	44.2	+0	02	8.3	Small reflection nebula in Orion
79	1904	5	22.1	−24	34	7.5	Globular cluster in Lepus
80	6093	16	14.0	−22	52	7.5	Globular cluster in Scorpius
81	3031	9	51.7	+69	18	7.9	Spiral galaxy in Ursa Major
82	3034	9	51.9	+69	56	8.4	Irregular galaxy in Ursa Major
83	5236	13	34.2	−29	37	10.1	Spiral galaxy in Hydra
84	4374	12	22.6	+13	10	9.4	S0 type galaxy in Virgo
85	4382	12	22.8	+18	28	9.3	S0 type galaxy in Coma Berenices
86	4406	12	23.6	+13	13	9.2	Elliptical galaxy in Virgo
87	4486	12	28.2	+12	40	8.7	Elliptical galaxy in Virgo
88	4501	12	29.4	+14	42	10.2	Spiral galaxy in Coma Berenices
89	4552	12	33.1	+12	50	9.5	Elliptical galaxy in Virgo

*New General Catalogue.

MESSIER CATALOGUE OF NEBULAE AND STAR CLUSTERS (Continued)

M	NGC*	Right Ascension (1950)		Declination (1950)		Apparent Visual Magnitude	Description
		(h)	(m)	(°)	(')		
90	4569	12	34.3	+13	26	9.6	Spiral galaxy in Virgo
91‡	4571(?)	—	—	—	—	—	
92	6341	17	15.6	+43	12	6.4	Globular cluster in Hercules
93	2447	7	42.4	−23	45	6.0	Open cluster in Puppis
94	4736	12	48.6	+41	24	8.3	Spiral galaxy in Canes Venatici
95	3351	10	41.3	+11	58	9.8	Barred spiral galaxy in Leo
96	3368	10	44.1	+12	05	9.3	Spiral galaxy in Leo
97	3587	11	12.0	+55	17	12.0	Owl Nebula; planetary nebula in Ursa Major
98	4192	12	11.2	+15	11	10.2	Spiral galaxy in Coma Berenices
99	4254	12	16.3	+14	42	9.9	Spiral galaxy in Coma Berenices
100	4321	12	20.4	+16	06	10.6	Spiral galaxy in Coma Berenices
101	5457	14	01.4	+54	36	9.6	Spiral galaxy in Ursa Major
102‡	5866(?)	—	—	—	—	—	
103	581	1	29.9	+60	26	7.4	Open cluster in Cassiopeia
104	4594	12	37.4	−11	21	8.3	Spiral galaxy in Virgo
105	3379	10	45.2	+13	01	9.7	Elliptical galaxy in Leo
106	4258	12	16.5	+47	35	8.4	Spiral galaxy in Canes Venatici
107	6171	16	29.7	−12	57	9.2	Globular cluster in Ophiuchus

*New General Catalogue.
‡Items of doubtful identification.

APPENDIX 6 IDENTIFYING STARS AND CONSTELLATIONS

Part of the joy of studying astronomy is in the ability to locate and recognize the important stars. For the uninitiated, the night sky appears as a confusing array of individual stars. However, with a little study it is possible to see that the stars occur in irregular groups that may be formed into a variety of geometric patterns. The ancient astronomers recognized this and used these patterns to form the constellations which in turn provided them with a convenient method for identifying notable stars. We shall do the same by first learning to recognize the important constellations.

The constellations, representing figures that are figments of the human imagination, have been handy points of reference not only for identifying stars but for locating nebulae and galaxies as well. For the professional astronomer attempting to locate a particular celestial object, more detailed data is required, but for the amateur, the constellations serve the purpose very well.

In any study of constellations the Big Dipper is the constellation first considered. The Big Dipper is an "asterism," that is, it resembles the object for which it was named. Unfortunately, the Big Dipper is not always visible. For example, in the October star chart the Big Dipper is on the northern horizon, and if the horizon is not clear, this constellation cannot be seen. As the months pass (see subsequent star charts), the Big Dipper will appear to swing counterclockwise around the polar star (Polaris) until, in April, the Big Dipper will be positioned between the polar star and the point directly overhead.

There are other signposts in the sky that work as well as the Big Dipper. Let us again consider the October chart. Notice that immediately to the west of the overhead point three prominent stars are indicated. These stars form an easily discernible triangle from which three constellations may be identified. From this point attempt to identify other prominent constellations in the October sky. Learn the name of the constellation and its English name and description. Then learn the names of the prominent stars associated with each constellation.

The star charts are designed for approximately 34°N latitude but may be used anywhere in the United States. For more northerly locations the overhead position will be closer to the north celestial pole and the constellations will be shifted to the south on the chart. The stars are as they appear at 10:00 P.M. at the beginning of the month, 9:00 P.M. during the middle of the month, and 8:00 P.M. toward the end of the month. To use the charts select the appropriate month, hold the chart vertical, and turn it so the direction you are facing is on the bottom of the chart.

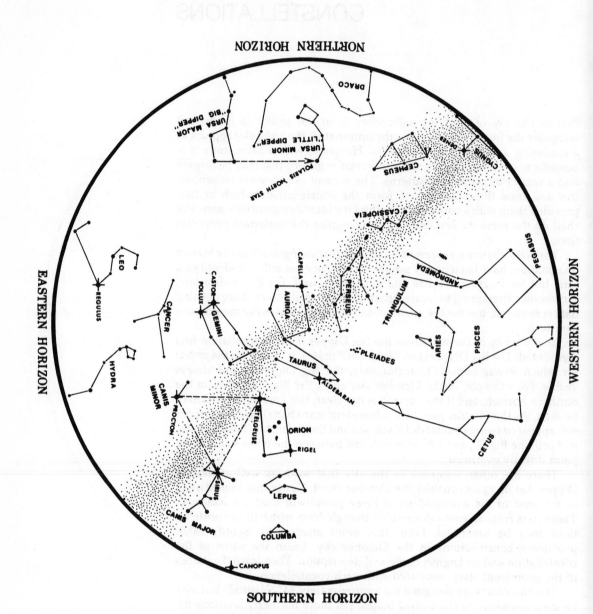

THE NIGHT SKY IN JANUARY

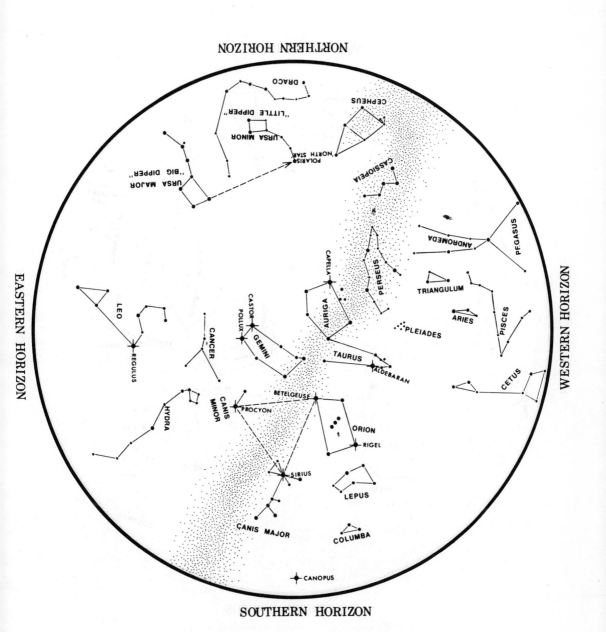

THE NIGHT SKY IN FEBRUARY

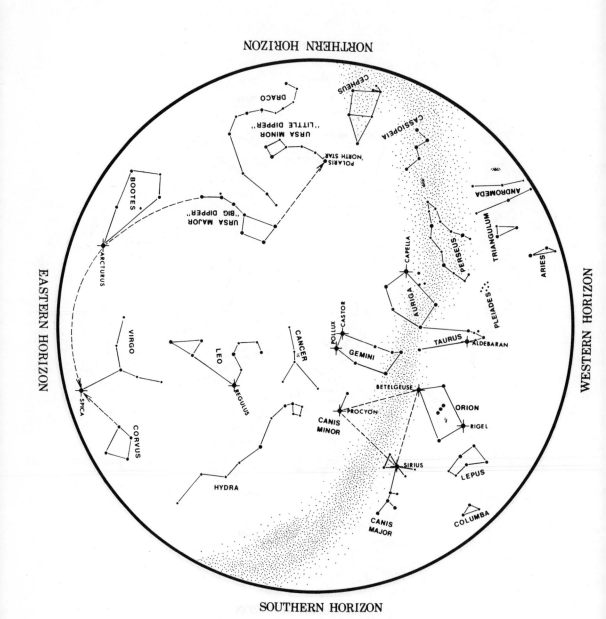

THE NIGHT SKY IN MARCH

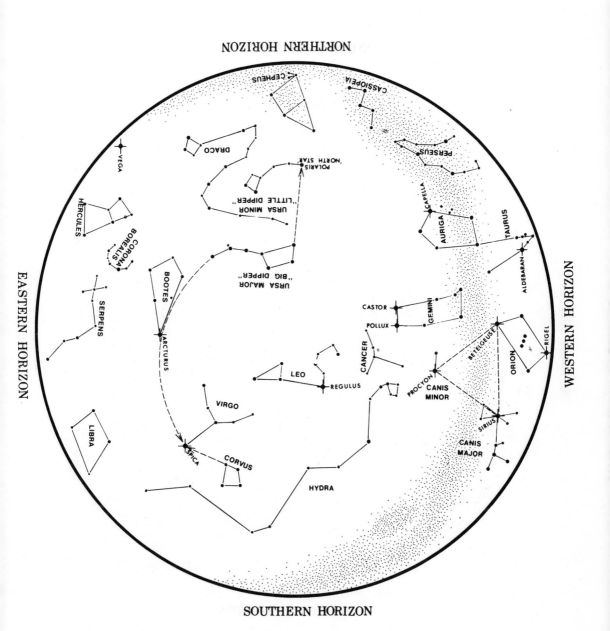

THE NIGHT SKY IN APRIL

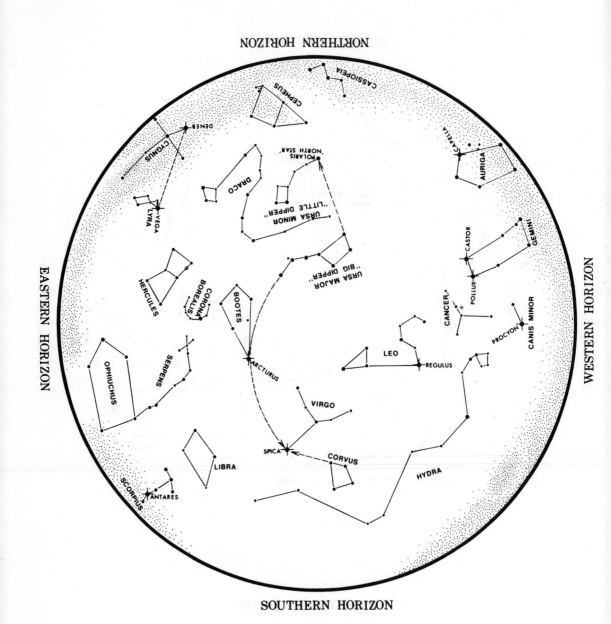

THE NIGHT SKY IN MAY

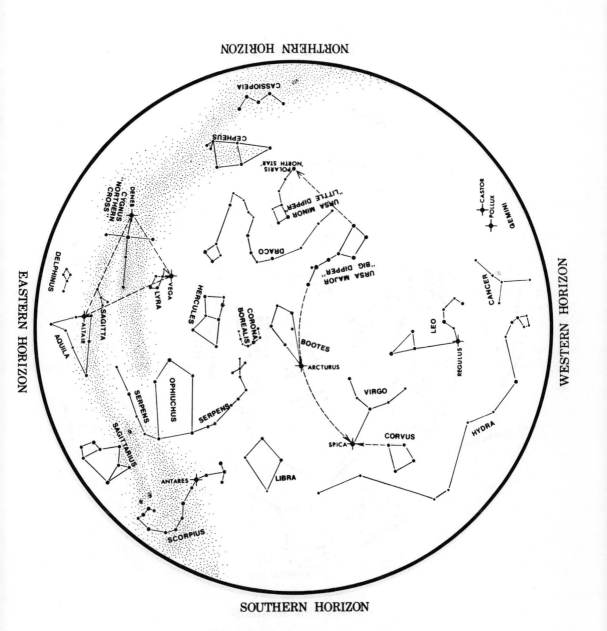

THE NIGHT SKY IN JUNE

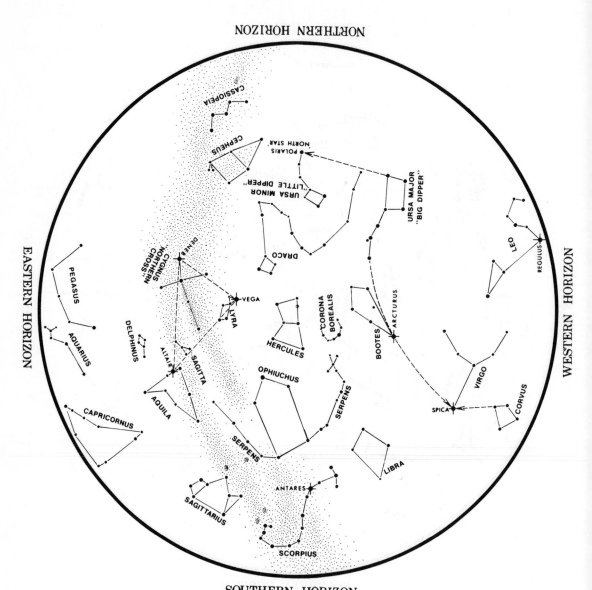

THE NIGHT SKY IN JULY

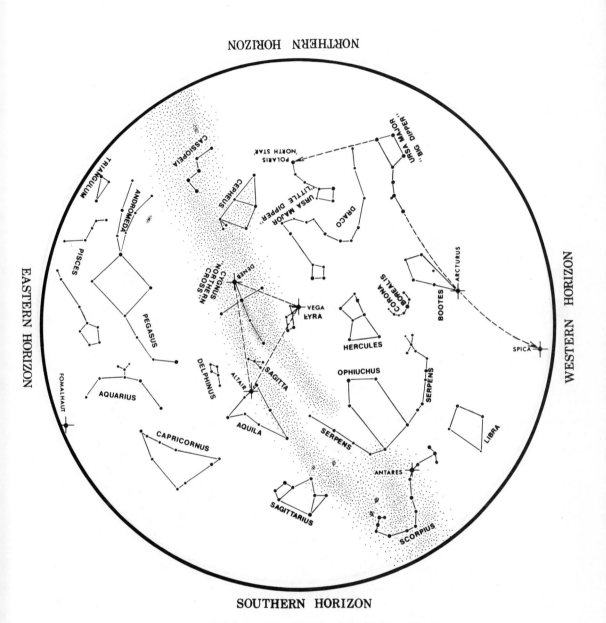

THE NIGHT SKY IN AUGUST

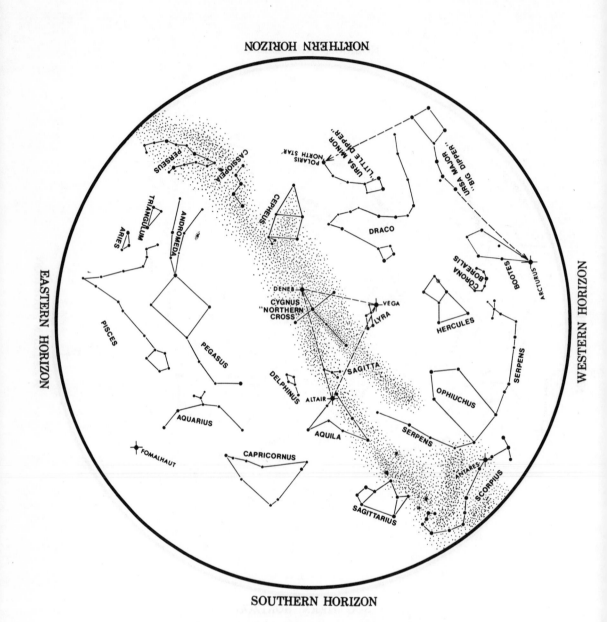

THE NIGHT SKY IN SEPTEMBER

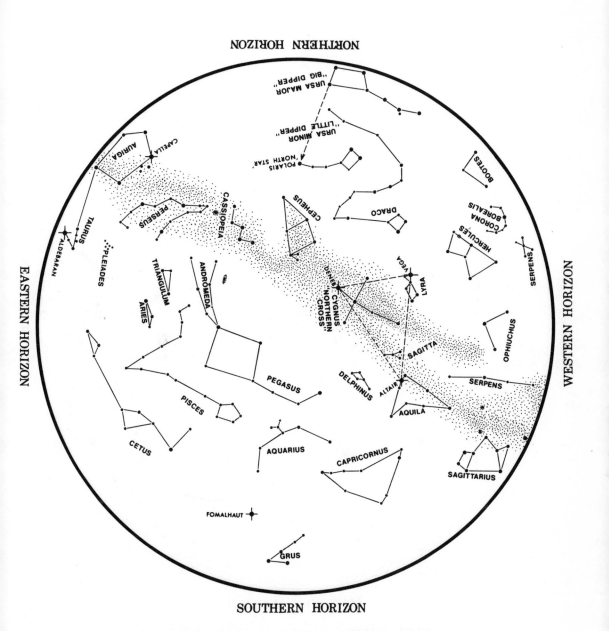

THE NIGHT SKY IN OCTOBER

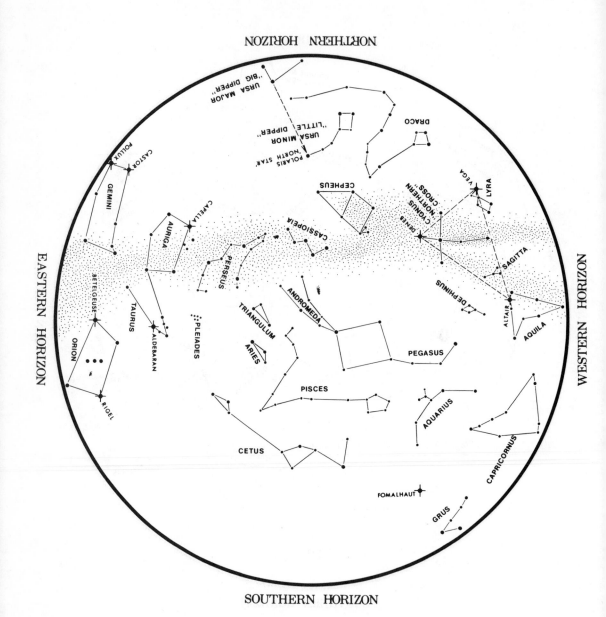

THE NIGHT SKY IN NOVEMBER

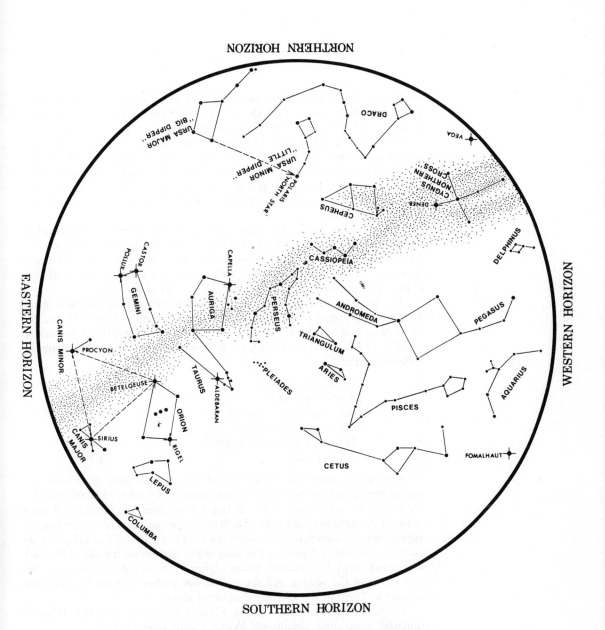

THE NIGHT SKY IN DECEMBER

APPENDIX 7   FINDING THE PLANETS

The two charts and the accompanying table make it possible to determine the approximate positions of the naked-eye planets with respect to Earth. In Chart 1 the orbits of the terrestrial planets are drawn to scale and, of necessity, in the same plane. The planets, you will recall, are not all exactly on the same plane (ecliptic), so the position of their respective orbits north of the ecliptic are indicated by the heavier arc. The sun ⊙ is shown in the center. On the outer circle are the symbols of the constellations in the zodiac. The same applies to Chart 2 which includes orbits for Earth, Jupiter, and Saturn.

In both charts Earth's progress is shown by the months while the orbits of the other planets are subdivided into tenths. Perihelion is indicated by 0.0 and aphelion by 0.5 and the numerical order designates the direction of revolution. The arc between two points measures the progress made by a planet during 10 percent of the planet's period of revolution.

The position of a planet is determined by the simple addition of the numbers associated with the day of the month, month of the year, and the units, tens, and hundreds number of the year. For example, let us find the position of Mars on January 20, 1978. From the table we would obtain the following data:

Day of month—20	0.03
Month—January	0.90
Units no. of the year—8	0.25
Tens no. of the year—7	0.21
Hundreds no. of the year—19	0.00
Total	1.39

We ignore anything to the left of the decimal and thus have the remaining value 0.39. Next locate this value on the orbit of Mars and also the position of Earth in its orbit on the desired date. We would find Mars to be on the opposite side of the Earth from the sun or directly overhead at approximately midnight and a little north of the ecliptic. The ecliptic may be shown for this purpose as the path of the sun across the sky. This path will change with the seasons, being higher in the sky during the summer and lower in the winter. All the naked-eye planets can be located with respect to Earth in the manner described above.

Calculate the positions of the planets for April 10, 1982. What is unusual about their alignment? Which planet does not fit?

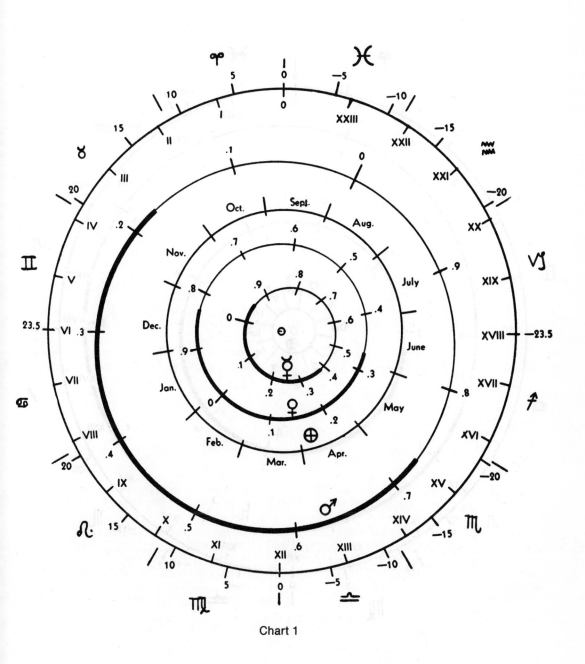

Chart 1

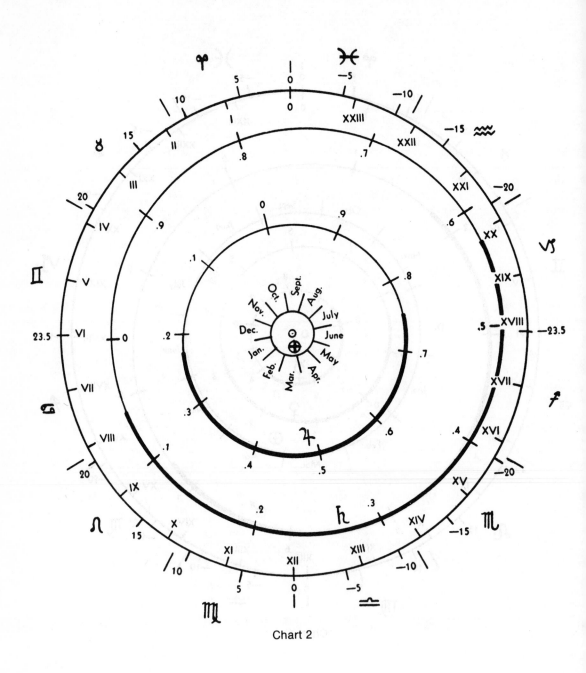

Chart 2

"DECIMALS" FOR PLANETS
Day of Month (Gregorian Calendar)

Day of month	1	4	7	10	13	16	19	22	25	28	31
Mercury	0.00	0.03	0.07	0.10	0.14	0.17	0.20	0.23	0.27	0.31	0.34
Venus	.00	.01	.03	.04	.05	.07	.08	.09	.11	.12	.13
Mars	0.00	0.00	0.01	0.01	0.02	0.02	0.03	0.03	0.04	0.04	0.04

Month

	Jan.	Feb.	Mar.	Apr.	May	June	July	Aug.	Sept.	Oct.	Nov.	Dec.
Mercury	0.29	0.64	0.96	0.31	0.65	0.00	0.35	0.70	0.06	0.40	0.75	0.09
Venus	.58	.72	.85	.99	.12	.26	.39	.53	.67	.80	.94	.07
Mars	.90	.95	.99	.03	.07	.11	.15	.20	.25	.29	.34	.38
Jupiter	.61	.62	.62	.63	.64	.65	.65	.66	.66	.67	.68	.68
Saturn	0.49	0.49	0.49	0.50	0.50	0.50	0.50	0.51	0.51	0.51	0.52	0.52

Year—Units

	0	1	2	3	4	5	6	7	8	9
Mercury	0.00	0.15	0.30	0.46	0.61	0.76	0.91	0.06	0.22	0.37
Venus	.00	.63	.25	.88	.50	.13	.75	.38	.00	.63
Mars	.00	.53	.06	.59	.13	.66	.19	.72	.25	.78
Jupiter	.00	.08	.17	.25	.34	.42	.51	.59	.67	.76
Saturn	0.00	0.03	0.07	0.10	0.14	0.17	0.20	0.24	0.27	0.31

Year—Tens

	0	1	2	3	4	5	6	7	8	9
Mercury	0.00	0.52	0.04	0.56	0.08	0.60	0.12	0.64	0.16	0.68
Venus	.00	.25	.51	.76	.01	.27	.53	.78	.04	.29
Mars	.00	.31	.63	.95	.27	.58	.90	.21	.53	.85
Jupiter	.00	.84	.69	.53	.37	.22	.06	.90	.74	.59
Saturn	0.00	0.33	0.68	0.02	0.35	0.70	0.04	0.38	0.72	0.06

Year—Hundreds

	14	15	16	17	18	19	20	21	22	23
Mercury	0.02	0.22	0.41	0.61	0.80	0.00	0.20	0.39	0.59	0.78
Venus	.29	.82	.36	.91	.45	.00	.55	.09	.64	.18
Mars	.18	.34	.51	.67	.84	.00	.16	.33	.49	.66
Jupiter	.85	.28	.71	.14	.57	.00	.43	.86	.29	.72
Saturn	0.02	0.42	0.82	0.21	0.61	0.00	0.39	0.79	0.18	0.58

Reprinted from *Highlights of Astronomy* by Walter Bartky by permission of The University of Chicago Press. Copyright 1935 by Walter Bartky.

APPENDIX 8   THE GREEK ALPHABET

A	α	alpha		N	ν	nu
B	β	beta		Ξ	ξ	xi
Γ	γ	gamma		O	o	omicron
Δ	δ	delta		Π	π	pi
E	ε	epsilon		P	ρ	rho
Z	ζ	zeta		Σ	σ	sigma
H	η	eta		T	τ	tau
Θ	θ	theta		Υ	υ	upsilon
I	ι	iota		Φ	ϕ	phi
K	κ	kappa		X	χ	chi
Λ	λ	lambda		Ψ	ψ	psi
M	μ	mu		Ω	ω	omega

GLOSSARY

A

Aberration of starlight Apparent displacement of a star on the celestial sphere due to the Earth's orbital motion.

Absolute magnitude The computed apparent magnitude of a star at a standard distance of 10 pc.

Absolute zero The theoretical temperature at which all molecular motion ceases: $0°K$, $-273°C$, $-460°F$.

Absorption spectrum Dark lines superimposed on a continuous spectrum.

Acceleration A change in speed with respect to time or a change in direction as in circular motion.

Acceleration of gravity Acceleration of an object in response to the pull of gravity at the surface of Earth or any other celestial body. On Earth's surface the value is $9.80 \, m/sec^2$.

Achromatic An optical system free of chromatic aberration.

Aerolite A stony meteorite.

Albedo The ratio of the light reflected by the moon or a planet to the total amount of incident light.

Alpha particle The nucleus of a helium atom containing two protons and two neutrons (^4_2He).

Altitude The angle of an object above or below the horizon measured along a vertical circle.

Angstrom ($\overset{\circ}{A}$) A unit of length equal to 10^{-8} centimetres, used in measuring the wavelength of light.

Angular diameter An angle subtended by the diameter of some object such as a planet.

Angular momentum A measure of the momentum of a body in motion about its axis or about a fixed point.

Annular eclipse A solar eclipse where the moon, directly between Earth and the sun, is a greater distance from Earth than the length of the lunar shadow.

Anomalistic month The period of the moon's revolution from perigee to following perigee: 27d 13h 24m 29s.

Anomalistic year The period of the Earth's revolution from perihelion to following perihelion: 365d 6h 13m 53s.

Antapex Position on the celestial sphere from which the sun appears to be receding relative to its stellar neighbors.

Apastron Point of greatest separation of the two components in a binary system.

Apex Position on the celestial sphere toward which the sun is moving with respect to its stellar neighbors.

Aphelion Position in the orbit of a planet where it is farthest from the sun.

Apogee Position in the orbit of an object orbiting the Earth where the object is farthest from the Earth.

Apparent magnitude The apparent brightness of a star or other celestial body without regard for distance as seen from Earth.

Apsides (line of) Major axis of an elliptical orbit.

Ascending node Point where the orbit of an object revolving around the sun crosses the ecliptic from south to north.

Association A loose grouping of stars with common characteristics indicating a common origin.

Asteroids A group of small objects, the majority of which are found revolving around the sun in orbit between Mars and Jupiter.

Astrometric binary A binary pair with one component invisible but detectable by means of its influence on the visible component.

Astronomical unit (AU) Formerly the semimajor axis of the Earth's orbit. Now considered to be the semimajor axis of a hypothetical body with mass and period assumed by Gauss for Earth. On this basis, semimajor axis of Earth is 1.000 000 230 AU. Generally rounded off to 1.0 AU or 1.50 $\times 10^8$ km.

Astronomy The branch of science dealing with all phenomena in space beyond the Earth's atmosphere.

Atom The smallest unit of an element that retains all the properties of the element.

Atomic mass unit (AMU) One-twelfth the mass of the most common isotope of carbon. The AMU is approximately that of a hydrogen atom, 1.67×10^{-24} grams.

Atomic number The number of protons in the atomic nucleus.

Atomic weight The mean mass of an atom of a particular element in AMU.

Aurora Light radiation from atomic particles in the Earth's ionosphere. Generally seen most prominently over the polar regions.

Autumnal equinox The point where the sun crosses the celestial equator apparently moving south. Signals beginning of the fall season.

Azimuth Angle measured eastward on the celestial horizon from the north point to where a verticle circle passing through a celestial object intersects the horizon.

B

Baily's beads Bright rays of sunlight seen through valleys along the limb of the moon at the instant before and immediately after a total solar eclipse.

Barred spiral Galaxy with bar composed of stars passing through the nucleus and tips of bar coiled to varying degrees.

Barycenter The center of mass of two mutually orbiting objects.

Base line A line of known length representing the base of a triangle used in triangulation. The line is between two including angles that are known or can be measured.

Beta particle A negatively charged particle; an electron.

Big bang theory A cosmological theory which assumes that the expansion of the universe was initiated by a primeval explosion.

Binary star A pair of stars revolving around a mutual center of gravity.

Black body An ideal radiator capable of absorbing all radiation falling upon it and reradiating all the energy without loss.

Black dwarf The final state of a star which has exhausted all energy sources and which no longer emits any radiation.

Black hole An object whose gravitational force is so powerful that its escape velocity exceeds the speed of light.

Blink comparator A device in which two photographs taken at different times of the same region in space are rapidly shifted back and forth in order to detect changes in brightness or position of celestial bodies.

Bode's law Not a law but rather a sequence of numbers that yield approximate distances in astronomical units of all planets except Neptune and Pluto.

Bolide A bright fireball passing through the Earth's atmosphere with a loud sound.

Bright-line spectrum An emission spectrum of bright lines seen only in the spectrum of an incandescent gas at low pressure.

Brightness Intensity of radiant energy, usually in the visible portion of the spectrum, where the sun's apparent luminosity is one unit of intensity at one pc.

C

Carbon-nitrogen cycle Stellar nuclear reactions involving carbon as a catalyst, in which hydrogen is transformed to helium with the emission of energy.

Cardinal points The four basic directions: N, E, S, W.

Cassegrain focus An optical system where light reflected from the primary mirror to a secondary mirror is in turn reflected through a hole in the primary to focus behind the primary.

Celestial equator A great circle representing an extension of Earth's equator projected to the celestial sphere.

Celestial horizon A great circle projected from the horizon to the celestial sphere. It is 90° from zenith and nadir.

Celestial meridian A great circle through the celestial poles and the zenith.

Celestial poles Points of apparent rotation of the celestial sphere. The celestial poles are extensions of the Earth's poles to the celestial sphere.

Celestial sphere The sphere or dome of the sky with an infinite radius centered on the observer.

Centrifugal reaction force The inertial reaction to centripetal force.

Centripetal force Force directed toward the center of a curve necessary to divert an object from a straight path into a curved path.

Cepheid varible A pulsating star, generally a giant or supergiant with a period of from one to 50 days.

Chromatic aberration An optical defect in lenses where light of different wavelengths (color) are focused at slightly different distances from the objective, yielding a slightly out-of-focus image.

Chromosphere A region of the solar atmosphere lying immediately above the photosphere.

Circumpolar stars Stars near the celestial poles that are always above the horizon.

Color index Difference in magnitudes of a star usually measured in wavelengths of yellow and blue light.

Coma A defect in an optical system in which off-axis rays reflected from the objective do not focus in the same place, resulting in a noncircular image.

Coma (comet) Gaseous portion surrounding the nucleus of the head of a comet.

Comet A diffuse body within the solar system orbiting the sun in elongated orbits.

Comparison spectrum Spectrum of some element (such as iron) photographed by the same equipment used to obtain the spectrum of a star, used for comparison purposes.

Conjunction Two celestial bodies with the same celestial longitude or right ascension. See *inferior conjunction* and *superior conjunction*.

Constellation A group of stars, or the area of the sky occupied by a group of stars, named for a mythical figure, animal, or object.

Continuous spectrum A continuous band of colors representing various wavelengths in the visible spectrum.

Coordinates A set of numbers that locate a point in space.

Corona The outer portion of the sun's atmosphere, generally seen during a total eclipse.

Coronagraph A special telescope designed for the purpose of observing and photographing the solar atmosphere without need of a total eclipse.

Cosmic rays High energy particles consisting mainly of the nuclei of hydrogen (protons) striking the Earth at near the speed of light.

Cosmogany The branch of astronomy that deals with the origin of the universe.

Cosmology The study of the evolution and structure of the universe.

Crape ring The faint innermost ring of Saturn.

Crater A depression observed on the moon and terrestrial planets resulting from volcanic activity or impact by a large meteorite.

Crescent moon Phase of the moon between new moon and first quarter and between third quarter and new moon, when moon is less than half illuminated.

D

Dark-line spectrum See *absorption spectrum*.

Dark nebula A cloud of interstellar dust and gas obscuring light from more distant stars and therefore appearing dark.

Declination The angular displacement of an object along the hour circle from the celestial equator. It is measured as + or north and – or south declination.

Deferent In the Ptolemaic system a circle around the Earth around which a planet and its epicycle move.

Degenerate gas A gas at tremendous high pressure and temperature that no longer acts according to the perfect gas laws. The electrons are extremely limited as to the energy states they may occupy.

Density The mass of a substance per unit of volume, generally expressed as g/cm^3.

Descending node A point where the orbit of a body crosses some reference plane (ecliptic or orbital plane) going from north to south.

Deuterium Heavy hydrogen composed of one proton and one neutron.

Diamond ring effect The flash of sunlight just prior to and just following a total solar eclipse during which the corona is visible.

Differential galactic rotation The galaxy rotating at different rates with distance from the center, and not as a solid wheel.

Diffraction The bending and spreading of light rays on passing through a narrow opening.

Diffraction grating An optical device of fine, closely placed slits used to produce a spectrum.

Disk (planet or other object) Circular shape of a celestial object seen through a telescope.

Disk (galactic) The central wheel-like portion of a spiral galaxy.

Dispersion The spreading of white light into different colors (wavelengths of light) in a spectroscope.

Distance modulus Difference in the apparent and absolute magnitude of an object. It is derived from $m-M = 5\log r-5$.

Diurnal Daily.

Diurnal libration Apparent motion of the moon in which a slight amount of the lunar surface can be seen around the limb due to the rotation of the Earth.

Doppler shift Apparent change in wavelength of radiation from an object due to its relative motion along the line of sight.

Double star An apparent single star to the unaided eye which is revealed as two stars with a telescope or spectroscope.

Dwarf star Generally a main sequence star as compared to a giant or supergiant. The term is also used for stars of low luminosity and high density that are considered to be approaching the end of stellar evolution.

Dyne A metric unit of force required to accelerate a mass of 1 gram 1 cm/sec^2.

E

Earthshine Dim light visible on the dark portion of the moon. Light is reflected from Earth to moon and back to Earth.

Eccentricity Ratio of the length of the major axis to distance between foci of an ellipse. Values are less than one for an ellipse; equal to zero for a circle; equal to one for a parabola; greater than one for a hyperbola.

Eclipse Interference of part or all of the light of one body such as the sun by another body such as the moon as it passes between the sun and Earth.

Eclipsing binary A pair of stars in a double-star system where the plane of their orbit is such that one star passes in front of the other as seen from Earth.

Ecliptic The apparent annual path of the sun against the background of distant stars. It is an extension of the Earth's orbit onto the celestial sphere.

Einstein-Rosen bridge A theoretical connection in spacetime between distinct regions of the universe, such as between a black hole at one point to a white hole at another through a "wormhole."

Electromagnetic spectrum The range of radiation of different wavelengths including gamma rays, X rays, ultraviolet light, visible light, infrared radiation, and long wave radiation.

Electron A negatively charged subatomic particle with small mass that revolves around the nucleus of the atom.

Element A substance that cannot be altered to simpler substances by ordinary chemical means.

Element of an orbit One of several quantities used to calculate the size, shape and orientation of the orbit of an object as well as the position of the object in orbit.

Ellipse A conic section where every point on the circumference is the same total distance from the fixed foci.

Elliptical galaxy A galaxy whose outline in space resembles an ellipse.

Ellipticity Ratio of the major axis to the difference between the major and minor axes in an ellipse.

Elongation The angle between the sun and a planet as seen from the Earth.

Emission line A bright line in the visible portion of the electromagnetic spectrum.

Emission nebula A bright nebula that receives its visible light from the fluorescence of ultraviolet light from stars in the nebula.

Emission spectrum A spectrum of bright lines from an incandescent gas at low pressure.

Energy The ability to accomplish work. Also the part of the universe that is not matter.

Epicycle A smaller circle whose center moves along the deferent in the Ptolemaic system.

Equant A point offset from the center of a circular orbit about which an object revolves with uniform angular velocity.

Equator A great circle on the Earth or the celestial sphere midway between the poles.

Equinox Point at which the ecliptic crosses the celestial equator. Also designates the time the sun crosses one of these points. See *vernal equinox* and *autumnal equinox*.

Erg A metric unit of energy representing the amount of work accomplished by a force of one dyne acting through a distance of one centimetre.

Eruptive variable A star characterized by sudden explosive or erratic outbursts of energy.

Escape velocity The velocity required of a body to escape from the gravitational pull of another body and move into space.

Ether The hypothetical medium thought to be necessary for the transmission of light through space.

Event horizon A hypothetical surface through which an object disappears from view as it approaches a black hole. It is the point at which the escape velocity at a black hole equals the speed of light.

Evolutionary cosmology The expanding model of the universe in which the universe has evolved from a small volume to its present state.

Exobiology The study of extraterrestrial life.

Expanding universe The view of the universe based on the Red Shift law which states that the universe is expanding in all directions uniformly. This is the basis for evolutionary cosmology.

Exploding galaxy A galaxy experiencing violent internal activity.

Extragalactic Beyond the Milky Way galaxy.

Extraterrestrial life Life thought to exist beyond the confines of the Earth.

F

Faculae Bright areas near the sun's limb.

Fireball A spectacularly bright meteor.

Fission Splitting the nucleus of the atom of a heavy element into two or more lighter elements.

Flare An eruptive outburst on the sun's surface.

Flare star A type of star that occasionally and unpredictably increases in brightness.

Flash spectrum The spectrum obtained of the sun's limb at the instant before totality during a solar eclipse.

Flocculi A bright region on the solar surface seen in monochromatic light. These are now generally called plages.

Fluorescence Absorption of light at one wavelength and its reemission at another wavelength, usually ultraviolet to visible light.

Focal length Distance from the center of a lens or surface of a mirror used as an objective to the point of focus of converging rays of light.

Focal plane The area where converging rays from the objective lens or mirror intersect to form an image.

Focus Point on the focal plane where converging rays meet.

Forbidden lines Spectral lines not usually seen under laboratory conditions. They originate from a gas at exceedingly high temperatures and low pressures.

Force Application of energy which can change the momentum or overcome the inertia of a body.

Foucault pendulum Device used by Jean Foucault to demonstrate the rotation of the Earth.

Fraunhofer lines Absorption lines in the spectrum of the sun or a star.

Frequency The number of waves passing a given point per unit of time. Also the number of vibrations per unit of time.

Full moon Phase of the moon where Earth is between the sun and the moon and the moon's disk is fully illuminated.

Fusion The buildup of heavy atomic nuclei from light ones; for example, hydrogen to helium.

G

Galactic center A region around which the stars in the Milky Way revolve, located in the direction of the constellation Sagittarius.

Galactic cluster A loose grouping of several dozen to several thousand stars having a common origin and common motions. Clusters are usually found in the spiral arms or disk of the galaxy.

Galactic equator A plane that bisects the principal disk of the Milky Way.

Galactic latitude Angular distance north and south of the galactic equator along a circle passing through the galactic poles.

Galactic longitude Angular distance measured eastward from the galactic center on the galactic plane.

Galactic plane The central plane through the Milky Way or galaxy.

Galactic poles North and south poles on a line through the center of the galaxy and perpendicular to the galactic plane. The poles are 90° from the galactic equator.

Galactic rotation Movement of the stars of a galaxy around the galactic center.

Galaxy An assemblage of millions or billions of stars. The term also applies to the Milky Way system.

Gamma rays High energy radiation emitted from a radioactive atom, a photon.

Gegenschein A faint glow seen in the night sky coming from sunlight reflected from particles located in space in the direction opposite the sun.

Geocentric Earth centered.

Giant star A large star of at least zero absolute magnitude.

Gibbous moon Phase of the moon between the first or third quarter and full moon.

Globular cluster A system of up to a million stars closely grouped, found mainly in the halo of the galaxy.

Granulation The "rice grain"-like structure of the solar surface as seen through the telescope.

Gravity The force whereby masses attract each other.

Gravitational constant (G) Constant of proportionality in Newton's law of universal gravitation; 6.668×10^{-8} dyne-cm^2/g^2.

Gravitational red shift Red shift in the visible spectrum caused by gravity.

Great circle Circle on the surface of a sphere representing the circumference of a plane passing through the center of the sphere.

Greatest (east or west) elongation The maximum angular separation between the sun and Mercury or Venus.

Greenhouse effect The warming effect resulting from the trapping of infrared radiation between a planet's surface and atmosphere. The heat is prevented from escaping into space by the presence of carbon dioxide in the atmosphere, which acts like the glass in a greenhouse.

Greenwich meridian The meridian of longitude 0° that passes from the north pole to the south pole through the site of the Royal Naval Observatory in Greenwich, England.

Gregorian calendar The calendar now in common use introduced by Pope Gregory XIII in 1582.

H

H I region Region in interstellar space containing neutral hydrogen.

H II region Region in interstellar space containing ionized hydrogen.

Half life Length of time required for half the radioactive atoms in a given sample to disintegrate.

Halo Distribution of globular clusters, stars, and hydrogen gas around the galaxy.

Harmonic law Kepler's third law.

Harvest moon Full moon nearest to the time of the autumnal equinox.

Hayashi track Rapid downward movement on the H-R diagram of a developing star before it reaches the main sequence. The star decreases in luminosity but maintains constant temperature.

Head The coma and nucleus of a comet.

Heliocentric Sun centered.

Heliocentric theory The idea that the sun is located in the center of the universe.

Helium flash The explosive formation of carbon from helium by the triple alpha process. This occurs at the core of a red giant.

Hertzsprung-Russell (H-R) diagram Plot of absolute magnitude against temperature, color index, or spectral class for any group of stars.

Horizon (astronomical) A great circle 90° from zenith and nadir.

Horizon system A coordinate system in which the horizon serves as the fundamental circle.

Hour angle An angle between the celestial meridian measured westward along the celestial equator to the hour circle passing through the object being observed.

Hour circle Any of a number of great circles on the celestial sphere passing through the celestial poles.

H-R diagram See Hertzsprung-Russell diagram.

Hubble constant A number relating the distance of external galaxies to their velocity of recession. Believed to be equal to 55 km/sec/Mpc.

Hubble's law Red shift law. Relation of the shift to longer wavelengths of light from distant galaxies to their velocity of recession. The shift is presumed to result from the Doppler shift.

Hyperbola A conic section with eccentricity greater than 1.0.

I

Image The likeness of an object produced by light rays from the object reflected by a mirror or refracted by a lens onto the focal plane.

Image tube A device in which electrons activated by light on the surface of a photocathode tube are focused electronically.

Inclination (orbital) The angle between the orbital plane of a revolving body and some fundamental plane used for reference. The reference may be the ecliptic or celestial equator.

Index of refraction Represents the ratio of the speed of light in a vacuum to the speed of light through a transparent substance.

Inertia The tendency of a body to preserve its state of rest or uniform motion in a straight line. Newton's first law of motion.

Inferior conjunction A state where a planet between the Earth and the sun has the same longitude as the sun. Applies only to Mercury and Venus.

Inferior planet A planet—Mercury or Venus—whose distance from the sun is less than that of the Earth's distance from the sun.

Infrared radiation Electromagnetic radiation longer in wavelength than the longest visible red radiation and shorter than radio waves.

Inner planets The innermost four planets in the solar system—the terrestrial planets, including Mercury, Venus, Earth, and Mars.

Insolation The amount of solar radiation received per unit of area at the Earth's surface.

International date line Line at 180° longitude at which the date changes ahead 24 hours when crossing to the west and back 24 hours when crossing to the east.

Interplanetary medium Dust and gas found in the spaces between the planets of the solar system.

Interstellar dust Microscopic particles in the vast spaces between the stars.

Interstellar gas Diffuse gas in the vast spaces between stars.

Interstellar line Dark lines (absorption) on the spectrum produced by interstellar gas.

Interstellar matter Dust and gas found in the vast spaces between the stars.

Ion An electrically charged atom that has reached this state by the loss or gain of an electron(s).

Ionization The process of an atom gaining or losing an electron.

Ionosphere The upper portion of the atmosphere extending upward from an elevation of 50 km. In this region of the atmosphere the gases are ionized by X rays and ultraviolet rays from the sun.

Irregular galaxy An asymmetrically shaped galaxy.

Island universe A name formerly applied to an external galaxy.

Isotope An atom having the same atomic number as other atoms of the same element but different atomic mass.

J

Jovian planet A name sometimes used in referring to the giant planets; Jupiter, Saturn, Uranus, and Neptune.

Julian calendar The calendar first introduced by Julius Caesar in 45 B.C.

K

Kepler's laws Laws of planetary motion introduced by J. Kepler in the early seventeenth century.

Kiloparsec (kpc) One thousand parsecs or approximately 3260 light years.

Kinetic energy The energy of motion which equals half the product of the mass of the object and the square of its velocity.

Kirchhoff's laws Three laws pertaining to the formation of spectra.

Kirkwood's gaps The gaps occurring in the asteroid belts resulting from the gravitational influence of the large planets.

L

Lagrangian points The five positions in the orbital plane of two bodies revolving around a barycenter, where a third body of negligible mass remains in equilibrium with respect to the larger bodies.

Latitude Imaginary lines on the Earth's surface that run parallel to the equator and measure angular distance along a meridian north and south of the equator. May be extended to the celestial sphere or may be measured from the ecliptic or galactic equator.

Latitudinal libration Libration resulting from the fact that the lunar axis of rotation is not perpendicular to the lunar orbital plane.

Law of areas Kepler's second law.

Law of conservation of energy and mass Energy can be neither created nor destroyed.

Law of gravitation Every particle in the universe attracts every other particle in the universe with a force proportional to the product of their masses and inversely proportional to the square of the distance between them.

Leap year Year with 366 days occurring every fourth year to adjust for the residual quarter day in each revolution of the Earth about the sun.

Libration Slight apparent (not real) movements of the moon which permit more than 50 percent of the lunar surface to be visible from the Earth.

Light Range of the electromagnetic spectrum to which the eye is sensitive.

Light curve Variations with time in the brightness of a variable star or binary system.

Light year The distance light travels in one year at 3×10^5 km per second. Equal to 9.46×10^{12} km or 5.88×10^{12} miles.

Limb Edge of the sun or moon as seen from Earth.

Limb darkening Apparent darkening of the sun or moon near the limb.

Line of apsides The line along the major axis of an ellipse joining the points farthest and nearest to the focus.

Line of nodes Line connecting the points where the orbital plane intersects a reference plane such as the ecliptic.

Local Group The small cluster of galaxies to which the Milky Way belongs.

Longitude Angular distance measured east or west on the equator from the Greenwich meridian.

Longitude of the ascending node An angle obtained by measuring eastward along the ecliptic from the vernal equinox to the ascending node of the moon or planet.

Longitudinal libration A libration occurring due to the variable speed of the moon in its orbit.

Long-period variable A star whose period of variability is not uniform and extends up to 100 days or more.

Luminosity The rate at which a star radiates energy compared with the sun.

Lunar Any reference to the moon.

Lyman alpha line First spectral line of the lyman series produced by electrons moving between ground state and second level of the hydrogen atom. It is located at 1216 angstroms.

Lyman lines Absorption or emission lines in the ultraviolet portion of the spectrum that arise from the transition of hydrogen atoms to and from the ground state.

M

Magellanic clouds Two neighboring irregular galaxies visible to the unaided eye in the southern hemisphere.

Magnetic field Region in space near a magnetized body where a magnetic force can be detected.

Magnetic pole One of two points of a magnet or a body (Earth) having a magnetic field where magnetic flux density is greatest. It is the direction toward which a compass needle points.

Magnetosphere Region in the near vicinity of Earth or other planet occupied by the Earth's magnetic field.

Magnitude A measure of the amount of light received from any luminous body such as a star.

Main sequence The major distribution of stars on the H-R diagram running diagonally from the upper left to the lower righthand part of the diagram.

Major axis The long diameter of an ellipse. See *line of apsides.*

Major planet A large (Jovian) planet.

Mantle That portion of the Earth between the crust and core.

Mare Latin term used by Galileo for sea, pl., *maria.*

Mascon A mass concentration beneath the lunar surface found mainly under the lunar *maria.*

Mass (gravitational) A measure of the amount of matter, which is proportional to the weight of the object and the gravitational force acting on it.

Mass (inertial) Measure of the resistance of a body to a change in motion. Inertial mass is m in the equation $F = ma$.

Mass-luminosity relationship The relationship between the mass and the luminosity of a star. Usually applies to main sequence stars.

Matter Anything that occupies space and has mass.

Maximum elongation See *greatest elongation.*

Mean solar day Interval between the successive passages of the mean sun past the celestial meridian.

Mean solar time The local hour angle of the mean sun plus 12 hours.

Mean sun A ficticious body moving eastward at uniform angular velocity along the celestial equator, making one complete circuit of the sky from vernal equinox to next vernal equinox.

Meridian See *celestial meridian.* Also any imaginary great circle on the Earth's surface passing through a particular site and the north and south poles.

Messier catalogue Catalogue of "nebulae" compiled by Charles Messier in the eighteenth century.

Meteor A bright streak in the sky resulting from a particle entering the Earth's atmosphere. Often called a "shooting star."

Meteorite A particle or meteoroid from space that survives the passage through the atmosphere and strikes the Earth's surface.

Meteoroid A small solid particle in interplanetary space.

Meteor shower Many meteors seeming to come from a common point in the sky and originating from a meteor stream or swarm.

Meteor stream Countless meteoroids revolving around the sun in a common orbit. The meteoroids are distributed over the entire orbit and may have originated from a comet.

Meteor swarm A cluster of meteoroids orbiting the sun as a group.

Micro- One millionth $= 10^{-6}$.

Micrometeoroid A microscopic meteoroid that does not burn up in the atmosphere but filters down to Earth's surface with dust.

Micron A metric unit of measure equal to one-millionth of a meter $= 10^{-6}$ meter.

Milky Way A faint and irregular band of light encircling the sky composed of billions of stars and diffuse nebulae lying near the plane of the galaxy of which the solar system is a part.

Minor axis The axis perpendicular to the line of apsides in an ellipse—the smallest diameter of the ellipse.

Minor planet See *asteroids*.

Molecule The combination of two or more atoms; a molecule represents the smallest particle of a substance that exhibits the chemical properties of that substance.

Momentum The measure of a body's inertia or state of motion; it is the product of the body's mass and its velocity: $p = mv$.

Monochromatic Light or color of a single wavelength.

N

Nadir Point on the celestial sphere directly below the observer; it is opposite the zenith at altitude $-90°$.

Neap tide The low tide that occurs during the quarter phase of the moon.

Nebula An interstellar dust and gas cloud.

Nebular hypothesis The theory suggested by LaPlace that the solar system was formed from a nebula.

Neutron A subatomic particle with no charge (neutral) and mass about equal to the proton.

Neutron star A gravitationally collapsed star thought to be composed primarily of neutrons and having an extremely high density.

New general catalogue (NGC) A catalogue of nebulae, star clusters, and galaxies compiled by J. L. E. Dreyer in 1888. It is the successor to the Messier Catalogue.

New moon Phase of the moon when the moon is in conjunction (between Earth and sun) with the sun.

Newtonian focus An arrangement in the reflecting telescope in which a flat secondary mirror is used to reflect the light beam at right angles from the objective out one side of the telescope tube.

Node One of two points in the orbit of a body where its orbit intersects some reference plane such as the ecliptic.

Normal spiral A galaxy from which arms extending from the nucleus are coiled, giving the stellar structure the appearance of a pinwheel.

Nova A star erupting with explosive energy, increasing in brightness to many thousands of times the normal degree of brightness.

Nucleus The central portion of an atom, comet, or galaxy.

O

Objective The principal image-forming lens or mirror of a telescope.

Oblate spheroid A sphere slightly flattened at the poles and bulging at the equator like Earth.

Obliquity Angle between the celestial equator and the orbital plane of a planet. This is the same angle as that between the axis of rotation and a perpendicular to the orbital plane of the planet.

Occultation The eclipsing of a star or planet by the moon or other planet.

Ocular Eyepiece of a telescope.

Open cluster See *galactic cluster.*

Opposition The position of a superior planet or the moon when it is most nearly opposite the Earth from the sun. Elongation is 180°.

Optical binary Two stars positioned in the line of sight causing them to appear close to each other when actually they may be quite distant.

Orbit The path of one body as it revolves around another or around a point in space.

Orbital plane The plane formed by a celestial body revolving around another body or a point in space.

Organic compounds Substances, of which carbon is a principal component, that are biologically essential to living organisms.

Outer planet One of the planets beyond Mars in the solar system.

Outgassing The means by which gases escape the solid planet and become a part of its atmosphere. This is generally accomplished by volcanic activity.

P

Parabola A conic section with an eccentricity of 1.0.

Parallax The apparent displacement of an object against a more distant background when observed from two different positions.

Parsec Distance at which a star's parallactic displacement would equal 1″ of arc. One parsec equals 3.26 light years.

Partial eclipse An eclipse in which the sun or moon is not totally covered.

Peculiar galaxy A galaxy that emits nonthermal radiation.

Peculiar velocity The velocity of a star with respect to the local standard of rest. It is the space motion of the star corrected for the motion of the sun with respect to neighboring stars.

Penumbra (eclipse) That part of the shadow resulting when only a portion of the light source is blocked during an eclipse.

Penumbra (sunspot) The outer, less-darkened portion of a sunspot.

Perfect cosmological principle The idea that the universe is everywhere the same at all times.

Periastron The point in the orbit of a binary pair where the stars are closest together.

Perigee The closest approach to the Earth of the moon or artificial satellites.

Perihelion The closest approach of the planets or other bodies in the solar system to the sun.

Period Time required for an object to complete one rotation on its axis or one revolution of its orbit.

Period-luminosity relationship The relationship between the period and the absolute magnitude of Cepheid variable stars.

Perturbation A small deviation in the normal movement of a body in orbit, usually arising from an outside force such as gravity.

Photographic magnitude Magnitude of celestial bodies as measured on blue and violet-sensitive photographic plates.

Photometer An instrument used to measure light intensity.

Photon An electromagnetic particle (e.g. light).

Photosphere The visible surface of the sun; the solar disk.

Photosynthesis The process whereby carbohydrates are formed in plants by combining carbon dioxide and water, using sunlight as a source of energy.

Plage A bright region on the solar surface seen in monochromatic light.

Planetary nebula An envelope of gas that is slowly expanding, surrounding a hot star.

Planetoid Alternate name for asteroid.

Plasma A hot, highly ionized gas.

Polar axis Earth's axis of rotation. Also applies to axis of a telescope mounting that is parallel to the Earth's axis.

Population I, II Two classes of stars grouped according to age, location in the galaxy, chemical properties, and spectral classification.

Positron A positively charged electron; an antielectron.

Precession of Earth Slow conical motion of the Earth's axis induced by the gravitational effect of the moon and sun on Earth's bulging equator.

Precession of the equinoxes Slow westward drift of the equinoxes due to Earth's precession.

Prime focus The point of focus of the objective in a telescope.

Prime meridian The terrestrial meridian or longitude line that goes from north pole to south pole through the site of the Royal Naval Observatory in Greenwich, England. It is identified as $0°$ longitude.

Primeval atom The original mass from which all matter in the universe presumably formed, according to the LeMaitre theory of the evolutionary universe.

Prominence A flamelike protuberence extending above the solar limb.

Proper motion The angular displacement per year of a star against the background of more distant stars.

Protogalaxy The original material from which a galaxy, planet, or star is formed.

Proton A positively charged subatomic particle which, along with the neutron, is found in the nucleus of atoms.

Proton-proton reaction A sequence of thermonuclear reactions in which helium is formed from hydrogen with a release of energy within the body of a star.

Pulsar A star from which bursts of energy are detected at very short periods. This is thought to result from the rapid rotation of a neutron star.

Pulsating star Stars varying in size and luminosity. See *Cepheid variables*.

Q

Quadrature The position of a planet with respect to Earth and the sun where the planet's elongation is 90°.

Quarter moon The two phases of the moon when the moon is at quadrature. The moon appears half full in these positions.

Quasar A term used to refer to quasi-stellar radio sources.

Quasi-stellar radio source (QSS) A starlike object suspected of being extragalactic that emits strong radio signals and is highly luminous. It has a very large red shift.

R

Radar A technique for sending out electromagnetic signals and measuring the period between transmission and reception of the reflected signal from a distant object for the purpose of determining its distance.

Radial velocity The velocity of an object along the line of sight determined from the Doppler shift of the lines in the spectrum of the object.

Radiant Point in the sky from which the meteors in a meteor shower seem to originate.

Radiation A means of transmission of electromagnetic energy through a vacuum. Also the transmitted electromagnetic waves.

Radioactivity The disintegration of the atomic nucleus of certain elements into lighter nuclei with the emission of subatomic particles and gamma rays.

Radio galaxy A galaxy that emits strong signals in the radio portion of the spectrum.

Ray Bright streaks on the lunar surface, sometimes appearing to radiate from a crater.

Red giant A huge, relatively cool star appearing in the upper righthand portion of the H-R diagram.

Red shift A shift of the lines in the visible portion of the spectrum toward the red end of the spectrum. The shift is thought to result from the Doppler shift.

Reflecting telescope A telescope in which the objective is a concave mirror.

Reflection The return of light rays by a surface.

Refracting telescope A telescope in which the objective is a lens.

Refraction The bending of light rays as they pass from one transparent medium to another.

Regression of nodes The westward motion of the nodes of the orbit of a planet or satellite along a reference plane such as the ecliptic or equatorial plane.

Relative orbit The orbit of one of two mutually revolving bodies referred to the other body.

Resolving power The ability of a telescope to separate two closely spaced bodies into separate entities. This ability is usually measured in seconds of arc.

Retrograde motion An apparent motion in which a planet seems to temporarily reverse its movement in orbit against the background of more distant stars. A motion opposite to the normal.

Revolution The movement of a body around a point.

Right ascension Angular distance measured in hours eastward from the vernal equinox along the ecliptic.

Rill(e) A narrow canyon or large crevice in the surface of the moon or a planet.

Roche's limit The closest distance to a body that a satellite of the same density can hold itself together by gravity. Inside this critical distance the gravitational force of the body would cause the satellite to break up.

Rotation The turning of a body on its own axis.

S

Saros The cycle of similar eclipses recurring at intervals of a little more than 18 years.

Satellite Any body, artificial or natural, that revolves around a larger body and is under the influence of the larger body's force of gravity.

Schwarzschild radius See *event horizon.*

Secondary mirror A mirror used to redirect the beam of light from the objective mirror in a Newtonian focus (reflecting telescope).

Semimajor axis Half the major axis of an ellipse.

Separation The angular displacement between two components of a visual binary star.

Seyfert galaxy A particular type of spiral galaxy that emits intense radiation from the nucleus on an irregular basis.

Short-period comet Comets with a period of less than 200 years.

Sidereal day Time between two successive meridian passages of the vernal equinox. Represents a $360°$ turn of a body (like Earth) on its axis with respect to some distant star.

Sidereal month Time required for the moon to complete one $360°$ revolution around the Earth with respect to a distant star.

Sidereal year Time required for the Earth to complete one $360°$ revolution around the sun with respect to a distant star.

Siderite A meteorite composed primarily of iron and nickel—essentially a metallic meteorite.

Siderolite A meteorite composed of a mixture of metal and stone.

Small circle Any circle on the surface of a sphere where the plane of the circle does not pass through the center of the sphere.

Solar constant Amount of solar radiation received per unit of area per unit of time at the surface of Earth's atmosphere in a direction perpendicular to the sun. The value of the solar constant is given as 1.37×10^6 ergs/cm^2/sec or 1.96 cal/cm^2/min.

Solar day Period of time between successive transits of the sun across a given meridian. Represents the 24-hour day.

Solar system The planets, satellites, comets, asteroids, meteoroids, and dust and gas controlled by the gravitational force of the sun.

Solar wind The radial flow of radiation and particulate matter from the sun.

Solstice The two points on the celestial sphere where the sun reaches the maximum distance north and south of the celestial equator.

Space motion Velocity of a star with respect to the sun.

Specific gravity The ratio of the density of some substance to that of water.

Spectral class or type A category in a system of stellar classification based on the appearance of the lines of the spectrum of the star.

Spectrogram Photograph of the spectrum.

Spectrograph A device for photographing the spectrum.

Spectroheliogram Photograph of the solar atmosphere in the light of a single spectral line.

Spectroheliograph A device for photographing the sun in the light of a single spectral line.

Spectroscope A device for viewing the spectrum of a light source.

Spectroscopic binary A double-star system in which the components can only be revealed by the Doppler shift of the spectral lines.

Spectroscopy The study of the spectra.

Spectrum The range of colors obtained from spreading light by passing it through a prism or special grid.

Speed The rate at which distance is covered without regard to direction.

Spherical aberration The inability of spherical lenses or mirrors to focus all light from an object passing through the lens or reflected from the mirror to a specific point.

Spicule A jet of material rising in the solar chromosphere.

Spiral arms Dust, gas, and stars that are coiled more or less tightly around the central nucleus of a galaxy usually in a single plane.

Spiral galaxy A flat, rotating galaxy having spiral arms.

Sporadic meteor An occasional meteor—one that is not part of a shower.

Spring tide High tides occurring when the sun and moon are on the same or opposite sides of the Earth from each other.

Star A self-luminous mass of gas.

Star cluster A group of stars acting as a unit by traveling in the same direction at the same speed and held together in a group by their mutual gravitational force.

Steady state A cosmological theory that holds that the density and form of the universe is always the same in space and time and that new matter is constantly being formed.

Stellar evolution The process of change in brightness, size, and structure of a star as it ages.

Stellar parallax The angle in seconds of arc subtended by one astronomical unit at the distance of a particular star.

Stratosphere A layer of the Earth's atmosphere between the troposphere and the ionosphere.

Subdwarf A star with a lower luminosity than a main sequence star of the same spectral class.

Subgiant A star intermediate between a main sequence star and a giant star of the same spectral class.

Summer solstice The point on the celestial sphere where the sun is at its northernmost point above the celestial equator.

Sun The nearest star around which the Earth and other planets revolve.

Sunspot A temporary spot on the solar surface that is cooler than the surrounding photosphere.

Sunspot cycle The more-or-less regular 11-year period during which the frequency and size of sunspots fluctuate.

Supergiant A huge star that has an absolute magnitude in excess of –1.

Superior conjunction The position of a planet in line with the Earth and sun, with the sun between the Earth and the planet.

Superior planet A planet that is a greater distance from the sun than the Earth.

Supernova A tremendous burst of energy from a star during which luminosity may increase to more than hundreds of thousands of times the normal. Star may be destroyed in the process.

Synodic month The period of the lunar phases or the period of time between successive conjunctions of the moon with the sun.

Synodic period The period of time between successive oppositions of a planet.

Syzygy The lining up of a planet or moon with the Earth and Sun.

T

Tachyon A hypothetical particle that is described as moving faster than the speed of light.

Tangential velocity The linear velocity of a star perpendicular to the line of sight.

Telescope A device for magnifying and increasing the visibility of distant objects. See *reflecting* and *refracting telescopes*.

Terminator The line between the illuminated and dark portions of the moon or planet.

Terrestrial planet Planets near the sun including Mercury, Venus, Earth, and Mars. Occasionally Pluto is included in this group.

Thermonuclear reaction A fusion reaction in which interacting nuclei are combined as a result of the kinetic energy imparted to them by heat. Hydrogen nuclei (protons) are combined to form helium by this process in stars. See *proton-proton reaction*.

Thermosphere A layer of the Earth's atmosphere in which the highest atmospheric temperatures are detected lying between the mesosphere and the exosphere.

Tidal force An unequal gravitational force that tends to deform a body.

Tide The deformation of a body by an unequal gravitational force exerted by another body, as, for example, the effect of the moon's and sun's gravitational force on Earth's oceans.

Total eclipse A solar eclipse in which the sun's disk is entirely covered by the moon. A lunar eclipse in which the moon passes entirely into the umbra of the Earth's shadow.

Train A temporary luminous trail left by the passage of a meteoroid through Earth's atmosphere.

Transit The passage of a body across the face of a larger body as viewed from Earth, as, for example, Mercury or Venus across the face of the sun. Also applies to a body moving across the meridian. An instrument used by surveyors for measuring vertical and horizontal angles.

Triangulation A surveying technique used to determine the distance to an inaccessible object from a known baseline.

Triple-alpha process A series of reactions whereby helium is formed into carbon in a star.

Tropical year Period of Earth's revolution from one vernal equinox to the next.

Troposphere Lowest level of Earth's atmosphere where most weather occurs.

Twinkling Variation in color and brightness of stars due to turbulence in Earth's atmosphere.

Tychonic system An earth-centered system suggested by Tycho Brahe in the sixteenth century according to which the sun revolved around a stationary Earth and the planets revolved around the sun.

U

UBV system A standardized color and magnitude system comparing radiation in the ultraviolet, blue, and visible (yellow) spectral regions.

Ultraviolet That portion of the electromagnetic spectrum that extends in wavelength from 100 angstroms to the visible portion of the spectrum—about 3900 angstroms.

Umbra The central dark portion of a shadow in which the light source is not visible. In a sunspot it is the central dark portion of the sunspot.

Universe The totality of all matter, energy, and space.

V

Van Allen radiation belts A region around the Earth where rapidly moving particles are trapped in the Earth's magnetic field.

Variable star A star that varies in luminosity on a periodic or nonperiodic basis.

Vector A quantity that has both magnitude and direction.

Velocity A vector denoting both the speed and direction in which a body is moving.

Vernal equinox The point on the celestial sphere where the sun crosses the celestial equator apparently moving north. Signals the beginning of the spring season.

Visual binary star A double star in which the components can be observed through a telescope.

W

Watt A unit of power equal to 10^7 ergs/sec.

Wavelength The distance of one point on a wave to a comparable point on the following wave.

Weight A measure of the force of gravity between the Earth or other body and an object on the surface.

White dwarf A star that has exhausted its nuclear fuel, occurring below the main sequence in the H-R diagram. It has collapsed to a small volume and high density due to its gravitational force.

White hole A hypothetical reversal of a black hole from which great energy and matter are supposedly emitted.

Widmanstatten figures The crystalline structure seen on the cut and polished surface of a meteorite.

Wein's law Equation that relates temperature emitted by an ideal black body to the wavelength at which it emits its greatest intensity of radiation.

Winter solstice Position on the celestial sphere where the sun is at its greatest distance south of the celestial equator.

Wormhole See *Einstein-Rosen bridge.*

X

X rays Electromagnetic radiation with photons of wavelength between one and 100 angstroms.

X-ray stars Stars other than the sun that emit detectable amounts of radiation in the X-ray portion of the spectrum.

Y

Year Measure of the period of revolution of the Earth around the sun.

Z

Zeeman effect The splitting or broadening of spectral lines under the influence of a magnetic field.

Zenith A point on the celestial sphere directly overhead at an altitude of $90°$.

Zero-age main sequence (ZAMS) The principal grouping of stars on the H-R diagram where the stars have achieved a degree of stability and are deriving their energy from thermonuclear reactions.

Zodiac A belt around the sky centered on the ecliptic.

Zodiacal light Faint band of light tapering upward from the Earth along the ecliptic. It is believed to come from sunlight reflected and scattered by interplanetary dust.

Zone of avoidance A region near the Milky Way where interstellar dust makes it difficult or impossible to see external galaxies in that region of the sky.

INDEX